Student Solutions Manual

C Trimble & Associates

Prealgebra

Fifth Edition

Jamie Blair

Orange Coast College

John Tobey

North Shore Community College

Jeffrey Slater

Orange Coast College

Jennifer Crawford

Normandale Community College

PEARSON

Boston Columbus Indianapolis New York San Francisco Upper Saddle River
Amsterdam Cape Town Dubai London Madrid Milan Munich Paris Montreal Toronto
Delhi Mexico City Sao Paulo Sydney Hong Kong Seoul Singapore Taipei Tokyo

The author and publisher of this book have used their best efforts in preparing this book. These efforts include the development, research, and testing of the theories and programs to determine their effectiveness. The author and publisher make no warranty of any kind, expressed or implied, with regard to these programs or the documentation contained in this book. The author and publisher shall not be liable in any event for incidental or consequential damages in connection with, or arising out of, the furnishing, performance, or use of these programs.

Reproduced by Pearson from electronic files supplied by the author.

Copyright © 2012, 2010, 2006, 2002 Pearson Education, Inc.
Publishing as Pearson, 75 Arlington Street, Boston, MA 02116.

All rights reserved. No part of this publication may be reproduced, stored in a retrieval system, or transmitted, in any form or by any means, electronic, mechanical, photocopying, recording, or otherwise, without the prior written permission of the publisher. Printed in the United States of America.

ISBN-13: 978-0-321-77353-1
ISBN-10: 0-321-77353-5

1 2 3 4 5 6 OPM 16 15 14 13 12

www.pearsonhighered.com

PEARSON

Contents

Chapter 1

1.1 Exercises

1. a. 8002
The number begins with 8 in the thousands place. The word name is eight thousand two.

b. 802
The number begins with 8 in the hundreds place. The word name is eight hundred two.

c. 82
The number begins with 8 in the tens place. The word name is eighty-two.

d. The number eight hundred twenty is written 820. the digit 0 is in the ones place. Its place value is one.

3. a. 9865
↑
hundreds

b. 9865
↑
ones

5. a. 754,310
↑
thousands

b. 754,310
↑
hundred thousands

7. a. 1,284,073
↑
millions

b. 1,284,073
↑
hundreds

9. $5876 = 5000 + 800 + 70 + 6$

11. $4921 = 4000 + 900 + 20 + 1$

13. $867{,}301 = 800{,}000 + 60{,}000 + 7000 + 300 + 1$

15. $562
$562 = 500 + 60 + 2$
5 hundred-dollar bills, 6 ten-dollar bills, and 2 one-dollar bills

17. $46

a. $46 = 40 + 6$
4 ten-dollar bills and 6 one-dollar bills; answers may vary.

b. $46 = 40 + 5 + 1$
4 ten-dollar bills, 1 five-dollar bill, and 1 one-dollar bill; answers may vary.

19. 6079
The number begins with 6 in the thousands place. The word name is six thousand, seventy-nine.

21. 86,491
The number begins with 8 in the ten thousands place. The word name is eighty-six thousand, four hundred ninety-one.

23. $672
Write 672.00 in the box following $. Write "Six hundred seventy-two and 00/100" on the line preceding DOLLARS.

25. 5 ? 7
5 is less than 7.
$5 < 7$

27. 6 ? 8
6 is less than 8.
$6 < 8$

29. 13 ? 10
13 is greater than 10.
$13 > 10$

31. 9 ? 0
9 is greater than 0.
$9 > 0$

33. 2131 ? 1909
2131 is greater than 1909.
$2131 > 1909$

35. 52,647 ? 616,000
52,647 is less than 616,000.
$52{,}647 < 616{,}000$

37. Five *is greater than* two.
↓ ↓ ↓
$5 \quad > \quad 2$

Copyright © 2012 Pearson Education, Inc.

39. Two *is less than* five.
↓ ↓ ↓
2 < 5

41. 45
Identify the round-off place digit: 45.
The digit to the right is 5 or more. Increase the round-off place digit by 1. Replace the digit to the right with a zero.
50

43. 661
Identify the round-off place digit: 661.
The digit to the right is less than 5. Do not change the round-off place digit. Replace the digit to the right with a zero.
660

45. 63,854
Identify the round-off place digit: 63,854.
The digit to the right is 5 or more. Increase the round-off place digit by 1. Replace all digits to the right with zeros.
63,900

47. 823,042
Identify the round-off place digit: 823,042.
The digit to the right is less than 5. Do not change the round-off place digit. Replace all digits to the right with zeros.
823,000

49. 38,431
Identify the round-off place digit: 38,431.
The digit to the right is less than 5. Do not change the round-off place digit. Replace all digits to the right with zeros.
38,000

51. 143,526
Identify the round-off place digit: 143,526.
The digit to the right is 5 or more. Increase the round-off place digit by 1. Replace all digits to the right with zeros.
144,000

53. 5,254,423
Identify the round-off place digit: 5,254,423.
The digit to the right is 5 or more. Increase the round-off place digit by 1. Replace all digits to the right with zeros.
5,300,000

55. 9,007,601
Identify the round-off place digit: 9,007, 601.
The digit to the right is less than 5. Do not change the round-off place digit. Replace all digits to the right with zeros.
9,000,000

57. 865,000 miles
Identify the round-off place digit: 865,000.
The digit to the right is 5 or more. Increase the round-off place digit by 1. Replace all digits to the right with zeros.
870,000 miles

59. Expedition XLT ? Supercab XLT
$35,010 ? $29,605
35,010 is greater than 29,605.
$35,010 > $29,605
Expedition XLT > Supercab XLT

61. $27,395
Identify the round-off place digit: 27,395.
The digit to the right is less than 5. Do not change the round-off place digit. Replace all digits to the right with zeros.
$27,000

63. 16,962
Identify the round-off place digit: 16,962.
The digit to the right is 5 or more. Increase the round-off place digit by 1. Since the round-off place digit is 9, place a zero in the round-off place and increase the digit in the next place to the left by 1. Replace all digits to the right with zeros.
17,000

65. 5,311,192,809,000
The number begins with 5 in the trillions place. The word name is five trillion, three hundred eleven billion, one hundred ninety-two million, eight hundred nine thousand.

67. 3 hours and 50 minutes
Since 50 minutes is more than one-half hour, we round up.
4 hours

69. 123 feet, 5 inches
Since 5 inches is less than one-half foot, we round down.
123 feet

Quick Quiz 1.1

1. 6402 = 6000 + 400 + 2

Copyright © 2012 Pearson Education, Inc.

2. **a.** 0 ? 10
0 is less than 10.
$0 < 10$

b. 15 ? 10
15 is greater than 10.
$15 > 10$

3. 154,572

a. Identify the round-off place digit: 1<u>5</u>4,572. The digit to the right is less than 5. Do not change the round-off place digit. Replace all digits to the right with zeros.
150,000

b. Identify the round-off place digit: 154,<u>5</u>72. The digit to the right is 5 or more. Increase the round-off place digit by 1. Replace all digits to the right with zeros.
154,600

4. Answers may vary. One possible solution follows. To round 8937 to the nearest hundred, first identify the round-off place digit: 8<u>9</u>37. The digit to the right is less than 5. Do not change the round-off place digit. Replace all digits to the right with zeros. 8900

1.2 Understanding the Concept Addition Facts Made Simple

1. $8 + 5 = (3 + 5) + 5 = 3 + (5 + 5) = 3 + 10 = 13$

2. $6 + 8 = 6 + (6 + 2) = (6 + 6) + 2 = 12 + 2 = 14$

1.2 Understanding the Concept Using Inductive Reasoning to Reach a Conclusion

1. 8, 14, 20, 26, 32, 38, ...
We observe a pattern that each number is 6 more than the preceding number: $14 = 8 + 6$, $20 = 14 + 6$, and so on. Therefore, if we add 6 to 38, we conclude that the next number in the sequence is 44.

2. 17, 28, 39, 50, 61, ...
We observe a pattern that each number is 11 more than the preceding number: $28 = 17 + 11$, $39 = 28 + 11$, and so on. Therefore, if we add 11 to 61, we conclude that the next number in the sequence is 72.

1.2 Exercises

1. 10 + x
↓ ↓ ↓
Ten plus a number
Answers may vary.

3. To evaluate $x + 6$ if x is equal to 9, replace x with 9 and then add 9 and 6.

5. $(2 + 3) + 4 = 2 + (3 + 4)$
The grouping of the addition is changed. This is the associative property of addition.

7. A number plus 2: $m + 2$

9. The sum of 5 and y: $5 + y$ or $y + 5$

11. Some number added to 12: $12 + m$ or $m + 12$

13. A number increased by 7: $m + 7$

15. $5 + a = a + 5$

17. $3 + x = x + 3$

19. By the commutative property of addition, $3542 + 216 = 216 + 3542$, so $216 + 3542 = 3758$.

21. By the commutative property of addition, $5 + n = n + 5$, so $n + 5 = 12$.

23. $x + 4 + 2 = x + (4 + 2) = x + 6$

25. $9 + 3 + n = (9 + 3) + n = 12 + n$ or $n + 12$

27. $x + 0 + 2 = x + (0 + 2) = x + 2$

29. $(x + 2) + 1 = x + (2 + 1) = x + 3$

31. $9 + (3 + n) = (9 + 3) + n = 12 + n$

33. $(n + 3) + 8 = n + (3 + 8) = n + 11$

35. $(x + 4) + 11 = x + (4 + 11) = x + 15$

37. $(2 + n) + 5 = (n + 2) + 5 = n + (2 + 5) = n + 7$

39. $8 + (1 + x) = (8 + 1) + x = 9 + x = x + 9$

41.
$$\begin{aligned} 2+(3+n)+4 &= (2+3)+(n+4) \\ &= 5+(4+n) \\ &= (5+4)+n \\ &= 9+n \\ &= n+9 \end{aligned}$$

Copyright © 2012 Pearson Education, Inc.

43. $\begin{aligned}(3+a+2)+8 &= (a+3+2)+8\\ &= (a+5)+8\\ &= a+(5+8)\\ &= a+13\end{aligned}$

45. $\begin{aligned}(5+x+7)+4 &= (x+5+7)+4\\ &= (x+12)+4\\ &= x+(12+4)\\ &= x+16\end{aligned}$

47. a. Replace y with 3.
$y + 7 = 3 + 7 = 10$
When y is equal to 3, $y + 7$ is equal to 10.

b. Replace y with 8.
$y + 7 = 8 + 7 = 15$
When y is equal to 8, $y + 7$ is equal to 15.

49. Replace x with 6 and y with 13.
$x + y = 6 + 13 = 19$
When x is 6 and y is 3, $x + y$ is 19.

51. Replace a with 9, b with 15, and c with 12.
$a + b + c = 9 + 15 + 12 = 36$
When a is 9, b is 15, and c is 12, $a + b + c$ is 36.

53. Replace n with 26 and m with 44.
$n + m + 13 = 26 + 44 + 13 = 83$
When n is 26 and m is 44, $n + m = 13$ is 83.

55. a. $\begin{aligned}\text{Bonus} &= x+y+250\\ &= 180+12+250\\ &= \$442\end{aligned}$

b. $\begin{aligned}\text{Bonus} &= x+y+250\\ &= 175+10+250\\ &= \$435\end{aligned}$

57. $\begin{array}{r} 15\\ +\ 23\\ \hline 38\end{array}$

59. $\begin{array}{r} 236\\ +\ \ 43\\ \hline 279\end{array}$

61. $\begin{array}{r} 32\\ 11\\ 20\\ +\ \ 7\\ \hline 70\end{array}$

63. $\begin{array}{r} 105\\ 8\\ 133\\ +\ \ 98\\ \hline 344\end{array}$

65. $\begin{array}{r} 236\\ 467\\ +\ \ 26\\ \hline 729\end{array}$

67. $\begin{array}{r} 281\\ 64\\ +539\\ \hline 884\end{array}$

69. $\begin{array}{r} 7287\\ 732\\ +\ \ 423\\ \hline 8442\end{array}$

71. $\begin{array}{r} 922{,}876\\ 54\\ 1{,}287\\ +\ \ \ 5{,}000\\ \hline 929{,}217\end{array}$

73. $\begin{array}{r} 3{,}107\\ 9{,}063\\ 54\\ +\ 379{,}626\\ \hline 391{,}850\end{array}$

75. a. $\begin{array}{r} 159\\ +\ 241\\ \hline 400\end{array}$

Total deposits were $400.

b. $\begin{array}{r} 63\\ 121\\ +\ \ 44\\ \hline 228\end{array}$

Total debits were $228.

Copyright © 2012 Pearson Education, Inc.

77.
$$\begin{array}{r} 875 \\ 875 \\ 500 \\ 24 \\ +\ \ 35 \\ \hline 2309 \end{array}$$
Charles and Vincent needed \$2309 to move into the apartment.

79. 13 in. + 5 in. + 13 in. + 5 in. = 36 in.
The perimeter is 36 inches.

81. 3 ft + 3 ft + 3 ft + 3 ft = 12 ft
The perimeter is 12 feet.

83. 4 in. + 4 in. + 6 in. = 14 in.
The perimeter is 14 inches.

85. The length of the side not labeled is 12 ft + 6 ft or 18 ft.
7 ft + 12 ft + 4 ft + 6 ft + 11 ft + 18 ft = 58 ft
The perimeter is 58 feet.

87. The length of the unlabeled side on the left is 150 in., and the length of the right side of the figure is 100 in. + 150 in. or 250 in.
150 in. + 120 in. + 100 in. + 190 in. + 250 in.
+ 310 in.
= 1120 in.
The perimeter is 1120 inches.

89. 1, 3, 5, 7, 9, 11, 13, ...
Each number is 2 more than the preceding number. The next number is 13 + 2 or 15.

91. 0, 5, 10, 15, 20, 25, ...
Each number is 5 more than the preceding number. The next number is 25 + 5 or 30.

93. 7, 16, 25, 34, 43, ...
Each number is 9 more than the preceding number. The next number is 43 + 9 or 52.

Quick Quiz 1.2

1. a.
$$\begin{aligned}(4+a)+9 &= (a+4)+9 \\ &= a+(4+9) \\ &= a+13\end{aligned}$$

b.
$$\begin{aligned}2+(1+x+7) &= 2+(1+7+x) \\ &= 2+(8+x) \\ &= (2+8)+x \\ &= 10+x \text{ or } x+10\end{aligned}$$

2. Replace m with 25 and n with 8.
$m + n + 13 = 25 + 8 + 13 = 46$
When m is 25 and n is 8, $n + m + 13$ is 46.

3. The length of the unlabeled side on the left is 40 in., and the length of the right side of the figure is 40 in. + 30 in. or 70 in.
40 in. + 40 in. + 30 in. + 135 in. + 70 in.
+ 175 in.
= 490 in.
The perimeter is 490 inches.

4. a. The 1 that is placed above the 9 is in the tens place. Its value is 1×10 or 10.

b. The 1 that is placed above the 3 is in the hundreds place. Its value is 1×100 or 100.

1.3 Understanding the Concept
Money and Borrowing

1. We can borrow only from a place value that has a nonzero whole number. In \$400 there are only 100-dollar bills to borrow from.

2. When we change the ten-dollar bill to 10 one-dollar bills, we have 0 ten-dollar bills and 10 one-dollar bills which is similar to borrowing in subtraction.

1.3 Exercises

1.
$$\begin{array}{ccc} 6 & - & x \\ \downarrow & \downarrow & \downarrow \end{array}$$
Six minus x
Answers may vary.

3. The key phrase "how many more" indicates the operation <u>subtraction</u>.

5. $7 - 4 = 3$

7. $6 - 2 = 4$

9. $9 - 3 = 6$

11. $8 - 7 = 1$

13. $15 - 0 = 15$

15. $20 - 20 = 0$

17. $700 - 600 = 100$
$700 - 601 = 99$
$700 - 602 = 98$
$700 - 603 = 97$

Copyright © 2012 Pearson Education, Inc.

19. $300 - 200 = 100$
$300 - 201 = 99$
$300 - 202 = 98$
$300 - 203 = 97$
$300 - 204 = 96$
$300 - 205 = 95$

21. Nine minus two: $9 - 2$

23. The difference of 8 and y: $8 - y$

25. Ten subtracted from seventeen: $17 - 10$

27. A number decreased by 1: $n - 1$

29. Two less than some number: $a - 2$

31. Replace n with 4.
$9 - n = 9 - 4 = 5$
If n is equal to 4, $9 - n$ is equal to 5.

33. Replace n with 9.
$9 - n = 9 - 9 = 0$
If n is equal to 9, $9 - n$ is equal to 0.

35. Replace x with 9.
$x - 2 = 9 - 2 = 7$
If x is equal to 9, $x - 2$ is equal to 7.

37. Replace x with 3.
$x - 2 = 3 - 2 = 1$
If x is equal to 3, $x - 2$ is equal to 1.

39. $\begin{array}{r} 97 \\ -\ 35 \\ \hline 62 \end{array}$ Check: $\begin{array}{r} 62 \\ +\ 35 \\ \hline 97 \end{array}$

41. $\begin{array}{r} 56 \\ -\ 23 \\ \hline 33 \end{array}$ Check: $\begin{array}{r} 23 \\ +\ 33 \\ \hline 56 \end{array}$

43. $\begin{array}{r} 83 \\ -\ 67 \\ \hline 16 \end{array}$ Check: $\begin{array}{r} 67 \\ +\ 16 \\ \hline 83 \end{array}$

45. $\begin{array}{r} 72 \\ -\ 18 \\ \hline 54 \end{array}$ Check: $\begin{array}{r} 18 \\ +\ 54 \\ \hline 72 \end{array}$

47. $\begin{array}{r} 873 \\ -\ 195 \\ \hline 678 \end{array}$ Check: $\begin{array}{r} 195 \\ +\ 678 \\ \hline 873 \end{array}$

49. $\begin{array}{r} 500 \\ -\ 43 \\ \hline 457 \end{array}$ Check: $\begin{array}{r} 43 \\ +\ 457 \\ \hline 500 \end{array}$

51. $\begin{array}{r} 8912 \\ -\ 3847 \\ \hline 5065 \end{array}$ Check: $\begin{array}{r} 3847 \\ +\ 5065 \\ \hline 8912 \end{array}$

53. $\begin{array}{r} 5301 \\ -\ 185 \\ \hline 5116 \end{array}$ Check: $\begin{array}{r} 185 \\ +\ 5116 \\ \hline 5301 \end{array}$

55. $\begin{array}{r} 15,107 \\ -\ 6,428 \\ \hline 8,679 \end{array}$ Check: $\begin{array}{r} 6,428 \\ +\ 8,679 \\ \hline 15,107 \end{array}$

57. $\begin{array}{r} 164,300 \\ -\ 58,923 \\ \hline 105,377 \end{array}$ Check: $\begin{array}{r} 58,923 \\ +\ 105,377 \\ \hline 164,300 \end{array}$

59. The length of the unlabeled side on the left is 7 ft − 2 ft or 5 ft, and the length of the unlabeled top side of the figure is 11 ft − 3 ft or 8 ft.
5 ft + 3 ft + 2 ft + 8 ft + 7 ft + 11 ft = 36 ft
The perimeter is 36 ft.

61. The length of the unlabeled side on the left is 17 ft − 6 ft or 11 ft, and the length of the unlabeled top side of the figure is 25 ft − 9 ft or 16 ft.
11 ft + 16 ft + 6 ft + 9 ft + 17 ft + 25 ft = 84 ft
The perimeter is 84 ft.

63.

Check #	Amount	Balance $1364
123	$238	$1126
124	$137	$989
125	$69	$920
126	$98	$822
127	$369	$453

65. $\begin{array}{r} 100 \\ -\ 85 \\ \hline 15 \end{array}$
The top speed of Superman the Escape is 15 miles per hour faster than the top speed of Goliath.

Copyright © 2012 Pearson Education, Inc.

67. $\begin{array}{r} 300 \\ -\ 195 \\ \hline 105 \end{array}$

The maximum drop of Magnum XL-200 is 105 feet less than that of Millennium Force.

69. $\begin{array}{r} 865{,}000 \\ -\quad 2{,}160 \\ \hline 862{,}840 \end{array}$

The diameter of the sun is 862,840 more miles than the diameter of the moon.

71. $x - y = y - x$ when the values of x and y are equal.

Cumulative Review

73. 5,117,206 > 13,842

74. 2,386,702 > 117,401

75. $\begin{array}{r} 120 \\ 135 \\ +\ 105 \\ \hline 360 \end{array}$

Edward worked 360 hours in the three-month period.

76. $\begin{array}{r} 430 \\ 32 \\ 12 \\ 28 \\ +\quad 6 \\ \hline 508 \end{array}$

Drew paid $508 for the dog and all the supplies.

Quick Quiz 1.3

1. a. Five subtracted from a number: $n - 5$

b. A number decreased by seven: $n - 7$

c. Eight less than a number: $n - 8$

2. a. $\begin{array}{r} 14{,}062 \\ -\ 7{,}283 \\ \hline 6{,}779 \end{array}$ Check: $\begin{array}{r} 7{,}283 \\ +\ 6{,}779 \\ \hline 14{,}062 \end{array}$

b. $\begin{array}{r} 601{,}307 \\ -\ 192{,}512 \\ \hline 408{,}795 \end{array}$ Check: $\begin{array}{r} 192{,}512 \\ +\ 408{,}795 \\ \hline 601{,}307 \end{array}$

3. $\begin{array}{r} 3270 \\ -\ 2860 \\ \hline 410 \end{array}$

Jose will earn $410 more per month at his new job.

Copyright © 2012 Pearson Education, Inc.

4. Answers may vary. One possible solution follows. We can borrow only from a place value that has a nonzero whole number. When we subtract 35 from 800 we cannot subtract 5 from 0, so we must change 800 to 790 and 10 ones.

1.4 Understanding the Concept Memorizing Multiplication Facts

1. a. $3(7) = 2(7) + 7 = 14 + 7 = 21$

b. $4(8) = 5(8) - 8 = 40 - 8 = 32$

c. $6(8) = 5(8) + 8 = 40 + 8 = 48$

d. $9(8) = 10(8) - 8 = 80 - 8 = 72$

1.4 Exercises

1. a. $4x$: Four times a number

b. ab: The product of a and b

3. 2 times 3:

★★★ ★★
★★★ Shapes may vary. ★★
★★

5. $3(6 \cdot 5) = (3 \cdot 6)5$
The grouping of the multiplication is changed. This is the associative property of multiplication.

7. $3 \cdot 4(2y) = (3 \cdot 4 \cdot \boxed{2}) \cdot y = \boxed{24}y$

9. $(3a) \cdot 4 \cdot 2 = 3 \cdot \boxed{a} \cdot 4 \cdot 2 = (3 \cdot 4 \cdot 2) \cdot \boxed{a} = \boxed{24}a$

11. a.

	White	Pink	Blue
Brown	Brown White	Brown Pink	Brown Blue
Black	Black White	Black Pink	Black Blue
Gray	Gray White	Gray Pink	Gray Blue
Dark Blue	Dark Blue White	Dark Blue Pink	Dark Blue Blue

b. $4(3) = 12$ different outfits

13. $8(5) = 40$ different ice cream dishes

15. $6(3) = 18$
The factors are 6 and 3. The product is 18.

17. $22x = 88$
The factors are 22 and x. The product is 88.

19. A number times 7: $x(7) = 7x$

Copyright © 2012 Pearson Education, Inc.

21. Triple a number: $3x$

23. The product of six and a number: $6x$

25. If $x \cdot y = 0$ and $x = 6$, then $y = 0$.

27. By the associative property of multiplication, $x(y \cdot z) = (x \cdot y)z$, so $(x \cdot y)z = 40$.

29. $(3)(6)(2)(5) = (3 \cdot 6)(2 \cdot 5) = 18(10) = 180$

31.
$$\begin{aligned}(2)(3)(8)(5) &= (3)(8)(2)(5)\\ &= (3 \cdot 8)(2 \cdot 5)\\ &= 24(10)\\ &= 240\end{aligned}$$

33. $2 \cdot 4 \cdot 6 \cdot 0 = 0$

35. $4 \cdot 2 \cdot 4 \cdot 5 = (4 \cdot 4) \cdot (2 \cdot 5) = 16 \cdot 10 = 160$

37. $8(6b) = (8 \cdot 6)b = 48b$

39. $5(z \cdot 8) = 5(8 \cdot z) = (5 \cdot 8)z = 40z$

41. $8(a \cdot 7) = 8(7 \cdot a) = (8 \cdot 7)a = 56a$

43. $2(7 \cdot c) = (2 \cdot 7)c = 14c$

45.
$$\begin{aligned}9(2)(x \cdot 5) &= 9(2)(5 \cdot x)\\ &= 9(2 \cdot 5)x\\ &= 9(10)x\\ &= 90x\end{aligned}$$

47. $9(2)(0 \cdot y) = 0$

49. $6(3)(1 \cdot b) = 18(b) = 18b$

51. $2 \cdot 3(5y) = 6(5y) = (6 \cdot 5)y = 30y$

53. $(6x)3 \cdot 7 = (6x) \cdot 21 = (6 \cdot 21)x = 126x$

55. $3(5y) \cdot 6 = 15y \cdot 6 = 15 \cdot 6y = 90y$

57.
$$\begin{array}{r} 637 \\ \times \quad 9 \\ \hline 5733 \end{array}$$

59.
$$\begin{array}{r} 602 \\ \times \quad 7 \\ \hline 4214 \end{array}$$

61.
$$\begin{array}{r} 398 \\ \times \quad 300 \\ \hline 119,400 \end{array}$$

63.
$$\begin{array}{r} 793 \\ \times \quad 600 \\ \hline 475,800 \end{array}$$

65.
$$\begin{array}{r} 76 \\ \times\ 68 \\ \hline 608 \\ 456 \\ \hline 5168 \end{array}$$

67.
$$\begin{array}{r} 32 \\ \times\ 59 \\ \hline 288 \\ 160 \\ \hline 1888 \end{array}$$

69.
$$\begin{array}{r} 847 \\ \times \quad 56 \\ \hline 5\ 082 \\ 42\ 35 \\ \hline 47,432 \end{array}$$

71.
$$\begin{array}{r} 455 \\ \times \quad 86 \\ \hline 2\ 730 \\ 36\ 400 \\ \hline 39,130 \end{array}$$

73.
$$\begin{array}{r} 354 \\ \times \quad 702 \\ \hline 708 \\ 247\ 80 \\ \hline 248,508 \end{array}$$

75.
$$\begin{array}{r} 432 \\ \times \quad 409 \\ \hline 3\ 888 \\ 172\ 80 \\ \hline 176,688 \end{array}$$

77.
$$\begin{array}{r} 8324 \\ \times \quad 922 \\ \hline 16\ 648 \\ 166\ 48 \\ 7\ 491\ 6 \\ \hline 7,674,728 \end{array}$$

Copyright © 2012 Pearson Education, Inc.

79.
$$\begin{array}{r} 3006 \\ \times \quad 837 \\ \hline 21\,042 \\ 90\;\;18 \\ 2\,404\,8 \\ \hline 2,516,022 \end{array}$$

81.
$$\begin{array}{r} 12,107 \\ \times \quad 808 \\ \hline 96\,856 \\ 9\,685\,60 \\ \hline 9,782,456 \end{array}$$

83.
$$\begin{array}{r} 61,711 \\ \times \quad 1000 \\ \hline 61,711,000 \end{array}$$

85.
$$\begin{array}{r} 40 \\ \times \;\; 8 \\ \hline 320 \end{array}$$
The cook earns $320 for the week.

87.
$$\begin{array}{r} 25 \\ \times 15 \\ \hline 125 \\ 25 \\ \hline 375 \end{array}$$
There are 375 orange trees in the grove.

89.
$$\begin{array}{r} 327 \\ \times \;\; 12 \\ \hline 654 \\ 327 \\ \hline 3924 \end{array}$$
The workbooks cost $3924.

91. $5 \times 40 = 200$
There are 200 rooms.
$25 \times 10 = 250$
There are 250 sets of curtains.
Yes, the owners can replace one set of curtains in each room because there are 200 rooms and 250 sets of curtains.

93. **a.** The bar for Portland is labeled 30. The high temperature was 30°F.

b. The temperature in Albany was 37%F.
$$\begin{array}{r} 37 \\ \times \;\; 2 \\ \hline 74 \end{array}$$
The temperature in Honolulu was 74°F.

Copyright © 2012 Pearson Education, Inc.

95. $2a(4b)(5c) = (2 \cdot 4 \cdot 5)(a \cdot b \cdot c) = 40abc$

97. $4(3a)(2b)(10c) = (4 \cdot 3 \cdot 2 \cdot 10)(a \cdot b \cdot c)$
$= 240abc$

99. $x(4y)7z = (4 \cdot 7)(x \cdot y \cdot z) = 28xyz$

101. a–d. See table.

e. 36

f. There are only a few blank spaces left in the table, so there are not many multiplication facts to learn.

	0	1	2	3	4	5	6	7	8	9
0	0	0	0	0	0	0	0	0	0	0
1	0	1	2	3	4	5	6	7	8	9
2	0	2	4	6	8	10	12	14	16	18
3	0	3	6			15				
4	0	4	8			20				
5	0	5	10	15	20	25	30	35	40	45
6	0	6	12			30				
7	0	7	14			35				
8	0	8	16			40				
9	0	9	18			45				

Cumulative Review

102.
$$\begin{array}{r} 426,862 \\ +\ \ 2,128 \\ \hline 428,990 \end{array}$$

103.
$$\begin{array}{r} 7000 \\ -\ 142 \\ \hline 6858 \end{array}$$

104. 826,540
Identify the round-off place digit: 82$\underline{6}$,540.
The digit to the right is 5 or more. Increase the round-off place digit by 1. Replace all digits to the right with zeros.
827,000

Copyright © 2012 Pearson Education, Inc.

105. 168,406,000
Identify the round-off place digit: 168,4$\underline{0}$6,000.
The digit to the right is 5 or more. Increase the round-off place digit by 1. Replace all digits to the right with zeros.
168,410,000

106.
$$\begin{array}{r} 120 \\ -\ 97 \\ \hline 23 \end{array}$$
The bill was $23 less than the budget allotment.

107.
$$\begin{array}{r} 920 \\ -\ 455 \\ \hline 465 \end{array}$$
Mary Ann must drive 465 miles the second day.

Quick Quiz 1.4

1. The product of six and a number: $6n$

2. $$\begin{array}{r} 1610 \\ \times\ \ 105 \\ \hline 8\ 050 \\ 161\ 00\ \ \\ \hline 169{,}050 \end{array}$$

3. $3a(2b)(5) = (3 \cdot 2 \cdot 5)(a \cdot b) = 30ab$

4. Answers may vary. One possible solution follows. Before we multiply 546 by 2000, we first separate the 2000 into its nonzero digits (2) and trailing zeros (000). We then multiply the nonzero digits by 546, or 2 × 546. To this product we then add to the right side the number of trailing zeros.

1.5 Understanding the Concept The Commutative Property and Division

1. $a \div b = b \div a$ when a and b are equal.

1.5 Understanding the Concept Conclusions and Inductive Reasoning

1. 1, 1, 2, ...
Notice that 1 + 0 = 1 and 1 + 1 = 2. If we follow a pattern of adding consecutive whole numbers (0, 1, 2, ...) to the preceding number, the next number is 2 + 2 = 4.
Notice that 1 · 1 = 1 and 1 · 2 = 2. If we follow a pattern of multiplying the preceding number by consecutive counting numbers (1, 2, 3, ...), the next number is 2 · 3 = 6.

1.5 Exercises

1. There are 220 ÷ 4 rows.

3. Each person will receive $225 \div n$ tickets.

5. (b) and (c) are correct.

7. Twenty-seven divided by a number: $27 \div x$

9. Forty-two dollars divided equally among six people: 42 ÷ 6

11. The quotient of thirty-six and six: 36 ÷ 6

13. The quotient of three and thirty-six: 3 ÷ 36

15. 42 ÷ 42 = 1

17. $\frac{0}{5} = 0$

19. 17 ÷ 0 undefined

21. 58 ÷ 9 = 6 R 4
$$\begin{array}{r} 6 \\ 9\overline{)58} \\ \underline{54} \\ 4 \end{array}$$
Check: 6(9) + 4 = 54 + 4 = 58

23. 2597 ÷ 7 = 371
$$\begin{array}{r} 371 \\ 7\overline{)2597} \\ \underline{21}\ \ \ \\ 49\ \ \\ \underline{49}\ \ \\ 7 \\ \underline{7} \\ 0 \end{array}$$
Check: 7(371) = 2597

25. 1346 ÷ 3 = 448 R 2
$$\begin{array}{r} 448 \\ 3\overline{)1346} \\ \underline{12}\ \ \ \\ 14\ \ \\ \underline{12}\ \ \\ 26 \\ \underline{24} \\ 2 \end{array}$$
Check: 3(448) + 2 = 1344 + 2 = 1346

 Copyright © 2012 Pearson Education, Inc.

27. $1268 \div 30 = 42 \text{ R } 8$

$$\begin{array}{r} 42 \\ 30\overline{)1268} \\ \underline{120} \\ 68 \\ \underline{60} \\ 8 \end{array}$$

Check: $30(42) + 8 = 1260 + 8 = 1268$

29. $632 \div 30 = 21 \text{ R } 2$

$$\begin{array}{r} 21 \\ 30\overline{)632} \\ \underline{60} \\ 32 \\ \underline{30} \\ 2 \end{array}$$

Check: $30(21) + 2 = 630 + 2 = 632$

31. $5817 \div 19 = 306 \text{ R } 3$

$$\begin{array}{r} 306 \\ 19\overline{)5817} \\ \underline{57} \\ 117 \\ \underline{114} \\ 3 \end{array}$$

Check: $19(306) + 3 = 5814 + 3 = 5817$

33. $1403 \div 29 = 48 \text{ R } 11$

$$\begin{array}{r} 48 \\ 29\overline{)1403} \\ \underline{116} \\ 243 \\ \underline{232} \\ 11 \end{array}$$

Check: $29(48) + 11 = 1392 + 11 = 1403$

35. $1369 \div 19 = 72 \text{ R } 1$

$$\begin{array}{r} 72 \\ 19\overline{)1369} \\ \underline{133} \\ 39 \\ \underline{38} \\ 1 \end{array}$$

Check: $19(27) + 1 = 1368 + 1 = 1369$

37. $18{,}985 \div 27 = 703 \text{ R } 4$

$$\begin{array}{r} 703 \\ 27\overline{)18985} \\ \underline{189} \\ 85 \\ \underline{81} \\ 4 \end{array}$$

Check: $27(703) + 4 = 18{,}981 + 4 = 18{,}985$

39. $11{,}571 \div 34 = 340 \text{ R } 11$

$$\begin{array}{r} 340 \\ 34\overline{)11571} \\ \underline{102} \\ 137 \\ \underline{136} \\ 11 \end{array}$$

Check: $34(340) + 11 = 11{,}560 + 11 = 11{,}571$

41. $113{,}317 \div 223 = 508 \text{ R } 33$

$$\begin{array}{r} 508 \\ 223\overline{)113317} \\ \underline{1115} \\ 1817 \\ \underline{1784} \\ 33 \end{array}$$

Check: $223(508) + 33 = 113{,}284 + 33 = 113{,}317$

43. $70{,}141 \div 136 = 515 \text{ R } 101$

$$\begin{array}{r} 515 \\ 136\overline{)70141} \\ \underline{680} \\ 214 \\ \underline{136} \\ 781 \\ \underline{680} \\ 101 \end{array}$$

Check: $136(515) + 101 = 70{,}040 + 101 = 70{,}141$

45.

$$\begin{array}{r} 4 \\ 14\overline{)60} \\ \underline{56} \\ 4 \end{array}$$

The remainder is 4, so 4 tickets were donated to the PTA.

Copyright © 2012 Pearson Education, Inc.

47.
$$\begin{array}{r} 17 \\ 5\overline{)85} \\ \underline{5} \\ 35 \\ \underline{35} \\ 0 \end{array}$$
Each person paid \$17.

49.
$$\begin{array}{r} 175 \\ 6\overline{)1050} \\ \underline{6} \\ 45 \\ \underline{42} \\ 30 \\ \underline{30} \\ 0 \end{array}$$
JoAnn should budget \$175 per day.

51.
$$\begin{array}{r} 62 \\ 2\overline{)124} \\ \underline{12} \\ 04 \\ \underline{4} \\ 0 \end{array}$$
The photographer would have to be 62 feet away with a regular lens.

53.
$$\begin{array}{r} 31 \\ 41\overline{)1300} \\ \underline{123} \\ 70 \\ \underline{41} \\ 29 \end{array}$$
31 stamps can be bought.

55. 5, 15, 45, 135, ...
Each number after the first is the preceding number multiplied by 3. The next number is 3×135 or 405.

57. 3, 4, 7, 12, 19, 28, 39, ...
$3 + 1 = 4$, $4 + 3 = 7$, $7 + 5 = 12$, $12 + 7 = 19$, $19 + 9 = 28$, $28 + 11 = 39$
The amount that is added each time is 2 more than what was added the previous time. Add 13 to find the next number. The next number is 39 + 13 or 52.

59. 7, 9, 10, 12, 13, 15, 16, ...
Alternate adding 2 and adding 1 to the preceding number. The next number is 16 + 2 or 18.

61. 0, 1, 4, ...
$0 + 1 = 1$, $1 + 3 = 4$
Using a pattern of alternating between adding 1 and adding 3, the next number would be 4 + 1 or 5.
$0 = 0 \times 0$, $1 \times 1 = 1$, $2 \times 2 = 4$
Using a pattern of multiplying the next consecutive whole number by itself, the next number is 3×3 or 9.

63. **a.** $(32 \div 4) \div 2 = 8 \div 2 = 4$

b. $32 \div (4 \div 2) = 32 \div 2 = 16$

c. Division is not associative.

Cumulative Review

65. Seven plus x equals eleven: $7 + x = 11$

66.
$$\begin{array}{r} 1060 \\ -\ 114 \\ \hline 946 \end{array}$$

67.
$$\begin{array}{r} 4031 \\ \times\ \ 202 \\ \hline 8\ 062 \\ 806\ 20 \\ \hline 814,262 \end{array}$$

68. 556,432
Identify the round-off place: 55<u>6</u>,432.
The digit to the right is less than 5. Do not change the round-off digit. Replace all digits to the right with zeros.
556,000

69.
$$\begin{array}{r} 1389 \\ -\ 430 \\ \hline 959 \end{array} \qquad \begin{array}{r} 959 \\ -\ 495 \\ \hline 464 \end{array}$$
Leo must drive 464 miles the third day.

70.
$$\begin{array}{r} 29,599 \\ -\ \ 6,200 \\ \hline 23,399 \end{array} \qquad \begin{array}{r} 23,399 \\ -\ \ 5,500 \\ \hline 17,899 \end{array}$$
The balance is \$17,899.

Quick Quiz 1.5

1. **a.** The quotient of 14 and 7: $14 \div 7$

b. The quotient of 7 and 14: $7 \div 14$

Copyright © 2012 Pearson Education, Inc.

2. $15,916 \div 39 = 408 \text{ R } 4$

$$\begin{array}{r} 408 \\ 39\overline{)15916} \\ \underline{156} \\ 316 \\ \underline{312} \\ 4 \end{array}$$

3.

$$\begin{array}{r} 1828000 \\ 3\overline{)5484000} \\ \underline{3} \\ 24 \\ \underline{24} \\ 08 \\ \underline{6} \\ 24 \\ \underline{24} \\ 0 \end{array}$$

Each college receives $1,828,000.

4. The following division problem has been partially solved.

$$\begin{array}{r} 2 \\ 13\overline{)2645} \\ \underline{26} \\ 04 \end{array}$$

The next step is to ask how many times the divisor (13) will go into the dividend (04). Because it doesn't, the next numeral in the quotient is 0.

$$\begin{array}{r} 20 \\ 13\overline{)2645} \\ \underline{26} \\ 04 \end{array}$$

Next we multiply the 0 in the quotient by the divisor (13) and subtract the product from the dividend.

$$\begin{array}{r} 20 \\ 13\overline{)2645} \\ \underline{26} \\ 04 \\ \underline{00} \\ 4 \end{array}$$

1.6 Exercises

1. What number squared is equal to 16?

3. $2 \cdot 2 \cdot 2 = 2^3$

5. $a \cdot a \cdot a \cdot a \cdot a = a^5$

7. $4 = 4^1$

9. $3 \cdot 3 \cdot 3 \cdot 3 = 3^4$

11. $5 \cdot 5 \cdot a \cdot a \cdot a = 5^2 a^3$

13. $2 \cdot 2 \cdot z \cdot z \cdot z \cdot z \cdot z = 2^2 z^5$

15. $5 \cdot 5 \cdot 5 \cdot y \cdot y \cdot x \cdot x = 5^3 y^2 x^2$

17. $n \cdot n \cdot n \cdot n \cdot n \cdot 9 \cdot 9 = n^5 \cdot 9^2$ or $9^2 n^5$

19. **a.** $7^3 = 7 \cdot 7 \cdot 7$

b. $y^5 = y \cdot y \cdot y \cdot y \cdot y$

21. $2^3 = 2 \cdot 2 \cdot 2 = 8$

23. $5^2 = 5 \cdot 5 = 25$

25. Repeated multiplication of 1 will always equal 1.
$1^6 = 1$

27. $7^2 = 7 \cdot 7 = 49$

29. $4^4 = 4 \cdot 4 \cdot 4 \cdot 4 = 256$

31. $10^1 = 10$

33. $5^3 = 5 \cdot 5 \cdot 5 = 125$

35. 10^6 is a 1 with 6 trailing zeros.
$10^6 = 1,000,000$

37. $x^2 = (5)^2 = 5 \cdot 5 = 25$
When $x = 5$, x^2 is equal to 25.

39. $a^4 = (1)^4 = 1$
When $a = 1$, a^4 is equal to 1.

41. Seven to the third power: 7^3

43. Nine squared: 9^2

45. $3 \cdot 4 - 7 = 12 - 7 = 5$

Copyright © 2012 Pearson Education, Inc.

47. $7^2+5-3=49+5-3=54-3=51$

49. $5\cdot 3^2=5\cdot 9=45$

51. $2\cdot 2^2=2\cdot 4=8$

53. $5^2-7+3=25-7+3=18+3=21$

55. $9+2\cdot 2=9+4=13$

57. $9+(6+2^2)=9+(6+4)=9+10=19$

59. $$\begin{aligned}40\div 5\times 2+3^2&=40\div 5\times 2+9\\&=8\times 2+9\\&=16+9\\&=25\end{aligned}$$

61. $2\times 15\div 5+10=30\div 5+10=6+10=16$

63. $2^2+8\div 4=4+8\div 4=4+2=6$

65. $$\begin{aligned}\frac{(8+4\div 2)}{(5-3)}&=(8+4\div 2)\div(5-3)\\&=(8+2)\div 2\\&=10\div 2=5\end{aligned}$$

67. $$\begin{aligned}\frac{(3+1)}{(12\div 6\times 2)}&=(3+1)\div(12\div 6\times 2)\\&=4\div(2\times 2)\\&=4\div 4\\&=1\end{aligned}$$

69. $$\begin{aligned}7+5(3\cdot 4+7)-2&=7+5(12+7)-2\\&=7+5(19)-2\\&=7+95-2\\&=102-2\\&=100\end{aligned}$$

71. $$\begin{aligned}59-4(1+5\cdot 2)+4&=59-4(1+10)+4\\&=59-4(11)+4\\&=59-44+4\\&=15+4\\&=19\end{aligned}$$

73. $$\begin{aligned}6+2(4\cdot 5+9)-11&=6+2(20+9)-11\\&=6+2(29)-11\\&=6+58-11\\&=64-11\\&=53\end{aligned}$$

75. $$\begin{aligned}32\cdot 6-4(4^3-5\cdot 2^2)+3&=192-4(64-5\cdot 4)+3\\&=192-4(64-20)+3\\&=192-4(44)+3\\&=192-176+3\\&=16+3\\&=19\end{aligned}$$

77. $$\begin{aligned}12\cdot 5-3(3^3-2\cdot 3^2)+1&=12\cdot 5-3(27-2\cdot 9)+1\\&=60-3(27-18)+1\\&=60-3(9)+1\\&=60-27+1\\&=33+1\\&=34\end{aligned}$$

79. He should have multiplied 3 times 2 first and then added 4 to get 10.

81. $21\cdot 10^1=210,\ 21\cdot 10^2=2100$

$21\cdot 10^3=21{,}000,\ 21\cdot 10^4=210{,}000$

The exponent on 10 determines the number of trailing zeros attached to the number.

Cumulative Review

83. $$\begin{array}{r}4079\\+\,2762\\\hline 6841\end{array}$$

84. $$\begin{array}{r}8900\\-\;\;477\\\hline 8423\end{array}$$

85. $$\begin{array}{r}387\\\times\;196\\\hline 2322\\34\,83\;\\38\,7\;\;\;\\\hline 75{,}852\end{array}$$

86. The product of two and some number: $2x$

Quick Quiz 1.6

1. a. $9\cdot 9\cdot 9\cdot x\cdot x=9^3x^2$

b. $5\cdot 5\cdot 5\cdot 5\cdot 5=5^5$

2. a. $2^4=2\cdot 2\cdot 2\cdot 2=16$

b. $1^5=1$

Copyright © 2012 Pearson Education, Inc.

3. $$\begin{aligned}2^2+2(10\div 2)-11&=2^2+2(5)-11\\&=4+2(5)-11\\&=4+10-11\\&=14-11\\&=3\end{aligned}$$

4. Answers may vary. One possible solution follows.
First, evaluate terms with exponents other than 1.
$50+3\times 5^2\div 25=50+3\times 25\div 25$
Next, evaluate division operations.
$50 + 3 \times 25 \div 25 = 50 + 3 \times 1$
Next, evaluate multiplication operations.
$50 + 3 \times 1 = 50 + 3$
Lastly, we evaluate addition operations.
$50 + 3 = 53$

Use Math to Save Money

1. $$\begin{array}{r}200.00\\150.50\\120.25\\50.00\\+\ 25.00\\\hline 545.75\end{array}$$
The total amount of her deposits is \$545.75.

2. $$\begin{array}{r}238.50\\75.00\\200.00\\28.56\\+\ 36.00\\\hline 578.06\end{array}$$
The total amount of her checks is \$578.06.

3. Since \$578.06 > \$545.75, she spent more than she deposited, but the \$300.50 would help her to cover her expenses.

4. 300.50 + 545.75 − 578.06 = 268.19
Assume her balance is \$268.19.

5. Eventually she will be in debt.

6. Answers will vary.

7. Answers will vary.

How Am I Doing? Sections 1.1–1.6

1. 9062 = 9000 + 60 + 2

2. $16 < 22$

3. 17,248,954 = 17,200,000 to the nearest hundred thousand.

4. a. $(6+a)+3=(a+6)+3=a+(6+3)=a+9$

b. $$\begin{aligned}(6+x+4)+2&=(x+6+4)+2\\&=(x+10)+2\\&=x+(10+2)\\&=x+12\end{aligned}$$

5. Replace x with 9 and y with 11.
$x + y = 9 + 11 = 20$

6. $$\begin{array}{r}9\ 532\\251\\+\ \ 322\\\hline 10{,}105\end{array}$$

7. The length of the bottom side is 8 in. + 6 in. or 14 in. The length of the other unlabeled side is 11 in. − 9 in. or 2 in.
9 in. + 8 in. + 2 in. + 6 in. + 11 in. + 14 in.
= 50 in.
The perimeter is 50 inches.

8. Eleven decreased by a number: $11 - x$

9. $$\begin{array}{r}39{,}204\\-\ \ 5{,}982\\\hline 33{,}222\end{array}$$ Check: $$\begin{array}{r}33{,}222\\+\ \ 5{,}982\\\hline 39{,}204\end{array}$$

10. Double a number: $2x$

11. $2(4)(y\cdot 5) = 8(5\cdot y) = (8\cdot 5)y = 40y$

12. $$\begin{array}{r}2371\\\times\ \ 126\\\hline 14\ 226\\47\ 42\ \ \\237\ 1\ \ \ \ \\\hline 298{,}746\end{array}$$

13. 6(12) = 72 rooms

14. The quotient of 144 and x: $144 \div x$

Copyright © 2012 Pearson Education, Inc.

15. $\frac{362,664}{721} = 503 \text{ R } 1$

$$\begin{array}{r} 503 \\ 721\overline{)362664} \\ \underline{3605} \\ 2164 \\ \underline{2163} \\ 1 \end{array}$$

16. $n \cdot n \cdot n \cdot n \cdot 3 \cdot 3 \cdot 3 = n^4 \cdot 3^3 = 3^3 n^4$

17. $4^3 = 4 \cdot 4 \cdot 4 = 64$

18. $2 \cdot 3^2 = 2 \cdot 9 = 18$

19. $$\begin{aligned}(2+10)+12 \div 6 - 3^2 &= (2+10)+12 \div 6 - 9 \\ &= 12 + 12 \div 6 - 9 \\ &= 12 + 2 - 9 \\ &= 14 - 9 \\ &= 5\end{aligned}$$

1.7 Exercises

1. $5(3 + 4) = 5 \cdot 3 + 5 \cdot 4$ represents the distributive property of multiplication over addition.

3. a. $8(3y) = 8 \cdot 3 \cdot 8 \cdot y$ is false because the distributive property is only used when the terms inside the parentheses are being added or subtracted.

b. $8(3 + y) = 8 \cdot 3 + 8y$ is true because the terms inside the parentheses are being added.

5. $2(x + 1) = 2 \cdot \boxed{x} + 2 \cdot \boxed{1}$

7. $6(y - 3) = 6 \cdot \boxed{y} - 6 \cdot \boxed{3}$

9. Six times y plus two: $6y + 2$

11. Seven times four minus one: $7 \cdot 4 - 1$

13. Four times the sum of three and nine: $4(3 + 9)$

15. Triple the sum of y and six: $3(y + 6)$

17. Eight times the difference of four and y: $8(4 - y)$

19. a. Four times two plus seven:
$4 \cdot 2 + 7 = 8 + 7 = 15$

b. Four times the sum of two and seven:
$4(2 + 7) = 4(9) = 36$

21. a. Four times three minus one:
$4 \cdot 3 - 1 = 12 - 1 = 11$

b. Four times the difference of three and one:
$4(3 - 1) = 4(2) = 8$

23. a. Twelve times one plus three:
$12 \cdot 1 + 3 = 12 + 3 = 15$

b. Twelve times the sum of one and three:
$12(1 + 3) = 12(4) = 48$

25. Replace a with 2 and b with 6.
$4a + 5b = 4 \cdot 2 + 5 \cdot 6 = 8 + 30 = 38$

27. Replace x with 9 and y with 2.
$8x - 6y = 8 \cdot 9 - 6 \cdot 2 = 72 - 12 = 60$

29. Replace x with 11.
$$\frac{(x+4)}{3} = \frac{(11+4)}{3} = \frac{15}{3} = 5$$

31. Replace a with 5 and b with 3.
$$\frac{(a^2-4)}{b} = \frac{(5^2-4)}{3} = \frac{(25-4)}{3} = \frac{21}{3} = 7$$

33. Replace x with 2 and y with 2.
$$\frac{(x^3+4)}{y} = \frac{(2^3+4)}{2} = \frac{(8+4)}{2} = \frac{12}{2} = 6$$

35. Replace a with 2 and b with 5.
$$\frac{(a^2+6)}{b} = \frac{(2^2+6)}{5} = \frac{(4+6)}{5} = \frac{10}{5} = 2$$

37. Replace y with 16.
$$\frac{(y-2)}{2} = \frac{(16-2)}{2} = \frac{14}{2} = 7$$

39. Replace m with 2 and n with 7.
$4m + 3n = 4 \cdot 2 + 3 \cdot 7 = 8 + 21 = 29$

41. Replace x with 5 and y with 4.
$$\frac{(x^2-5)}{y} = \frac{(5^2-5)}{4} = \frac{(25-5)}{4} = \frac{20}{4} = 5$$

43. $4(x + 1) = 4 \cdot x + 4 \cdot 1 = 4x + 4$

45. $3(n - 5) = 3 \cdot n - 3 \cdot 5 = 3n - 15$

Copyright © 2012 Pearson Education, Inc.

47. $3(x-6) = 3 \cdot x - 3 \cdot 6 = 3x - 18$

49. $4(x+4) = 4 \cdot x + 4 \cdot 4 = 4x + 16$

51. $$\begin{aligned} 2(x+6)+5 &= 2 \cdot x + 2 \cdot 6 + 5 \\ &= 2x + 12 + 5 \\ &= 2x + 17 \end{aligned}$$

53. $$\begin{aligned} 2(y+1)+5 &= 2 \cdot y + 2 \cdot 1 + 5 \\ &= 2y + 2 + 5 \\ &= 2y + 7 \end{aligned}$$

55. $$\begin{aligned} 4(x+3)+6 &= 4 \cdot x + 4 \cdot 3 + 6 \\ &= 4x + 12 + 6 \\ &= 4x + 18 \end{aligned}$$

57. $$\begin{aligned} 9(y+1)-3 &= 9 \cdot y + 9 \cdot 1 - 3 \\ &= 9y + 9 - 3 \\ &= 9y + 6 \end{aligned}$$

59. $$\begin{aligned} 3(x+1)-1 &= 3 \cdot x + 3 \cdot 1 - 1 \\ &= 3x + 3 - 1 \\ &= 3x + 2 \end{aligned}$$

61. Replace y with 6 and x with 2.
$yx^2 - 3 = 6 \cdot 2^2 - 3 = 6 \cdot 4 - 3 = 24 - 3 = 21$

63. Replace a with 5 and b with 2.
$$\begin{aligned} \frac{(a^2-3)+2^3}{b} &= \frac{(5^2-3)+2^3}{2} \\ &= \frac{(25-3)+8}{2} \\ &= \frac{22+8}{2} \\ &= \frac{30}{2} \\ &= 15 \end{aligned}$$

65. a. $(x+2)+(x+2)+(x+2)+(x+2) = 4x + 8$

b. $4(x+2) = 4x + 4(2) = 4x + 8$

c. The answers are the same.

Cumulative Review

67. $8(2)(x \cdot 4) = 16(4x) = 64x$

68. Replace x with 2.
$4 + x = 4 + 2 = 6$

69. Replace x with 1 and y with 3.
$x + y + 4 = 1 + 3 + 4 = 4 + 4 = 8$

70. $$\begin{array}{r} 2001 \\ -\ 463 \\ \hline 1538 \end{array}$$

Quick Quiz 1.7

1. Double the sum of n and five: $2(n+5)$

2. $$\begin{aligned} 6(y+1)+3 &= 6 \cdot y + 6 \cdot 1 + 3 \\ &= 6y + 6 + 3 \\ &= 6y + 9 \end{aligned}$$

3. a. Replace x with 3 and y with 4.
$3x + 2y = 3 \cdot 3 + 2 \cdot 4 = 9 + 8 = 17$

b. Replace x with 4 and y with 7.
$$\frac{x^2-2}{y} = \frac{(4^2-2)}{7} = \frac{(16-2)}{7} = \frac{14}{7} = 2$$

4. Answers may vary. One possible solution is to first simplify $5(x+1)$ using the distributive property.
$5(x+1) = 5x + 5$
Next, evaluate the simplified expression for $x = 2$, by substituting 2 for x.
$5x + 5 = 5(2) + 5 = 10 + 5 = 15$

1.8 Understanding the Concept Evaluate or Solve?

1. Answers will vary.

1.8 Exercises

1. $7x$: seven times x or the product of seven and x.

3. $8x = 40$: eight times what number equals 40?

5. $2x + 3y$ cannot be added because they are not like terms.

7. In the expression $6x + 5$, $6x$ is called a variable term and 5 is called a constant term.

9. The numerical part of $8x$ is 8 and is called the coefficient of the term.

11. $y = 1y$

13. $7x + 3\boxed{x} = 10x$

15. $3xy + \boxed{4xy} = 7xy$

17. $3x + 5xy + \boxed{4x} = 7x + 5xy$

Copyright © 2012 Pearson Education, Inc.

19. Three x's: $3x$

21. $a + a + a + a = 4a$

23. In $5x + 3y + 2x + 8m + 7y$, $5x$ and $2x$ are like terms; $3y$ and $7y$ are like terms.

25. In $2mn + 3y + 4mn + 6$, $2mn$ and $4mn$ are like terms.

27. $7x + 2x = (7 + 2)x = 9x$

29. $9y - y = 9y - 1y = (9 - 1)y = 8y$

31. $3x + 2x + 6x = (3 + 2 + 6)x = 11x$

33. $8x+4a+3x+a = 8x+3x+4a+1a$
$= (8+3)x+(4+1)a$
$= 11x+5a$

35. $6xy+4b+3xy = 6xy+3xy+4b$
$= (6+3)xy+4b$
$= 9xy+4b$

37. $6xy+3x+9+9xy = 6xy+9xy+3x+9$
$= (6+9)xy+3x+9$
$= 15xy+3x+9$

39. $12ab - 5ab + 9 = (12 - 5)ab + 9 = 7ab + 9$

41. $14xy+4+3xy+6 = 14xy+3xy+4+6$
$= (14+3)xy+(4+6)$
$= 17xy+10$

43. $(4x+7y)+5x+(4x+7y)+5x$
$= (4x+5x+4x+5x)+(7y+7y)$
$= 18x+14y$
The perimeter is $18x + 14y$.

45. $(8a + 2b) + (6a + 5b) + (8a + 2b) + (6a + 5b)$
$= (8a + 6a + 8a + 6a) + (2b + 5b + 2b + 5b)$
$= 28a + 14b$
The perimeter is $28a + 14b$.

47. $x+6y+(5x+2y) = (x+5x)+(6y+2y)$
$= 6x+8y$
The perimeter is $6x + 8y$.

49. Five plus what number equals sixteen?
$5 + x = 16$

51. What number times three equals thirty-six?
$3x = 36$

53. If a number is subtracted from forty-five the result is six.
$45 - x = 6$

55. Twenty-five divided by what number is equal to five?
$\frac{25}{n} = 5$ or $25 \div n = 5$

57. James' age, J, plus 12 years equals 25.
$J + 12 = 25$

59. Chuong's monthly salary, C, decreased by \$50 equals \$1480.
$C - 50 = 1480$

61. Replace the variable with 4.
$9 - x = 3$
$9 - 4 \stackrel{?}{=} 3$
$5 = 3$, false
No, 4 is not a solution.

63. Replace the variable with 15.
$x + 4 = 19$
$15 + 4 \stackrel{?}{=} 19$
$19 = 19$, true
Yes, 19 is a solution.

65. $x + 5 = 9$
What number plus five is equal to nine?
$4 + 5 = 9$
The solution is $x = 4$.
Check: $x + 5 = 9$
$4 + 5 \stackrel{?}{=} 9$
$9 = 9$ ✓

67. $11 - n = 3$
Eleven minus what number is equal to 3?
$11 - 8 = 3$
The solution is $n = 8$.
Check: $11 - n = 3$
$11 - 8 \stackrel{?}{=} 3$
$3 = 3$ ✓

69. $x - 6 = 0$
What number minus 6 is equal to 0?
$6 - 6 = 0$
The solution is $x = 6$.
Check: $x - 6 = 0$
$6 - 6 \stackrel{?}{=} 0$
$0 = 0$ ✓

Copyright © 2012 Pearson Education, Inc.

71. $2 + x = 13$
Two plus what number is equal to 13?
$2 + 11 = 13$
The solution is $x = 11$.
Check: $2+x=13$
$2+11 \stackrel{?}{=} 13$
$13=13$ ✓

73. $25 - x = 20$
Twenty-five minus what number is equal to 20?
$25 - 5 = 20$
The solution is $x = 5$.
Check: $25-x=20$
$25-5 \stackrel{?}{=} 20$
$20=20$ ✓

75. $8x = 16$
Eight times what number equals sixteen?
$8(2) = 16$
The solution is $x = 2$.
Check: $8x=16$
$8(2) \stackrel{?}{=} 16$
$16=16$ ✓

77. $4y = 12$
Four times what number equals twelve?
$4(3) = 12$
The solution is $y = 3$.
Check: $4y=12$
$4(3) \stackrel{?}{=} 12$
$12=12$ ✓

79. $8x = 56$
Eight times what number equals fifty-six?
$8(7) = 56$
The solution is $x = 7$.
Check: $8x=56$
$8(7) \stackrel{?}{=} 56$
$56=56$ ✓

81. $\frac{15}{y}=1$
Fifteen divided by what number is equal to 1?
$\frac{15}{15}=1$
The solution is $y = 15$.
Check: $\frac{15}{y}=1$
$\frac{15}{15} \stackrel{?}{=} 1$
$1=1$ ✓

83. $\frac{14}{x}=2$
Fourteen divided by what number is equal to 2?
$\frac{14}{7}=2$
The solution is $x = 7$.
Check: $\frac{14}{x}=2$
$\frac{14}{7} \stackrel{?}{=} 2$
$2=2$ ✓

85. $(x+1)+3=7$
$x+(1+3)=7$
$x+4=7$
What number plus four is equal to seven?
$3 + 4 = 7$
The solution is $x = 3$.
Check: $(x+1)+3=7$
$(3+1)+3 \stackrel{?}{=} 7$
$4+3 \stackrel{?}{=} 7$
$7=7$ ✓

87. $2+(7+y)=10$
$(2+7)+y=10$
$9+y=10$
Nine plus what number is equal to ten?
$9 + 1 = 10$
The solution is $y = 1$.
Check: $2+(7+y)=10$
$2+(7+1) \stackrel{?}{=} 10$
$2+8 \stackrel{?}{=} 10$
$10=10$ ✓

89. $3+(n+5)=10$
$3+(5+n)=10$
$(3+5)+n=10$
$8+n=10$
Eight plus what number is equal to ten?
$8 + 2 = 10$
The solution is $n = 2$.
Check: $3+(n+5)=10$
$3+(2+5) \stackrel{?}{=} 10$
$3+7 \stackrel{?}{=} 10$
$10=10$ ✓

91. $3n+n=12$
$3n+1n=12$
$(3+1)n=12$
$4n=12$
Four times what number is equal to 12?

Copyright © 2012 Pearson Education, Inc.

$4(3) = 12$
The solution is $n = 3$.
Check: $3n + n = 12$
$3 \cdot 3 + 3 \stackrel{?}{=} 12$
$9 + 3 \stackrel{?}{=} 12$
$12 = 12$ ✓

93. $7x - 2x - x = 4$
$7x - 2x - 1x = 4$
$(7 - 2 - 1)x = 4$
$(5 - 1)x = 4$
$4x = 4$
Four times what number is equal to 4?
$4(1) = 4$
The solution is $x = 1$.
Check: $7x - 2x - x = 4$
$7 \cdot 1 - 2 \cdot 1 - 1 \stackrel{?}{=} 4$
$7 - 2 - 1 \stackrel{?}{=} 4$
$4 = 4$ ✓

95. $5x = 20$
Five times what number equals 20?
$5(4) = 20$
The solution is $x = 4$.
Check: $5x = 20$
$5(4) \stackrel{?}{=} 20$
$20 = 20$ ✓

97. $16 - x = 1$
Sixteen minus what number is equal to one?
$16 - 15 = 1$
The solution is $x = 15$.
Check: $16 - x = 1$
$16 - 15 \stackrel{?}{=} 1$
$1 = 1$ ✓

99. $1 + (4 + a) = 15$
$(1 + 4) + a = 15$
$5 + a = 15$
Five plus what number is equal to 15?
$5 + 10 = 15$
The solution is $a = 10$.
Check: $1 + (4 + a) = 15$
$1 + (4 + 10) \stackrel{?}{=} 15$
$1 + 14 \stackrel{?}{=} 15$
$15 = 15$ ✓

101. $8x - 5x - x = 10$
$8x - 5x - 1x = 10$
$(8 - 5 - 1)x = 10$
$2x = 10$
Two times what number is equal to ten?
$2(5) = 10$
The solution is $x = 5$.
Check: $8x - 5x - x = 10$
$8(5) - 5(5) - 5 \stackrel{?}{=} 10$
$40 - 25 - 5 \stackrel{?}{=} 10$
$10 = 10$ ✓

103. Four plus what number equals eight?

a. $4 + x = 8$

b. $4 + 4 = 8$
The solution is $x = 4$.

105. Three times what number is equal to nine?

a. $3x = 9$

b. $3(3) = 9$
The solution is $x = 3$.

107. $50 + 50 + x = 170$
$100 + x = 170$
One hundred plus what number is equal to 170?
$100 + 70 = 170$
The solution is $x = 70$.
The length of the missing side is 70 feet.

109. $6 + (2x^2 + 5) + (7 + 3x^2) + x^2$
$= (2x^2 + 3x^2 + x^2) + (6 + 5 + 7)$
$= (2 + 3 + 1)x^2 + 18$
$= 6x^2 + 18$

111. a. $2x + 3x + 5y = (2 + 3)x + 5y = 5x + 5y$

b. $(2x)(5y) = (2 \cdot 5)(x \cdot y) = 10xy$

113. a. $5a + 6y + 2a = (5 + 2)a + 6y = 7a + 6y$

b. $(5a)(6y) = (5 \cdot 6)(a \cdot y) = 30ay$

115. a. From the graph, we see that a grizzly bear can run 30 miles per hour, so a domestic cat can also run 30 miles per hour.

b. From the graph, we see that a lion's speed is 50 miles per hour. Since this is twice the speed of an elephant, an elephant's speed is 25 miles per hour.

Cumulative Review

117. "Split equally between" describes division. The answer is (d).

Copyright © 2012 Pearson Education, Inc.

118. "Find the number of items in an array" describes multiplication. The answer is (c).

119. "Find the total" describes addition. The answer is (a).

120. "How much less" describes subtraction. The answer is (b).

Quick Quiz 1.8

1. $$\begin{aligned}2ab+4a+1+ab &= (2ab+ab)+4a+1\\ &= (2ab+1ab)+4a+1\\ &= (2+1)ab+4a+1\\ &= 3ab+4a+1\end{aligned}$$

2. a. $\frac{10}{a}=2$

Ten divided by what number is equal to two?

$\frac{10}{5}=2$

The solution is $a = 5$.

Check: $$\begin{aligned}\frac{10}{a}&=2\\ \frac{10}{5}&\stackrel{?}{=}2\\ 2&=2\ \checkmark\end{aligned}$$

b. $$\begin{aligned}4+(x+7)&=12\\ 4+(7+x)&=12\\ (4+7)+x&=12\\ 11+x&=12\end{aligned}$$

Eleven plus what number is equal to twelve?

$11 + 1 = 12$

The solution is $x = 1$.

Check: $$\begin{aligned}4+(x+7)&=12\\ 4+(1+7)&\stackrel{?}{=}12\\ 4+8&\stackrel{?}{=}12\\ 12&=12\ \checkmark\end{aligned}$$

c. $$\begin{aligned}8y+y&=72\\ 8y+1y&=72\\ (8+1)y&=72\\ 9y&=72\end{aligned}$$

Nine times what number equals seventy-two?

$9(8) = 72$

The solution is $y = 8$.

Check: $$\begin{aligned}8y+y&=72\\ 8\cdot 8+8&\stackrel{?}{=}72\\ 64+8&\stackrel{?}{=}72\\ 72&=72\ \checkmark\end{aligned}$$

3. a. The product of three and what number is equal to eighteen?

$3x = 18$

Three times what number equals eighteen?

$3(6) = 18$

The solution is $x = 6$.

b. Dave's age, D, increased by seven is equal to twenty-one.

$D + 7 = 21$

What number plus seven equals twenty-one?

$14 + 7 = 21$

The solution is $D = 14$.

Dave is 14 years old.

4. Answers may vary. One possible solution follows.

The first step in both processes is to combine like terms.

$3x + x + 2x = 3x + 1x + 2x = (3 + 1 + 2)x = 6x$

At this point the process for (a) is complete. Because (b) has an equals sign, it can be solved by isolating the x. To isolate, and therefore solve for x, divide both sides of the equation by the coefficient of x.

$$\begin{aligned}6x&=12\\ \frac{x}{6}&=\frac{12}{6}\\ x&=2\end{aligned}$$

1.9 Exercises

1. a. Rounded to the nearest ten, the costs are \$80, \$40, \$20, and \$10.

\$80 + \$40 + \$20 + \$10 = \$150

b. \$81 + \$36 + \$22 + \$14 = \$153

Emma spent \$153.

c. Yes, \$150 is a reasonable estimate for \$153.

3. Day 1: 597 miles rounds to 600 miles.

Day 2: 512 miles rounds to 500 miles.

Day 3: 389 miles rounds to 400 miles.

Day 4: 310 miles rounds to 300 miles.

600 mi + 500 mi = 1100 mi

They drove about 1100 miles the first two days.

400 mi + 300 mi = 700 mi

They drove about 700 miles the last two days.

Copyright © 2012 Pearson Education, Inc.

1100 mi − 700 mi = 400 mi
the Arismendi family drove about 400 more miles the first two days than the last two days.

5. Find the total cost.

7 tables = 7(475) = $3325
20 chairs = 20(65) = $1300
2 light fixtures = 2(650) = $1300
Total: $5925

Divide the total cost by the number of owners.
5925 ÷ 5 = 1185
Each owner paid $1185.

7. 2 adult tickets = 2(13) = $26
4 child tickets = 4(5) = $20
1 senior ticket = 1(7) = $ 7
Total: $53

Dave and his family spent $53 on tickets.

9. 20 + 25 + 80 + 150 + 80 + 25 + 20 = 400 feet
John must purchase 400 feet of fence.

11. a. 6 tagged players = 6(3) = 18 points
pulling the flag = 22 points
hanging the flag = 50 points
5 players left = 5(1) = 5 points
Total: 95 points

The Torches had 95 points at the end of the match.

b. 7 tagged players = 7(3) = 21 points
pulling the flag = 22 points
1 player left = 1(1) = 1 point
Total: 44 points

13. a.

Gather the facts	What am I asked to do?	How do I proceed?	Key points to remember
Cook earns: $8 per hour for 40 hr; $12 per hour for overtime Hours worked = 52	Calculate the cook's total pay.	**1.** Multiply $8 × 40 to find base pay. **2.** Multiply $12 × 12 to find overtime pay. **3.** Add results from steps 1 and 2.	Overtime hours are hours worked in addition to 40 hours per week.

b. $8×40 = $320
$12×12 = $144
Total: $464

The cook's total pay was $464.

Copyright © 2012 Pearson Education, Inc.

15. a. Subtract the expenses from the amount made to find the profit.

$$\begin{array}{r} 68,542 \\ -\ 14,372 \\ \hline 54,170 \end{array}$$

The company's profit was \$54,170.

b. Divide the profit by 2.
$54,170 \div 2 = 27,085$
Each owner received \$27,085.

17. Find the total expenses.

$$\begin{array}{rr} \text{boat rental} = 2(450) = & \$900 \\ \text{gasoline} = 2(50) = & \$100 \\ \text{food and bait} = & \underline{\$200} \\ \text{Total:} & \$1200 \end{array}$$

Divide the total by the number of people.
$1200 \div 4 = 300$
Each person's share will be \$300.

19. Justin rides the bus $5 + 2 = 7$ times per week. There are about 4 weeks per month, so Justin rides the bus about $7(4)(6) = 168$ times in 6 months. At \$2 per trip, his total would be $2(168) = \$336$. Since $\$336 < \$400$, it is cheaper for Justin to pay each time than to buy the six-month pass.

21. a. Divide the total inheritance by 2.
$6000 \div 2 = 3000$
Glenda will invest \$3000 in a CD.

b. The second part of the inheritance will also be \$3000. Divide this amount by the number of children.
$3000 \div 3 = 1000$
Each child will receive \$1000.

23. a. Find the total of Jesse's purchases.
$30 + 240 + 170 = 440$
Divide the total by 50.
$440 \div 50 = 8 \text{ R } 40$
Jesse earned $8(5) = 40$ points from total purchases. He earned an additional 25 points for having one purchase over \$200.
$40 + 25 = 65$
Jesse earned 65 points in June.

b. Divide the number of points by 10.
$65 \div 10 = 6 \text{ R } 5$
Jesse earned 6 discount dollars.

25. a. Find the total of Alyssa's purchases.
$170 + 260 = 430$
Divide the total by 50.
$430 \div 50 = 8 \text{ R } 30$
Alyssa earned $8(10) = 80$ points from total purchases. She earned an additional 50 points for having one purchase over \$200.
$80 + 50 = 130$
Alyssa earned 130 points in January.

b. Divide the number of points by 25.
$130 \div 25 = 5 \text{ R } 5$
Alyssa earned five \$5 discounts for a total of \$25 in discounts.

27. Find the total payments for 3 years at \$175 per month.
$3(12)(\$175) = \$6300$
Subtract the original debt from the total of the payments.
$\$6300 - \$5000 = \$1300$
The interest paid at the end of three years is \$1300.

29. 9, 25, 49, 81, 121, ...
Write the numbers in exponent form.
$3^2, 5^2, 7^2, 9^2, 11^2, \ldots$
The sequence consists of the squares of consecutive odd numbers. The next odd number is 13, so the next number in the sequence is $13^2 = 169$.

31. The sequence alternates between squares and circles, so the next figure is a square and one after that is a circle. The pattern of dots is rotated in each figure. In the next two figures, the dots will be back to their original positions. the next two figures are identical to the first two figures in the sequence.

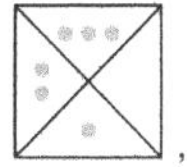, 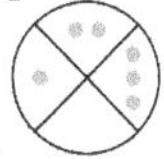

Cumulative Review

32. $4 \cdot 3 \cdot 2 \cdot 5 = 12 \cdot 10 = 120$

33. $6x = 30$
Six times what number equals 30?
$6(5) = 30$
The solution is $x = 5$.

Copyright © 2012 Pearson Education, Inc.

34. $x + 9 = 12$
What number plus 9 equals 12?
$3 + 9 = 12$
The solution is $x = 3$.

Quick Quiz 1.9

1. a.

350 adults = 350(20) =	\$7000
200 students = 200(15) =	\$3000
47 children = 47(7) =	\$329
Total:	\$10,329

The total income from tickets was \$10,329.

b. Subtract the expenses from the income.

$$\begin{array}{r} 10,329 \\ -\ 3,400 \\ \hline 6,929 \end{array}$$

The profit for the event was \$6929.

2. Multiply the number of cows by the amount of milk each cow produces per day.

$$\begin{array}{r} 35 \\ \times\ 7 \\ \hline 245 \end{array}$$

Multiply this amount by the number of days in a week.

$$\begin{array}{r} 245 \\ \times\ 7 \\ \hline 1715 \end{array}$$

The herd produces 245 gallons of milk in one day and 1715 gallons in one week.

3.

2 filing cabinets:	2(500) =	1000
1 desk:		800
1 office chair:		300
2 bookcases:	2(300) =	600
6 guest chairs:	6(200) =	1200
2 end tables:	2(300) =	600
1 coffee table:		400
	Total:	\$4900

The price of the furniture is about \$4900.

4. Answers may vary. One possible solution is to list all deposits to her vacation account over the six months.

Description of deposit	deposit amount
Initial balance	\$200
Monthly deposit × months $\$100 \times 6 = \$600$	\$600
Tax return deposit ÷ 2 $\$900 \div 2 = \$450$	\$450
Sum in vacation account	\$1250

Sahara will have \$1250 after six months in her vacation account. She will not have enough money to take a \$1500 vacation.

You Try It

1. In words, 23,327,414 is written as twenty-three million, three hundred twenty-seven thousand, four hundred fourteen.

2. 2 ? 11
2 is less than 11.
$2 < 11$

17 ? 13
17 is greater than 13.
$17 > 13$

3. 133,442
Identify the round-off place digit: 13<u>3</u>,442.
The digit to the right is less than 5. Do not change the round-off place digit. Replace all digits to the right with zeros.
133,000

4. $(x + 6) + 8 = x + (6 + 8) = x + 14$

5.

$$\begin{array}{r} 121 \\ 46 \\ 592 \\ +\ 3 \\ \hline 762 \end{array}$$

6. unlabeled vertical side: 9 m − 2 m = 7 m
unlabeled horizontal side: 18 m − 6 m = 12 m
$6 + 2 + 12 + 7 + 18 + 9 = 54$ m
The perimeter is 54 meters.

7.

$$\begin{array}{r} 47,621 \\ -\ 5,935 \\ \hline 41,686 \end{array}$$

8. a. Twice a number: $2n$

b. Five times a number: $5n$

c. A number times eight: $x \cdot 8$

d. The product of four and two: $4 \cdot 2$

9. $4(y \cdot 5) = 4(5 \cdot y) = (4 \cdot 5)y = 20y$

Copyright © 2012 Pearson Education, Inc.

10.
$$\begin{array}{r} 468 \\ \times\ 251 \\ \hline 468 \\ 2340 \\ 936 \\ \hline 117,468 \end{array}$$

11. a. The quotient of six and x: $6 \div x$

b. The quotient of x and six: $x \div 6$

c. A number divided by 3: $n \div 3$

12. $988 \div 21 = 47 \text{ R } 1$
$$\begin{array}{r} 47 \\ 21\overline{)988} \\ 84 \\ \hline 148 \\ 147 \\ \hline 1 \end{array}$$

13. a. $8 \cdot 8 \cdot 8 \cdot n \cdot n = 8^3 n^2$

b. $2^4 = 2 \cdot 2 \cdot 2 \cdot 2 = 16$

14.
$$\begin{aligned} 4+8 \div 2^2 \cdot 5 - 3^2 &= 4+8 \div 4 \cdot 5 - 9 \\ &= 4+2 \cdot 5 - 9 \\ &= 4+10-9 \\ &= 14-9 \\ &= 5 \end{aligned}$$

15. a. Four times the difference of x and 5: $4(x - 5)$

b. Four times x minus five: $4x - 5$

16. $7(n - 3) = 7 \cdot n - 7 \cdot 3 = 7n - 21$

17. $4mn + 2n + 6mn = 10mn + 2n$

18. a.
$$\begin{aligned} 4n &= 24 \\ 4 \cdot 6 &= 24 \\ n &= 6 \end{aligned}$$

b.
$$\begin{aligned} \frac{35}{x} &= 7 \\ \frac{35}{5} &= 7 \\ x &= 5 \end{aligned}$$

19.
$$\begin{aligned} 3+(x+2) &= 15 \\ 3+(2+x) &= 15 \\ (3+2)+x &= 15 \\ 5+x &= 15 \\ 5+10 &= 15 \\ x &= 10 \end{aligned}$$

20.
$$\begin{aligned} 10-x &= 2 \\ 10-8 &= 2 \\ x &= 8 \end{aligned}$$

21. a. Replace x with 3 and y with 2.
$5x + 3y = 5(3) + 3(2) = 15 + 6 = 21$
When $x = 3$ and $y = 2$, $5x + 3y = 21$.

b. Replace x with 10.
$$\frac{(x-4)}{3} = \frac{(10-4)}{3} = \frac{6}{3} = 2$$
When $x = 10$, $\frac{x-4}{3} = 2$.

22. \$2499 rounds to \$2500.
\$2130 rounds to \$2100.
\$2500 − \$2100 = \$400
Sara saved approximately \$400.

Chapter 1 Review Problems

1. A rectangle is a four-sided figure with adjoining sides that are perpendicular and opposite sides that are equal.
2. A square is a rectangle with all sides equal.
3. A right angle is an angle that measures 90°.
4. A triangle is a three-sided figure with three angles.
5. The perimeter is the distance around an object.
6. Factors are the numbers or variables that we multiply.
7. A term is a number, a variable, or a product of a number and one or more variables.
8. A constant term is a term that has no variable.
9. The coefficient is the number factor in a term.
10. Like terms are terms with identical variable parts.

Copyright © 2012 Pearson Education, Inc.

11. An equation is two expressions separated by an equals sign.

12. a. In the number 175,493, the digit 7 is in the ten thousands place.

b. In the number 175,493, the digit 5 is in the thousands place.

13. \$187
Write 187.00 in the box following \$. Write "One hundred eighty-seven and 00/100" on the line preceding DOLLARS.

14. $7694 = 7000 + 600 + 90 + 4$

15. $5831 = 5000 + 800 + 30 + 1$

16. 2 ? 8
2 is less than 8.
$2 < 8$

17. 12 ? 0
12 is greater than 0.
$12 > 0$

18. Six is greater than one: $6 > 1$

19. Three is less than five: $3 < 5$

20. 61,269
Identify the round-off place digit: 61,2̲69.
The digit to the right is 5 or more. Increase the round-off place digit by 1. Replace all digits to the right with zeros.
61,300

21. 382,240
Identify the round-off place digit: 382,2̲40.
The digit to the right is less than 5. Do not change the round off place digit. Replace all digits to the right with zeros.
382,200

22. 6,365,534
Identify the round-off place digit: 6,3̲65,534.
The digit to the right is 5 or more. Increase the round-off place digit by 1. Replace all digits to the right with zeros.
6,400,000

23. 8,118,701
Identify the round-off place digit: 8,1̲18,701.
The digit to the right is less than 5. Do not change the round-off place digit. Replace all digits to the right with zeros.
8,100,000

24. Seven more than a number: $x + 7$

25. The sum of some number and five: $n + 5$

26. $7 + (9 + x) = (7 + 9) + x = 16 + x$

27. $(2 + n) + 9 = (n + 2) + 9 = n + (2 + 9) = n + 11$

28.
$$\begin{aligned} 5+(n+2) &= (n+2)+5 \\ &= n+(2+5) \\ &= n+7 \text{ or } 7+n \end{aligned}$$

29.
$$\begin{aligned} (5+x+3)+2 &= (x+5+3)+2 \\ &= (x+8)+2 \\ &= x+10 \end{aligned}$$

30.
$$\begin{array}{r} 8398 \\ 372 \\ +\ 255 \\ \hline 9025 \end{array}$$

31.
$$\begin{array}{r} 17,456 \\ 213 \\ +\ \ 982 \\ \hline 18,651 \end{array}$$

32.
$$\begin{array}{r} 1434 \\ 1596 \\ 1423 \\ +\,1565 \\ \hline 6018 \end{array}$$
A total of 6018 students attend the college.

33. The length of the right side of the figure is $8 + 5 = 13$ meters, and the length of the bottom is $13 + 7 = 20$ meters.
8 m + 13 m + 5 m + 7 m + 13 m + 20 m = 66 m
The perimeter is 66 meters.

34. Eight decreased by a number: $8 - n$

35. The difference of a number and six: $n - 6$

36. Ten subtracted from a number: $x - 10$

37. Replace x with 3.
$8 - x = 8 - 3 = 5$
If x is equal to 3, then $8 - x$ is equal to 5.

38. Replace y with 15.
$y - 9 = 15 - 9 = 6$
If y is equal to 15, then $y - 9$ is equal to 6.

Copyright © 2012 Pearson Education, Inc.

39. $\begin{array}{r} 8502 \\ -\ 2957 \\ \hline 5545 \end{array}$ Check: $\begin{array}{r} 2957 \\ +\ 5545 \\ \hline 8502 \end{array}$

40. $\begin{array}{r} 9021 \\ -\ 5862 \\ \hline 3159 \end{array}$ Check: $\begin{array}{r} 5862 \\ +\ 3159 \\ \hline 9021 \end{array}$

41. $\begin{array}{r} 29,104 \\ -\ \ 4,988 \\ \hline 24,116 \end{array}$ Check: $\begin{array}{r} 24,116 \\ +\ \ 4,988 \\ \hline 29,104 \end{array}$

42. The player won 4900,000 in 2009 and \$522,000 in 2006.
$\begin{array}{r} 900,000 \\ -\ 522,000 \\ \hline 378,000 \end{array}$
The player won \$378,000 more in 2009.

43. The player won \$450,000 in 2005 and \$720,000 in 2008.
$\begin{array}{r} 720,000 \\ -\ 450,000 \\ \hline 270,000 \end{array}$
The player won \$270,000 less in 2005.

44. $4x = 32$
The factors are 4 and x.

45. Triple a number: $3x$

46. $7y = 63$
Seven times what number equals 63?

47. $7 \cdot 2 \cdot 3 \cdot 0 = 0$

48. $5 \cdot 3 \cdot 2 \cdot 2 = (5 \cdot 2) \cdot (3 \cdot 2) = 10 \cdot 6 = 60$

49. $6(y \cdot 7) = 6(7y) = (6 \cdot 7)y = 42y$

50. $3(5)(x \cdot 2) = 15(2x) = (15 \cdot 2)x = 30x$

51. $3(2)(x \cdot 4) = 6(4x) = (6 \cdot 4)x = 24x$

52. $\begin{array}{r} 416 \\ \times\ \ 2000 \\ \hline 832,000 \end{array}$

53. $\begin{array}{r} 4251 \\ \times\ \ \ \ 352 \\ \hline 8\ 502 \\ 212\ 55 \\ 1\ 275\ 3 \\ \hline 1,496,352 \end{array}$

54. $\begin{array}{r} 6424 \\ \times\ \ \ \ 903 \\ \hline 19\ 272 \\ 5\ 781\ 60 \\ \hline 5,800,872 \end{array}$

55. $\begin{array}{r} 17 \\ \times 18 \\ \hline 136 \\ 17 \\ \hline 306 \end{array}$
Lisa can travel 306 miles.

56. There are $6 \times 21 = 126$ apartments, so there are $4 \times 126 = 504$ doors.

57. There are $300 \div 20$ rows.

58. Each person will receive $500 \div n$.

59. Five divided by a number: $5 \div y$

60. The quotient of a number and thirteen: $n \div 13$

61. $10 \div 0$ undefined

62. $33 \div 33 = 1$

63. $1456 \div 29 = 50 \text{ R } 6$
$\begin{array}{r} 50 \\ 29\overline{)1456} \\ \underline{145} \\ 06 \\ \underline{0} \\ 6 \end{array}$

64. $369,757 \div 922 = 401 \text{ R } 35$
$\begin{array}{r} 401 \\ 922\overline{)369757} \\ \underline{3688} \\ 957 \\ \underline{922} \\ 35 \end{array}$

Copyright © 2012 Pearson Education, Inc.

65. $\frac{510,144}{846} = 603 \text{ R } 6$

$$\begin{array}{r} 603 \\ 846\overline{)510144} \\ \underline{5076} \\ 2544 \\ \underline{2538} \\ 6 \end{array}$$

66.
$$\begin{array}{r} 111 \\ 4\overline{)447} \\ \underline{4} \\ 04 \\ \underline{4} \\ 07 \\ \underline{4} \\ 3 \end{array}$$

The remainder is 3. The club deposited \$3.

67.
$$\begin{array}{r} 147 \\ 24\overline{)3528} \\ \underline{24} \\ 112 \\ \underline{96} \\ 168 \\ \underline{168} \\ 0 \end{array}$$

The payments will be \$147.

68. $2 \cdot 2 \cdot 2 \cdot n \cdot n = 2^3 n^2$

69. $z \cdot z \cdot z \cdot z \cdot 5 \cdot 5 \cdot 5 = z^4 \cdot 5^3$ or $5^3 z^4$

70. $x^3 = x \cdot x \cdot x$

71. $6^5 = 6 \cdot 6 \cdot 6 \cdot 6 \cdot 6$

72. $10^3 = 10 \cdot 10 \cdot 10 = 1000$

73. $2^4 = 2 \cdot 2 \cdot 2 \cdot 2 = 16$

74. Six cubed: 6^3

75. x to the fifth power: x^5

76.
$$\begin{aligned} 6 + 24 \div 8 - 2^2 &= 6 + 24 \div 8 - 4 \\ &= 6 + 3 - 4 \\ &= 9 - 4 \\ &= 5 \end{aligned}$$

77. $(15 + 25 \div 5) \div (8 - 4) = (15 + 5) \div 4 = 20 \div 4 = 5$

78. $5 \cdot 2^2 = 5 \cdot 4 = 20$

79. **a.** Three times x plus two: $3x + 2$

b. Three times the sum of x and two: $3(x + 2)$

80. **a.** Four times x minus five: $4x - 5$

b. Four times the difference of x and five: $4(x - 5)$

81. **a.** Three times seven plus one: $3 \cdot 7 + 1 = 21 + 1 = 22$

b. Three times the sum of seven and one: $3(7 + 1) = 3(8) = 24$

82. Replace x with 3 and y with 2.

$$\frac{x^3 - 1}{y} = \frac{3^3 - 1}{2} = \frac{27 - 1}{2} = \frac{26}{2} = 13$$

If x is equal to 3 and y is equal to 2, $\frac{x^3 - 1}{y}$ is equal to 13.

83. Replace m with 8 and n with 2.
$2m + 3n = 2 \cdot 8 + 3 \cdot 2 = 16 + 6 = 22$
If m is equal to 8 and n is equal to 2, $2m + 3n$ is equal to 22.

84. $5(x + 1) = 5x + 5(1) = 5x + 5$

85. $4(x - 1) = 4x - 4(1) = 4x - 4$

86. $3(x + 1) + 5 = 3 \cdot x + 3 \cdot 1 + 5 = 3x + 3 + 5 = 3x + 8$

87. $2x + x + 6x = 2x + 1x + 6x = (2 + 1 + 6)x = 9x$

88.
$$\begin{aligned} 5x + 6y + 6x &= (5x + 6x) + 6y \\ &= (5 + 6)x + 6y \\ &= 11x + 6y \end{aligned}$$

89.
$$\begin{aligned} 3xy + 5y + 2xy + 8y &= (3xy + 2xy) + (5y + 8y) \\ &= (3 + 2)xy + (5 + 8)y \\ &= 5xy + 13y \end{aligned}$$

 Copyright © 2012 Pearson Education, Inc.

90. $(2x+4y)+(3x+y)+(2x+4y)+(3x+y)$
$=(2x+3x+2x+3x)+(4y+y+4y+y)$
$=10x+10y$
The perimeter is $10x + 10y$.

91. $x + 2 = 9$
What number plus two is equal to nine?
$7 + 2 = 9$
The solution is $x = 7$.
Check: $x+7=9$
$2+7 \stackrel{?}{=} 9$
$9=9$ ✓

92. $10 - n = 6$
Ten minus what number is equal to six?
$10 - 4 = 6$
The solution is $n = 4$.
Check: $10-n=6$
$10-4 \stackrel{?}{=} 6$
$6=6$ ✓

93. $(3+x)+1=8$
$(x+3)+1=8$
$x+(3+1)=8$
$x+4=8$
What number plus four is equal to eight?
$4 + 4 = 8$
The solution is $x = 4$.
Check: $(3+x)+1=8$
$(3+4)+1 \stackrel{?}{=} 8$
$7+1 \stackrel{?}{=} 8$
$8=8$ ✓

94. $2+(n+7)=10$
$2+(7+n)=10$
$(2+7)+n=10$
$9+n=10$
Nine plus what number is equal to ten?
$9 + 1 = 10$
The solution is $n = 1$.
Check: $2+(n+7)=10$
$2+(1+7) \stackrel{?}{=} 10$
$2+8 \stackrel{?}{=} 10$
$10=10$ ✓

95. $9x = 27$
Nine times what number is equal to 27?
$9(3) = 27$
The solution is $x = 3$.
Check: $9x=27$
$9 \cdot 3 \stackrel{?}{=} 27$
$27=27$ ✓

96. $\frac{15}{x}=5$
Fifteen divided by what number is equal to five?
$15 \div 3 = 5$
The solution is $x = 3$.
Check: $\frac{15}{x}=5$
$\frac{15}{3} \stackrel{?}{=} 5$
$5=5$ ✓

97. $12n-n=22$
$12n-1n=22$
$(12-1)n=22$
$11n=22$
Eleven times what number equals 22?
$11(2) = 22$
The solution is $n = 2$.
Check: $12n-n=22$
$12 \cdot 2-2 \stackrel{?}{=} 22$
$24-2 \stackrel{?}{=} 22$
$22=22$ ✓

98. $y+3y+2y=12$
$1y+3y+2y=12$
$(1+3+2)y=12$
$6y=12$
Six times what number is equal to 12?
$6(2) = 12$
The solution is $y = 2$.

Check: $y+3y+2y=12$
$2+3 \cdot 2+2 \cdot 2 \stackrel{?}{=} 12$
$2+6+4 \stackrel{?}{=} 12$
$12=12$ ✓

99. What number subtracted from eighteen equals three?

a. $18 - x = 3$

b. $18 - 15 = 3$
The solution is $x = 15$.

100. What number increased by five equals eleven?

a. $x + 5 = 11$

b. $6 + 5 = 11$
The solution is $x = 6$.

Copyright © 2012 Pearson Education, Inc.

101. Triple what number is equal to twelve?

a. $3 \cdot x = 12$

b. $3 \cdot 4 = 12$
The solution is $x = 4$.

102. Rounded to the nearest ten, the costs are \$30, \$30, \$90, and \$160.
\$30 + \$30 + \$90 + \$160 = \$310
Joseph will pay about \$310.

103. Find the total deductions.

$$\begin{array}{r} 499 \\ 218 \\ +\ 97 \\ \hline 814 \end{array}$$

Subtract the amount of the deductions from the salary.

$$\begin{array}{r} 3560 \\ -\ 814 \\ \hline 2746 \end{array}$$

The check was \$2746 after deductions.

104. a. Balance & Deposits

$$\begin{array}{r} 5021 \\ 759 \\ 2534 \\ +\ 532 \\ \hline 8846 \end{array}$$

Withdrawals

$$\begin{array}{r} 799 \\ 533 \\ +\ 88 \\ \hline 1420 \end{array}$$

Subtract the withdrawals from the total of the balance and deposits.

$$\begin{array}{r} 8846 \\ -1420 \\ \hline 7426 \end{array}$$

Her ending balance was \$7426.

b. $7426 \div 2 = 3713$
Jean will have \$3713 in each account.

105. The perimeter of the living room is $20 + 25 + 20 + 25 = 90$ feet, and the perimeter of the dining room is $15 + 18 + 15 + 18 = 66$ feet. Ruth Ann needs to purchase a total of $90 + 66$ or 156 feet of crown molding. At \$3 per foot, the total cost is \$3 × 156 or \$468.

How Am I Doing? Chapter 1 Test

1. $1525 = 1000 + 500 + 20 + 5$

2. a. 7 ? 2
7 is greater than 2.
$7 > 2$

b. 5 ? 0
5 is greater than 0.
$5 > 0$

3. 2925

a. Identify the round-off place digit: $\underline{2}925$.
The digit to the right is 5 or more. Increase the round-off place digit by 1. Replace all digits to the right with zeros.
3000

b. Identify the round-off place digit: $2\underline{9}25$.
The digit to the right is less than 5. Do not change the round-off place digit. Replace all digits to the right with zeros.
2900

4. a.

$$\begin{aligned} 3+(8+x) &= (8+x)+3 \\ &= (x+8)+3 \\ &= x+(8+3) \\ &= x+11 \end{aligned}$$

b. $5 + y + 2 = y + 5 + 2 = y + 7$

c.

$$\begin{aligned} 1+(n+2)+4 &= (n+2)+1+4 \\ &= (n+2)+5 \\ &= n+(2+5) \\ &= n+7 \end{aligned}$$

5.

$$\begin{array}{r} 12,389 \\ 4 \\ +\ 2,302 \\ \hline 14,695 \end{array}$$

6.

$$\begin{array}{r} 244,869,201 \\ +\qquad 19,077 \\ \hline 244,888,278 \end{array}$$

7. a.

$$\begin{array}{r} 613 \\ -\ 75 \\ \hline 538 \end{array}$$

b.

$$\begin{array}{r} 20,105 \\ -\ 7,826 \\ \hline 12,279 \end{array}$$

8. The length of the unlabeled top side is $9 - 7 = 2$ feet, and the length of the right side of the figure is $6 - 1 = 5$ feet.
6 ft + 2 ft + 1 ft + 7 ft + 5 ft + 9 ft = 30 ft
The perimeter is 30 feet.

9. $2(4)(y \cdot 2) = 8(2y) = (8 \cdot 2)y = 16y$

Copyright © 2012 Pearson Education, Inc.

10. a.

$$\begin{array}{r} 432 \\ \times\ 312 \\ \hline 864 \\ 4\,32 \\ 129\,6 \\ \hline 134{,}784 \end{array}$$

b.

$$\begin{array}{r} 2031 \\ \times\ 129 \\ \hline 18\,279 \\ 40\,62 \\ 203\,1 \\ \hline 261{,}999 \end{array}$$

11. a. $492 \div 12 = 41$

$$\begin{array}{r} 41 \\ 12\overline{)492} \\ \underline{48} \\ 12 \\ \underline{12} \\ 0 \end{array}$$

b. $5523 \div 46 = 120 \text{ R } 3$

$$\begin{array}{r} 120 \\ 46\overline{)5523} \\ \underline{46} \\ 92 \\ \underline{92} \\ 03 \\ \underline{0} \\ 3 \end{array}$$

12. a. Seven subtracted from a number: $n - 7$

b. The product of ten and a number: $10n$

c. y to the fourth power: y^4

d. 7 cubed: 7^3

e. Six times the sum of x and nine: $6(x + 9)$

13. a.

$$\begin{aligned} 3xy + 2y + 4xy - 2 &= (3xy + 4xy) + 2y - 2 \\ &= (3 + 4)xy + 2y - 2 \\ &= 7xy + 2y - 2 \end{aligned}$$

b.

$$\begin{aligned} 2m + 5 + m + 6mn &= (2m + m) + 5 + 6mn \\ &= (2m + 1m) + 5 + 6mn \\ &= (2 + 1)m + 5 + 6mn \\ &= 3m + 5 + 6mn \end{aligned}$$

14. $3(y + 4) = 3 \cdot y + 3 \cdot 4 = 3y + 12$

15.

$$\begin{aligned} 8(x + 1) + 2 &= 8 \cdot x + 8 \cdot 1 + 2 \\ &= 8x + 8 + 2 \\ &= 8x + 10 \end{aligned}$$

16. a. Replace x with 16 and y with 4.
$2x - 3y = 2 \cdot 16 - 3 \cdot 4 = 32 - 12 = 20$
If x is equal to 16 and y is equal to 4, $2x - 3y$ is equal to 20.

b. Replace a with 9 and b with 7.

$$\frac{a^2 - 4}{b} = \frac{9^2 - 4}{7} = \frac{81 - 4}{7} = \frac{77}{7} = 11$$

If $a = 9$ and $b = 7$, then $\frac{a^2 - 4}{b} = 11$.

17. $6 \cdot 6 \cdot 6 \cdot 6 \cdot 6 \cdot n \cdot n \cdot n = 6^5 n^3$

18. a. $5^3 = 5 \cdot 5 \cdot 5 = 125$

b. $10^5 = 10 \cdot 10 \cdot 10 \cdot 10 \cdot 10 = 100{,}000$

19. $24 \div 4 - 2 \cdot 3 = 6 - 2 \cdot 3 = 6 - 6 = 0$

20.

$$\begin{aligned} 6^2 - 7 + 3 \cdot 4 &= 36 - 7 + 3 \cdot 4 \\ &= 36 - 7 + 12 \\ &= 29 + 12 \\ &= 41 \end{aligned}$$

21.

$$\begin{aligned} 3 \cdot 2 + 4(7 - 1) &= 3 \cdot 2 + 4(6) \\ &= 6 + 4(6) \\ &= 6 + 24 \\ &= 30 \end{aligned}$$

22. a. $7 + x = 13$
Seven plus what number is equal to thirteen?
$7 + 6 = 13$
The solution is $x = 6$.

b. $\frac{x}{4} = 2$
What number divided by four is equal to two?
$8 \div 4 = 2$
The solution is $x = 8$.

Copyright © 2012 Pearson Education, Inc.

c. $x+3x=36$
$1x+3x=36$
$(1+3)x=36$
$4x=36$
Four times what number is equal to 36?
$4(9) = 36$
The solution is $x = 9$.

d. $5+(b+2)=18$
$5+(2+b)=18$
$(5+2)+b=18$
$7+b=18$
Seven plus what number is equal to eighteen?
$7 + 11 = 18$
The solution is $b = 11$.

e. $9n-n=32$
$9n-1n=32$
$(9-1)n=32$
$8n=32$
Eight times what number is equal to 32?
$8(4) = 32$
The solution is $n = 4$.

23. Fred's checking account balance, B, decreased by \$155 equals \$275: $B - 155 = 275$.

24. What number divided by six equals two?

a. $x \div 6 = 2$

b. $12 \div 6 = 2$
The solution is $x = 12$.

25. Three subtracted from what number equals one.

a. $x - 3 = 1$

b. $4 - 3 = 1$
The solution is $x = 4$.

26. a. $412 \text{ adults} = 412(25) = \$10,300$
$280 \text{ children} = 280(18) = \$\ 5,040$
Total: $\$15,340$
The total income from tickets was \$15,340.

b. Subtract the expenses from the income.
$$\begin{array}{r} 15,340 \\ -\ 7,350 \\ \hline 7,990 \end{array}$$
The profit for the event was \$7990.

27. Beth has four choices of types of sandwiches and three choices of breads. Multiply to find the number of different sandwiches.
$4(3) = 12$
There are 12 different sandwiches possible.

28. Find the total deductions.
$$\begin{array}{r} 265 \\ 78 \\ +\ 57 \\ \hline 400 \end{array}$$
Subtract the amount of the deductions from the salary.
$$\begin{array}{r} 1540 \\ -\ 400 \\ \hline 1140 \end{array}$$
The check was \$1140 after deductions.

29.
$$\begin{array}{r} 525 \\ 525 \\ 200 \\ +\ 40 \\ \hline 1290 \end{array}$$
Fred needed \$1290 to move into the apartment.

30. a. Rounded to the nearest hundred, the expenses are \$800, \$200, \$100, \$200 and \$300.
\$800 + \$200 + \$100 + \$200 + \$300 = \$1600
Sylvia's expenses were about \$1600.

b. Rounded to the nearest hundred, Sylvia's income was \$1900.
$1900 - 1600 = 300$
Sylvia had about \$300 left.

31. Divide the total number of miles by 2 to find how many 3-point awards Elizabeth will accumulate.
$5000 \div 2 = 2500$
Multiply this number by 3 to obtain the total number of points.
$3(2500) = 7500$
Elizabeth will accumulate 7500 points.

Copyright © 2012 Pearson Education, Inc.

Chapter 2

2.1 Exercises

1. $-(-1)$ is the opposite of negative one.

3. Negative four minus two: $-4 - 2$

5. -9 is a <u>negative</u> number and 9 is a <u>positive</u> number.

7. To graph -6, start at zero and count six places in the negative direction.
To graph -4, start at zero and count four places in the negative direction.
To graph 3, start at zero and count three places in the positive direction.
To graph 4, start at zero and count four places in the positive direction.

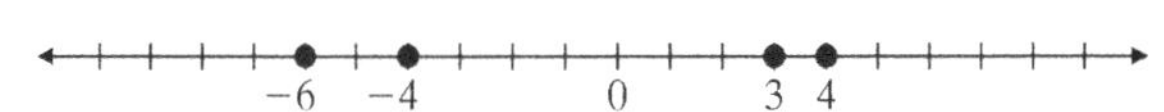

9. To graph -3, start at zero and count three places in the negative direction.
To graph -1, start at zero and count one place in the negative direction.
To graph 3, start at zero and count three places in the positive direction.
To graph 1, start at zero and count one place in the positive direction.

11. The dot labeled *A* represents the larger number since it lies to the right of the dot labeled *B* on the number line.

13. $-9 \; ? \; 4$
Negative numbers are always less than positive numbers.
$-9 < 4$

15. $4 \; ? \; -3$
Positive numbers are always greater than negative numbers.
$4 > -3$

17. $-5 \; ? \; 5$
Negative numbers are always less than positive numbers.
$-5 < 5$

19. $-9 \; ? \; -5$
-9 lies to the left of -5 on the number line.
$-9 < -5$

21. $-8 \; ? \; -6$
-8 lies to the left of -6 on the number line.
$-8 < -6$

23. $-298 \; ? \; -350$
-298 lies to the right of -350 on the number line.
$-298 > -350$

25. <u> − </u> A loss of \$100

27. <u> + </u> A raise of \$100

29. <u> − </u> A discount of \$10

31. <u> + </u> A plane ascends 1000 feet.

33. Start at -1. Locate the number that is the same distance from zero but lies on the opposite side of zero.

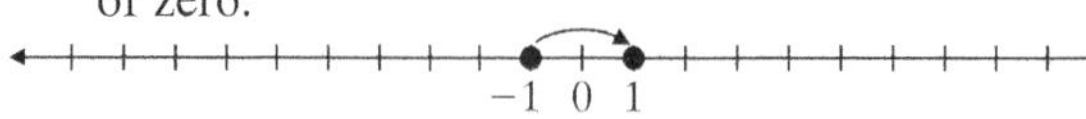

The opposite of -1 is 1.

35. The opposite of -3 is <u> 3 </u>.

37. The opposite of 16 is <u> −16 </u>.

39. The opposite of -4 is 4.
$-(-4) = 4$

41. The opposite of 8 is -8.
$-(8) = -8$

43. $-(-(-7)) = -(7) = -7$

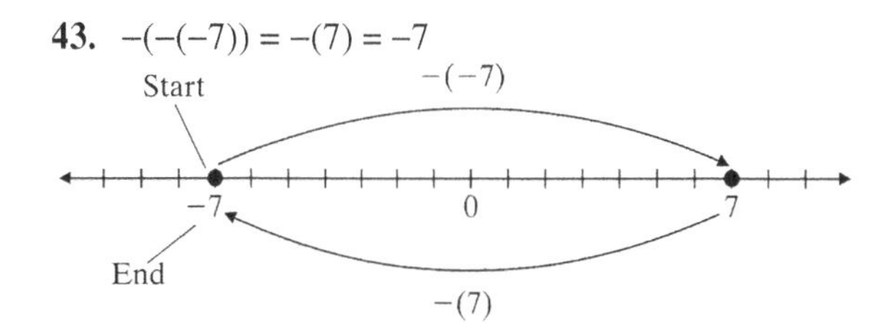

45. $-(-(13)) = -(-13) = 13$

47. $-(-(-(-1))) = -(-(1)) = -(-1) = 1$

49. Place parentheses around *a* and then replace *a* with 6.
$-(-a) = -(-(a)) = -(-(6)) = -(-6) = 6$

Copyright © 2012 Pearson Education, Inc.

51. Place parentheses around x and then replace x with -1.
$$\begin{aligned} -(-(-x)) &= -(-(-(x))) \\ &= -(-(-(-1))) \\ &= -(-(1)) \\ &= -(-1) \\ &= 1 \end{aligned}$$

53. Place parentheses around y and then replace y with -2.
$$\begin{aligned} -(-(-(-y))) &= -(-(-(-(y)))) \\ &= -(-(-(-(-2)))) \\ &= -(-(-(2))) \\ &= -(-(-2)) \\ &= -(2) \\ &= -2 \end{aligned}$$

55. The absolute value of a positive number is positive.
$|8| = 8$

57. The absolute value of a negative number is positive.
$|-5| = 5$

59. The absolute value of a negative number is positive.
$|-16| = 16$

61. The absolute value of zero is zero.
$|0| = 0$

63.
$$\begin{aligned} |-3| &\ ?\ |1| \\ 3 &\ ?\ 1 \\ 3 &> 1 \\ |-3| &> |1| \end{aligned}$$

65.
$$\begin{aligned} |8| &\ ?\ |-8| \\ 8 &\ ?\ 8 \\ 8 &= 8 \\ |8| &= |-8| \end{aligned}$$

67.
$$\begin{aligned} |-9| &\ ?\ |16| \\ 9 &\ ?\ 16 \\ 9 &< 16 \\ |-9| &< |16| \end{aligned}$$

69.
$$\begin{aligned} |-35| &\ ?\ |-8| \\ 35 &\ ?\ 8 \\ 35 &> 8 \\ |-35| &> |-8| \end{aligned}$$

71. $-|-3| = -(3) = -3$

73. $|-14| = 14$

75. a. The dot on the graph for Fargo is lower than the dot for Albany, so the temperature was colder in Fargo.

b. The dots for Boston and Cleveland are above the horizontal line indicating zero, so positive temperatures were recorded in Boston and Cleveland. The dots for Albany, Anchorage, and Fargo are below the horizontal line indicating zero, so negative temperatures were recorded in these three cities.

77.
$$\begin{aligned} -|-16| &\ ?\ -16 \\ -(16) &\ ?\ -16 \\ -16 &= -16 \\ -|-16| &= -16 \end{aligned}$$

79.
$$\begin{aligned} |-9| &\ ?\ -|-9| \\ 9 &\ ?\ -(9) \\ 9 &> -9 \\ |-9| &> -|-9| \end{aligned}$$

81. $-(-6) + |-5| = 6 + 5 = 11$

83. $-(-|8|) - |-1| = -(-8) - 1 = 8 - 1 = 7$

85. The number -33 has a larger absolute value than 20 because -33 is further from 0 on the number line.

87. The number 129 has a larger absolute value than -112 because 129 is further from 0 on the number line.

89. The statement is false. For example, $-1 > -2$ but -1 is not positive.

91. If m is a negative number, then $-m$ is a positive number.

Cumulative Review

93.
$$\begin{array}{r} 5009 \\ -\ 258 \\ \hline 4751 \end{array}$$

Copyright © 2012 Pearson Education, Inc.

94.
$$\begin{array}{r} 5699 \\ +\ 351 \\ \hline 6050 \end{array}$$

95.
$$\begin{array}{r} 256 \\ \times\ 91 \\ \hline 256 \\ 23\ 04\ \ \\ \hline 23,296 \end{array}$$

96. $456 \div 3 = 152$
$$\begin{array}{r} 152 \\ 3\overline{)456} \\ \underline{3}\ \ \\ 15\ \\ \underline{15}\ \\ 06 \\ \underline{6} \\ 0 \end{array}$$

97. Add the expenses.
$$\begin{array}{r} 480 \\ 1200 \\ +\ 350 \\ \hline \$2030 \end{array}$$
Subtract the expenses from the budget.
$$\begin{array}{r} 2600 \\ -\ 2030 \\ \hline \$570 \end{array}$$
Wanda will have $570 left to spend.

98. Find the total cost.
$$\begin{array}{r} 780 \\ 520 \\ 450 \\ 1150 \\ 203 \\ +\ 45 \\ \hline \$3148 \end{array}$$
Subtract the down payment.
$$\begin{array}{r} 3148 \\ -\ 800 \\ \hline \$2348 \end{array}$$
Tran has to finance $2348.

Quick Quiz 2.1

1. a. $-5 \ ?\ -6$
-5 lies to the right of -6 on the number line.
$-5 > -6$

b. $|-11| \ ?\ |13|$
$11 \ ?\ 13$
$11 < 13$
$|-11| < |13|$

2. $-(-(-7)) = -(7) = -7$

3. a. $|-3| = 3$

b. $-|-5| = -(5) = -5$

4. Answers may vary. The numbers are arranged left to right on the number line. They are arranged from smallest to largest.
$-10, -6, -4, -1, 0$

2.2 Exercises

1. Answers may vary. When two negative numbers are added the first move in the negative direction is followed by a second move in the negative direction resulting in a negative.

3. To add two numbers with different signs we keep the sign of the larger absolute value and subtract.

5. a. $-2 + (-3) = \boxed{-5}$
Rule: When adding two numbers with the same sign, we use the common sign in the answer and add the absolute values of the numbers.

b. $2 + (-3) = \boxed{-1}$
Rule: When adding two numbers with different signs we keep the sign of the larger absolute value and subtract the absolute values.

c. $-2 + 3 = \boxed{+1}$
Rule: When adding two numbers with different signs we keep the sign of the larger absolute value and subtract the absolute values.

Copyright © 2012 Pearson Education, Inc.

7. a.

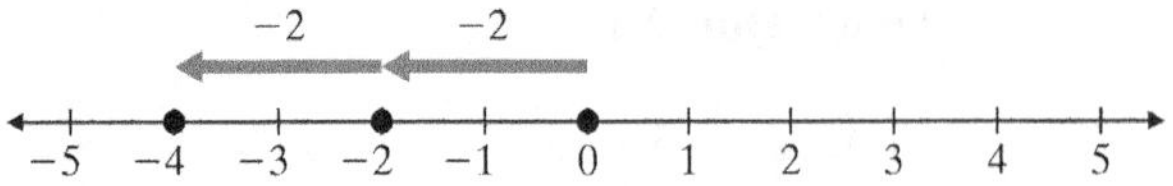

b. Negative

c. $-2 + (-2)$

d. From the number line, sum is -4.

9. a.

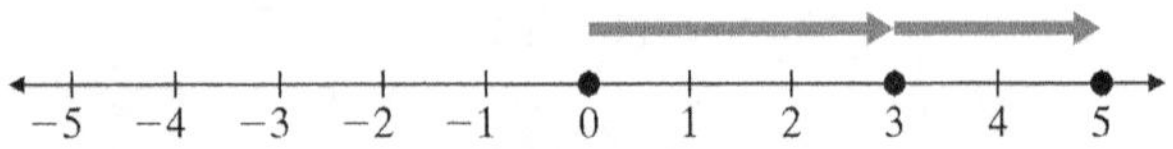

b. Positive

c. $3 + 2$

d. From the number line, sum is 5.

11. a. A decrease of 10°F followed by a decrease of 5°F results in a decrease of 15°F.

b. $-10°\text{F} + (-5°\text{F}) = -15°\text{F}$

13. a. A profit of \$100 followed by a profit of \$50 results in a profit of \$150.

b. \$100 + \$50 = \$150

15. We are adding two numbers with the same sign, so we keep the common sign and add the absolute values.

a. $-11 + (-13) = -24$

b. $11 + 13 = 24$

17. We are adding two numbers with the same sign, so we keep the common sign and add the absolute values.

a. $-29 + (-39) = -68$

b. $29 + 39 = 68$

19. We are adding two numbers with the same sign, so we keep the common sign and add the absolute values.

a. $-53 + (-18) = -71$

b. $53 + 18 = 71$

21. a. $-3 + 10$

b. Positive

c. $-3 + 10 = 7$

23. a. $2 + (-4)$

b. Negative

c. $2 + (-4) = -2$

Copyright © 2012 Pearson Education, Inc.

25. $+300 \text{ ft} + (-400 \text{ ft}) = -100 \text{ ft}$

27. $-\$400 + (+\$500) = \$100$

29. a. The answer is negative since the negative number has the larger absolute value.
$6 + (-8) = -2$

b. The answer is positive since the positive number has the larger absolute value.
$-6 + 8 = 2$

31. a. The answer is positive since the positive number has the larger absolute value.
$5 + (-1) = 4$

b. The answer is negative since the negative number has the larger absolute value.
$-5 + 1 = -4$

33. a. The answer is positive since the positive number has the larger absolute value.
$22 + (-16) = 6$

b. The answer is negative since the negative number has the larger absolute value.
$-22 + 16 = -6$

35. a. The numbers have different signs. The answer is positive since the positive number has the larger absolute value.
$3 + (-1) = 2$

b. The numbers have the same sign, so we keep the common sign.
$-3 + (-1) = -4$

c. The numbers have different signs. The answer is negative since the negative number has the larger absolute value.
$-3 + 1 = -2$

37. a. The numbers have the same sign, so we keep the common sign.
$-9 + (-11) = -20$

b. The numbers have different signs. The answer is negative since the negative number has the larger absolute value.
$9 + (-11) = -2$

c. The numbers have different signs. The answer is positive since the positive number has the larger absolute value.
$-9 + 11 = 2$

39. Since 2 and −2 are additive inverses, their sum is 0.
$2 + (-2) = 0$

41. Since −9 and 9 are additive inverses, their sum is 0.
$-9 + 9 = 0$

43. Since −360 and 360 are additive inverses, their sum is 0.
$-360 + 360 = 0$

45. Since 452 and −452 are additive inverses, their sum is 0.
$452 + (-452) = 0$

47. $x + 19 = 0$
The sum of additive inverses is 0. Thus $x = -19$ and $-19 + 19 = 0$.

49. $-12 + x = 0$
The sum of additive inverses is 0. Thus $x = 12$ since $-12 + 12 = 0$.

51. The answer is positive since the positive number has the larger absolute value.
$12 + (-11) = 1$

53. The answer is negative since the negative number has the larger absolute value.
$-10 + 4 = -6$

55. The answer is positive since the positive number has the larger absolute value.
$-7 + 14 = 7$

57. The answer is positive since the positive number has the larger absolute value.
$22 + (-10) = 12$

59. The numbers have the same sign, so we keep the common sign.
$-33 + (-5) = -38$

61. The numbers have the same sign, so we keep the common sign.
$-27 + (-12) = -39$

63. Since −15 and 15 are additive inverses, their sum is 0.
$-15 + 15 = 0$

65. The answer is positive since the positive number has the larger absolute value.
$-12 + 16 = 4$

Copyright © 2012 Pearson Education, Inc.

67. The answer is negative since the negative number has the larger absolute value.
$6 + (-8) = -2$

69. Since 15 and −15 are additive inverses, their sum is 0.
$15 + (-15) = 0$

71. $6+(-9)+1+(-3) = [6+1]+[(-9)+(-3)]$
$= 7+[(-9)+(-3)]$
$= 7+(-12)$
$= -5$

73. $-21+16+(-33) = [-21+(-33)]+16$
$= -54+16$
$= -38$

75. $57+(-29)+(-34)+23$
$= [57+23]+[(-29)+(-34)]$
$= 80+[(-29)+(-34)]$
$= 80+(-63)$
$= 17$

77. $-15+7+(-10)+3 = [-15+(-10)]+[7+3]$
$= -25+[7+3]$
$= -25+10$
$= -15$

79. $y + 5$

a. Replace y with −2.
$(y) + 5 = (-2) + 5 = 3$

b. Replace $y = -7$.
$(y) + 5 = (-7) + 5 = -2$

81. $a + (-8)$

a. Replace a with 4.
$(a) + (-8) = (4) + (-8) = -4$

b. Replace a with −5.
$(a) + (-8) = (-5) + (-8) = -13$

83. $-2 + x + y$

a. Replace x with −6 and y with 4.
$-2+(x)+(y) = -2+(-6)+(4)$
$= [-2+(-6)]+4$
$= -8+4$
$= -4$

b. Replace x with 5 and y with −9.
$-2+(x)+(y) = -2+(5)+(-9)$
$= [-2+(-9)]+5$
$= -11+5$
$= -6$

85. Replace x with −3 and y with −1.
$-x+y+6 = -(x)+(y)+6$
$= -(-3)+(-1)+6$
$= 3+(-1)+6$
$= [3+6]+(-1)$
$= 9+(-1)$
$= 8$

87. Replace a with −3 and b with −5.
$-a+b+(-1) = -(a)+(b)+(-1)$
$= -(-3)+(-5)+(-1)$
$= 3+[(-5)+(-1)]$
$= 3+(-6)$
$= -3$

89. 1st quarter gain: $30,000
2nd quarter loss: −$20,000
30,000 + (−20,000) = 10,000
At the end of the second quarter, the company had a profit of $10,000.

91. $-121 + 200 = 79$
The balance was $79.

93. $-75 + (-150) = -225$
The distance is −225 feet.

95. Replace x with −1, y with 9, and z with −5.
$3+x+y+(-1)+z = 3+(x)+(y)+(-1)+(z)$
$= 3+(-1)+(9)+(-1)+(-5)$
$= [3+9]+[(-1)+(-1)+(-5)]$
$= 12+(-7)$
$= 5$

97. $-33+24+(-38)+19+(-3)$
$= [-33+(-38)+(-3)]+[24+19]$
$= -74+43$
$= -31$

99. $12 + (-45) + (-9) + 5 + (-19)$
$= [12 + 5] + [(-45) + (-9) + (-19)]$
$= 17 + (-73)$
$= -56$

101. $-2 + \boxed{-3} = -5$

103. $3 + \boxed{-4} = -1$

Copyright © 2012 Pearson Education, Inc.

105. $-11 + 1 = -10$
1 can be added to -11 to obtain -10.

107. If $x + y + 30 = 0$, then $(x + y) + 30 = 0$.
The sum of additive inverses is 0. Thus $x + y$ is the additive inverse of 30 or $x + y = -30$ since $-30 + 30 = 0$. Since $-15 + (-15) = -30$, possible values for x and y are -15 and -15.

109. There are two possible solutions using six squares.
First solution:
$\boxed{-2} + \boxed{1} = -1$
$-1 + \boxed{-4} = -5$
$-5 + \boxed{4} = -1$
$-1 + \boxed{-1} = -2$
$-2 + \boxed{7} = 5$
The numbers are -2, 1, -4, 4, -1, and 7.
Second solution:
$\boxed{8} + \boxed{-5} = 3$
$3 + \boxed{2} = 5$
$5 + \boxed{-8} = -3$
$-3 + \boxed{-1} = -4$
$-4 + \boxed{7} = 3$
The numbers are 8, -5, 2, -8, -1, 7.

Cumulative Review

111. $4x + 6x = (4 + 6)x = 10x$

112. $2(3x) = (2 \cdot 3)x = 6x$

113. $8x - 3x = (8 - 3)x = 5x$

114. $3(x - 4) = 3 \cdot x - 3 \cdot 4 = 3x - 12$

115. 110 mi + 150 mi = 260 mi
Vu drove 260 miles each way.
260 mi + 260 mi = 520 mi
Vu drove a total of 520 miles.
23,566 mi + 520 mi = 24,086 mi
The reading on the odometer was 24,086 miles.

116. Numbers exiting: -4, -7
Numbers boarding: 12, 8, 11, 15
$[-4 + (-7)] + [12 + 8 + 11 + 15] = -11 + 46 = 35$
There were 35 people on the bus after the third stop.

Quick Quiz 2.2

1. a. The numbers have the same sign, so we keep the common sign.
$-6 + (-4) = -10$

b. The answer is negative since the negative number has the larger absolute value.
$-6 + 4 = -2$

c.
$$\begin{aligned} -9+5+(-3)+6 &= [-9+(-3)]+[5+6] \\ &= -12+11 \\ &= -1 \end{aligned}$$

2. Replace a with -1 and b with -7.
$$\begin{aligned} -a+b+3 &= -(a)+(b)+3 \\ &= -(-1)+(-7)+3 \\ &= 1+(-7)+3 \\ &= [1+3]+(-7) \\ &= 4+(-7) \\ &= -3 \end{aligned}$$

3. 1st quarter loss: $-\$3000$
2nd quarter loss: $-\$2500$
3rd quarter profit: \$8000
4th quarter loss: $-\$3500$
$-3000 + (-2500) + 8000 + (-3500)$
$= [-3000 + (-2500) + (-3500)] + 8000$
$= -9000 + 8000$
$= -1000$
At the end of the fourth quarter, the company had a loss of \$1000.

4. Answers may vary. Since the values given for x, y, and z are all negative, their sum is negative. However, it should be obvious that the absolute value of this sum is less than 132. So when the sum of x, y, and z is added to 132, we keep the sign of 132, which is positive.

2.3 Understanding the Concept
Another Approach to Subtracting Several Integers

1. a.
$$\begin{aligned} 2-6-8-11 &= 2+[(-6)+(-8)+(-11)] \\ &= 2+(-25) \\ &= -23 \end{aligned}$$

b.
$$\begin{aligned} 2-6-8-11 &= 2+(-6)-8-11 \\ &= -4-8-11 \\ &= -4+(-8)-11 \\ &= -12-11 \\ &= -12+(-11) \\ &= -23 \end{aligned}$$

c. Answers may vary.

Copyright © 2012 Pearson Education, Inc.

2.3 Exercises

1. To subtract two numbers, we change the subtraction sign to <u>addition</u> and take the <u>opposite</u> of the second number. Then we <u>add</u>.

3. We can think of this subtraction as measuring the distance between the two numbers. −10 is 10 units below 0 on a vertical number line, so we must add these 10 units to the 25 units above 0.

5. To subtract −3 we add $\boxed{3}$.

7. To subtract 5 we add $\boxed{-5}$.

9. $-7-6=-7+\boxed{-6}=\boxed{-13}$

11. $4-9=4+\boxed{-9}=\boxed{-5}$

13. $6-(-3)=6\boxed{+}3=\boxed{9}$

15. $9-(-5)=9\boxed{+}5=\boxed{14}$

17. **a.** $7-4=7+(-4)=3$

b. $15-7=15+(-7)=8$

c. $10-8=10+(-8)=2$

19. $20 − $35 = $20 + (−$35) = −$15

21. $6 − $7 = $6 + (−$7) = −$1

23. $-6-4=-6+(-4)=-10$

25. $-5-4=-5+(-4)=-9$

27. $5-(-2)=5+2=7$

29. $5-(-9)=5+9=14$

31. $-8-(-6)=-8+6=-2$

33. $-8-(-8)=-8+8=0$

35. $2-7=2+(-7)=-5$

37. $3-7=3+(-7)=-4$

39. $50-70=50+(-70)=-20$

41. $-85-(-20)=-85+20=-65$

43. $$\begin{aligned}12-9-5-8&=12+(-9)+(-5)+(-8)\\&=12+(-22)\\&=-10\end{aligned}$$

45. $$\begin{aligned}2-1-9-7&=2+(-1)+(-9)+(-7)\\&=2+(-17)\\&=-15\end{aligned}$$

47. $$\begin{aligned}9-10-2+3&=9+(-10)+(-2)+3\\&=9+3+(-10)+(-2)\\&=12+(-12)\\&=0\end{aligned}$$

49. $$\begin{aligned}-8-(-3)+(-10)&=-8+3+(-10)\\&=-8+(-10)+3\\&=-18+3\\&=-15\end{aligned}$$

51. $-7-(-2)-(-5)=-7+2+5=-7+7=0$

53. $$\begin{aligned}-3-(-8)+(-6)&=-3+8+(-6)\\&=-3+(-6)+8\\&=-9+8\\&=-1\end{aligned}$$

55. $7-21=7+(-21)=-14$

57. $-9-(-9)=-9+9=0$

59. $-13-18=-13+(-18)=-31$

61. $40-(-1)=40+1=41$

63. $$\begin{aligned}8-1-9-5&=8+(-1)+(-9)+(-5)\\&=8+(-15)\\&=-7\end{aligned}$$

65. $$\begin{aligned}7+8-6-11&=7+8+(-6)+(-11)\\&=15+(-17)\\&=-2\end{aligned}$$

67. $$\begin{aligned}9-10-2+3&=9+(-10)+(-2)+3\\&=9+3+(-10)+(-2)\\&=12+(-12)\\&=0\end{aligned}$$

69. Replace a with −8.
$a-9=(a)-9=(-8)-9=-8+(-9)=-17$

71. Replace x with −3.
$x-11=(x)-11=(-3)-11=-3+(-11)=-14$

73. Replace m with −5.
$14-m=14-(m)=14-(-5)=14+5=19$

Copyright © 2012 Pearson Education, Inc.

75. Replace y with –1 and x with 2.
$$\begin{aligned}21-y+x &= 21-(y)+(x)\\ &= 21-(-1)+(2)\\ &= 21+1+2\\ &= 24\end{aligned}$$

77. Replace x with –4 and y with 2.
$$\begin{aligned}-8-x-y &= -8-(x)-(y)\\ &= -8-(-4)-(2)\\ &= -8+4+(-2)\\ &= -8+(-2)+4\\ &= -10+4\\ &= -6\end{aligned}$$

79. Replace x with –6 and y with –5.
$$\begin{aligned}-1-x+y &= -1-(x)+(y)\\ &= -1-(-6)+(-5)\\ &= -1+6+(-5)\\ &= -1+(-5)+6\\ &= -6+6\\ &= 0\end{aligned}$$

81. 3556 – (–150) = 3556 + 150 = 3706
The difference in altitude is 3706 feet.

83. 326 – (–18) = 326 + 18 = 344
The roof is 344 feet above the floor of the basement.

85. a. The highest temperature in the chart is 88°F, which occurred in Brownsville, Texas.

b. 77 – (–9) = 77 + 9 = 86
The difference between the record high and record low on day 3 was 86°F.

87. –11 – (–15) = –11 + 15 = 4
The difference between Sandra's score and Tran's score was 4 points after the fourteenth hole.

89. After the eighteenth hole, Sandra's score was –15 and Tran's score was –13. Since -15 < –13 and the low score determines the winner, Sandra won the tournament.

91. –31 + 24 – 13 – 12 + (–14) – 3
= –31 + 24 + (–13) + (–12) + (–14) + (–3)
= –31 + (–13) + (–12) + (–14) + (–3) + 24
= –73 + 24
= –49

93. Replace a with –13, b = –4, and c with –9.
$$\begin{aligned}7-a-b+c+2 &= 7-(a)-(b)+(c)+2\\ &= 7-(-13)-(-4)+(-9)+2\\ &= 7+13+4+(-9)+2\\ &= 7+13+4+2+(-9)\\ &= 26+(-9)\\ &= 17\end{aligned}$$

95. –3009 – 893 = –3000 + (–893) = –3902

97. –2001 – (–987) = –2001 + 987 = –1014

99. a. –1 – 8 = n

b. –1 – 8 = –1 + (–8) = –9; n = –9

101. –4 is a solution to x – 3 = –7 since –4 – 3 = –4 + (–3) = –7.

103. –2 – $\boxed{2}$ = –4

105. –6, –3, 0, 3, 6, ...
Each number is 3 more than the preceding number. The next number is 6 + 3 = 9.

107. 4, –1, –6, –11, –16, ...
Each number is 5 less than the preceding number. The next number is
–16 – 5 = –16 + (–5) = –21.

Cumulative Review

108. 2 + 3(5) = 2 + 15 = 17

109. 12 – 3(4 – 1) = 12 – 3(3) = 12 – 9 = 3

110.
$$\begin{aligned}3^2+4(2)-5 &= 9+4(2)-5\\ &= 9+8-5\\ &= 17-5\\ &= 12\end{aligned}$$

111.
$$\begin{aligned}3+[3+2(8-6)] &= 3+[3+2(2)]\\ &= 3+[3+4]\\ &= 3+7\\ &= 10\end{aligned}$$

112.
$$\begin{array}{r} 45 \\ 12\overline{)550} \\ \underline{48} \\ 70 \\ \underline{60} \\ 10 \end{array}$$

They will need 45 full boxes plus 10 more pencils. The school should order 46 boxes. There will be two extra pencils.

Copyright © 2012 Pearson Education, Inc.

113. $8670 \div 85 = 102$ min = 1 hr 42 min

Quick Quiz 2.3

1. a. $-9 - 15 = -9 + (-15) = -24$

b. $-6 - (-8) = -6 + 8 = 2$

2. $$\begin{aligned}-8+6-(-5)-10 &= -8+6+5+(-10)\\ &= [-8+(-10)]+(6+5)\\ &= -18+11\\ &= -7\end{aligned}$$

3. $6300 - (-419) = 6300 + 419 = 6719$
The difference in altitude is 6719 feet.

4. Answers may vary. The problem was not completed correctly. First, subtraction is rewritten as addition of the opposite.
$-6 - (-3) + (-7) = -6 + 3 + (-7)$
Next, like signs are combined.
$= -13 + 3$
Next, unlike signs are combined.
$= -10$

Use Math to Save Money

1. $$\begin{aligned}1000+36\times 388.06 &= 1000+13,970.16\\ &= 14,970.16\end{aligned}$$
Louvy would pay \$14,970.16 over the entire length of the lease.

2. $$\begin{aligned}1000+36\times 669.28 &= 1000+24,094.08\\ &= 25,094.08\end{aligned}$$
Louvy would pay \$25,094.08 over the entire length of the loan.

3. $14,970.16 + 11,000.00 = 25,970.16$
This would bring the total cost up to \$25,970.16.

4. $25,970.16 - 25,094.08 = 876.08$
Louvy saves \$876.08 if he buys the car instead of leasing it.

5. $669.28 - 388.06 = 281.22$
He will save \$281.22 each month in car payments.

6. To get the best overall price, Louvy should buy the car since he will save \$876.08 on the total price. To get a lower monthly payment, Louvy should lease the car since he will save \$281.22 each month in car payments.

7. Answers will vary.

8. Answers will vary.

9. Answers will vary.

How Am I Doing? Sections 2.1–2.3

1. $-12 \; ? \; -7$
-12 lies to the left of -7 on the number line.
$-12 < -7$

2. $|-11| \; ? \; |8|$
$11 \; ? \; 8$
$11 > 8$
$|-11| > |8|$

3. $-|-8| = -(8) = -8$

4. $-(-(-3)) = -(3) = -3$

5. Replace x with -6.
$-(-x) = -(-(x)) = -(-(-6)) = -(6) = -6$

6. We are adding two numbers with the same sign, so we keep the common sign and add the absolute values.
$-2 + (-14) = -16$

7. $$\begin{aligned}-8+3+(-1)+4 &= [-8+(-1)]+[3+4]\\ &= -9+7\\ &= -2\end{aligned}$$

8. Replace a with -9 and b with -5.
$$\begin{aligned}a+b+12 &= (a)+(b)+12\\ &= (-9)+(-5)+12\\ &= -14+12\\ &= -2\end{aligned}$$

9. Replace x with -8 and y with -11.
$$\begin{aligned}-x+y+7 &= -(x)+(y)+7\\ &= -(-8)+(-11)+7\\ &= 8+(-11)+7\\ &= 8+7+(-11)\\ &= 15+(-11)\\ &= 4\end{aligned}$$

10. 1st quarter loss: $-\$20,000$
2nd quarter profit: \$20,000
3rd quarter loss: $-\$10,000$
4th quarter profit: \$30,000
$$\begin{aligned}&-20,000+20,000+(-10,000)+30,000\\ &= -20,000+(-10,000)+20,000+30,000\\ &= -30,000+50,000\\ &= 20,000\end{aligned}$$
At the end of the fourth quarter, the company's overall profit was \$20,000.

Copyright © 2012 Pearson Education, Inc.

11. $7 - 19 = 7 + (-19) = -12$

12. $-3 - (-5) = -3 + 5 = 2$

13. $-8 - (-2) - (-1) = -8 + 2 + 1 = -8 + 3 = -5$

14. $$\begin{aligned} -5-6+(-1)-(-7) &= -5+(-6)+(-1)+7 \\ &= -12+7 \\ &= -5 \end{aligned}$$

15. Replace x with -1 and y with -2.
$$\begin{aligned} -5-x-y &= -5-(x)-(y) \\ &= -5-(-1)-(-2) \\ &= -5+1+2 \\ &= -5+3 \\ &= -2 \end{aligned}$$

16. $7622 - (-161) = 7622 + 161 = 7783$
The difference in altitude is 7783 feet.

2.4 Exercises

1. When we add two negative numbers, the sum is negative. So the statement is only true for multiplication and division.

3. If you multiply 4 negative numbers, the product will be a <u>positive</u> number.

5. The quotient of a positive number and a <u>negative</u> number is negative.

7. $3(-4) = (-4) + (-4) + (-4) = -12$

9. $4(-6) = (-6) + (-6) + (-6) + (-6) = -24$

11. $2(-3) = (-3) + (-3) = -6$

13. a. $3 \cdot \boxed{3} = 9$

b. $3 \cdot \boxed{-3} = -9$

c. $-3 \cdot \boxed{3} = -9$

d. $-3 \cdot \boxed{-3} = 9$

15. a. $\dfrac{12}{\boxed{4}} = 3$

b. $\dfrac{-12}{\boxed{-4}} = 3$

c. $\dfrac{12}{\boxed{-4}} = -3$

d. $\dfrac{-12}{\boxed{4}} = -3$

17. a. The number of negative signs, 0, is even, so the answer is positive.
$9(2) = 18$

b. The number of negative signs, 1, is odd, so the answer is negative.
$9(-2) = -18$

c. The number of negative signs, 1, is odd, so the answer is negative.
$-9(2) = -18$

d. The number of negative signs, 2, is even, so the answer is positive.
$-9(-2) = 18$

19. a. The number of negative signs, 0, is even, so the answer is positive.
$5(2) = 10$

b. The number of negative signs, 2, is even, so the answer is positive.
$-5(-2) = 10$

c. The number of negative signs, 1, is odd, so the answer is negative.
$-5(2) = -10$

d. The number of negative signs, 1, is odd, so the answer is negative.
$5(-2) = -10$

21. The number of negative signs, 2, is even, so the answer is positive.
$-2(-9) = 18$

23. The number of negative signs, 2, is even, so the answer is positive.
$-1(-6) = 6$

25. The number of negative signs, 1, is odd, so the answer is negative.
$8(-7) = -56$

27. The number of negative signs, 1, is odd, so the answer is negative.
$-5(9) = -45$

Copyright © 2012 Pearson Education, Inc.

29. $(-1)(3)(-236)(42)(-16)(-90)$ is a positive number because it contains an even number of negative factors.

31. $(-943)(-721)(-816)(-96)(-51)$ is a negative number because it contains an odd number of negative factors.

33. $4(-5)(-2) = (-20)(-2) = 40$

35. $(-3)(-2)(-3)(-4) = (6)(12) = 72$

37. $2(-1)(5)(-7) = (-2)(-35) = 70$

39. $(-2)(-1)(4)(-5) = 2(-20) = -40$

41. $$\begin{aligned}(-5)(4)(-3)(2)(-1) &= (-20)(-6)(-1)\\ &= (120)(-1)\\ &= -120\end{aligned}$$

43. The value of $(-2)^{13}$ is a negative number because the exponent is odd.

45. The value of $(-96)^{52}$ is a positive number because the exponent is even.

47. -96^{52} is a negative number because -96^{52} is the opposite of 96^{52}, which is positive.

49. $(-10)^2 = (-10)(-10) = 100$

51. $(-5)^3 = (-5)(-5)(-5) = -125$

53. a. $(-4)^2 = (-4)(-4) = 16$

b. $(-4)^3 = (-4)(-4)(-4) = -64$

55. a. $(-1)^{13} = -1$, 13 is odd.

b. $(-1)^{24} = 1$, 24 is even.

57. a. $-4^2 = -(4)(4) = -16$

b. $(-4)^2 = (-4)(-4) = 16$

59. a. $-2^3 = -(2)(2)(2) = -8$

b. $(-2)^3 = (-2)(-2)(-2) = -8$

61. a. $(-4)^3 = (-4)(-4)(-4) = -64$

b. $-4^3 = -4 \cdot 4 \cdot 4 = -64$

63. a. $(-9)^2 = (-9)(-9) = 81$

b. $-9^2 = -9 \cdot 9 = -81$

65. a. $35 \div 7 = 5$

b. $35 \div (-7) = -5$

c. $-35 \div 7 = -5$

d. $-35 \div (-7) = 5$

67. a. $40 \div 8 = 5$

b. $40 \div (-8) = -5$

c. $-40 \div 8 = -5$

d. $-40 \div (-8) = 5$

69. $30 \div (-5) = -6$

71. $\frac{-45}{5} = -45 \div 5 = -9$

73. $-16 \div (-2) = 8$

75. $\frac{-49}{-7} = -49 \div (-7) = 7$

77. a. $22 \div (-2) = -11$

b. $22(-2) = -44$

79. a. $-4 \div (-2) = 2$

b. $-4(-2) = 8$

81. a. $-15 \div 3 = -5$

b. $-15(3) = -45$

83. a. $14 \div -7 = -2$

b. $-14(7) = -98$

85. Replace x with -1.
$x^2 = (x)^2 = (-1)^2 = 1$

Copyright © 2012 Pearson Education, Inc.

87. Replace x with −42 and y with −7.

$$\frac{-x}{y}=\frac{-(x)}{(y)}=\frac{-(-42)}{(-7)}=\frac{42}{-7}=-6$$

89. Replace m with −20 and n with 2.

$$\frac{-m}{-n}=\frac{-(m)}{-(n)}=\frac{-(-20)}{-(2)}=\frac{20}{-2}=-10$$

91. a. Replace y with −2.

$$-y^3=-(y)^3=-(-2)^3=-(-8)=8$$

b. Replace y with −2.

$$-y^4=-(y)^4=-(-2)^4=-(16)=-16$$

93. Multiply the rate by the time to find the distance.
$-30(3)=-90$
The projectile travels 90 meters to the left in 3 seconds. Since the projectile starts at zero, it is 90 meters to the left of zero.

95. The discount on a single glove can be represented by the integer −2. Multiply by 350 to find the total reduction for 350 gloves.
$350(-2)=-700$
The total reduction in price can be represented by the −$700.

97. $\frac{x}{2}=-12$

$\frac{8}{2}\stackrel{?}{=}-12$

$4=-12$, false

No, 8 is not a solution.

99. $\frac{x}{-3}=8$

What number divided by −3 is equal to eight?

$\frac{-24}{-3}=-24\div(-3)=8$

The value of x is −24.

Cumulative Review

101.
$$\begin{aligned}2^2+3(5)-1&=4+3(5)-1\\&=4+15-1\\&=19-1\\&=18\end{aligned}$$

102. $8+2(9\div 3)=8+2(3)=8+6=14$

103.
$$\begin{aligned}2^3+(4\div 2+6)&=2^3+(2+6)\\&=2^3+8\\&=8+8\\&=16\end{aligned}$$

104.
$$\begin{aligned}3^2+(6\div 2+8)&=3^2+(3+8)\\&=3^2+11\\&=9+11\\&=20\end{aligned}$$

105. To find the time, divide the distance by the speed of the sound.

$$\begin{array}{r}3\\1087\overline{)3261}\\\underline{3261}\\0\end{array}$$

It took the sound 3 seconds to reach Kristina.

106. $|-1|\ ?\ |-20|$

$|-1|=1, |-20|=20$

$1<20$

$|-1|<|-20|$

Quick Quiz 2.4

1. a. The number of negative signs, 1, is odd, so the answer is negative.
$4(-3)=-12$

b. $-45\div 5=-9$

c. $\frac{-48}{-8}=-48\div(-8)=6$

2. $(-2)(-6)(-1)(3)=-12(3)=-36$

3. a. Replace x with −1.

$$x^5=(x)^5=(-1)^5=-1$$

b. Replace a with 10 and b with −2.

$$\frac{-a}{b}=\frac{-(a)}{(b)}=\frac{-(10)}{(-2)}=\frac{-10}{-2}=5$$

4. Answers may vary. One possible solution is to recognize that an even number of negative signs occurs in the expression, the product of which must be positive.

2.5 Exercises

1. $2+3(-1)\neq 5(-1)=-5$ because we must multiply $3(-1)$ before we add.

Copyright © 2012 Pearson Education, Inc.

3. Yes, $-2^2 + 8 = -4 + 8 = 4$ because there are no parentheses around −2 we square 2 and not −2 and then take the opposite.

5. $-2 + 3 \cdot 4 = -2 + 12 = 10$

7. $1 + 7(2 - 6) = 1 + 7(-4) = 1 + (-28) = -27$

9. $-3 + 6(8 - 5) = -3 + 6(3) = -3 + 18 = 15$

11.
$$\begin{aligned} 12 - 5(2 - 6) &= 12 - 5(-4) \\ &= 12 - (-20) \\ &= 12 + 20 \\ &= 32 \end{aligned}$$

13. $5(-3)(4 - 7) + 9 = -15(-3) + 9 = 45 + 9 = 54$

15. $-3(6 \div 3) + 7 = -3(2) + 7 = -6 + 7 = 1$

17. $3(-2)(9 - 5) - 10 = -6(4) - 10 = -24 - 10 = -34$

19. $-24 \div 12 - 8 = -2 - 8 = -10$

21. $(-3)^2 + 5(-9) = 9 + 5(-9) = 9 + (-45) = -36$

23.
$$\begin{aligned} (-3)^3 - 7(8) &= -27 - 7(8) \\ &= -27 - 56 \\ &= -27 + (-56) \\ &= -83 \end{aligned}$$

25. $(-2)^3 + 2(-8) = -8 + 2(-8) = -8 + (-16) = -24$

27. $36 \div (-6) + (-6) = -6 + (-6) = -12$

29.
$$\begin{aligned} 12 - 20 \div 4(-4)^2 + 9 &= 12 - 20 \div 4(16) + 9 \\ &= 12 - 5(16) + 9 \\ &= 12 - 80 + 9 \\ &= -68 + 9 \\ &= -59 \end{aligned}$$

31.
$$\begin{aligned} 8 - 2(5 - 2^2) + 6 &= 8 - 2(5 - 4) + 6 \\ &= 8 - 2(1) + 6 \\ &= 8 - 2 + 6 \\ &= 6 + 6 \\ &= 12 \end{aligned}$$

33. $\dfrac{(-50 \div 2 + 3)}{(20 - 9)} = \dfrac{(-25 + 3)}{(20 - 9)} = \dfrac{-22}{(20 - 9)} = \dfrac{-22}{11} = -2$

35.
$$\begin{aligned} \frac{[3^2 + 4(-6)]}{[-3 + (-2)]} &= \frac{[9 + 4(-6)]}{[-3 + (-2)]} \\ &= \frac{[9 + (-24)]}{[-3 + (-2)]} \\ &= \frac{-15}{[-3 + (-2)]} \\ &= \frac{-15}{-5} \\ &= 3 \end{aligned}$$

37.
$$\begin{aligned} \frac{[-12 - 3(-2)]}{(15 - 17)} &= \frac{[-12 - (-6)]}{(15 - 17)} \\ &= \frac{(-12 + 6)}{(15 - 17)} \\ &= \frac{-6}{(15 - 17)} \\ &= \frac{-6}{[15 + (-17)]} \\ &= \frac{-6}{-2} \\ &= 3 \end{aligned}$$

39.
$$\begin{aligned} -16 \div \{-4 \cdot [8 \div (-2)]\} &= -16 \div [-4 \cdot (-4)] \\ &= -16 \div 16 \\ &= -1 \end{aligned}$$

41.
$$\begin{aligned} -60 \div \{5 \cdot [-2 \cdot (-12 \div 4)]\} &= -60 \div \{5 \cdot [-2 \cdot (-3)]\} \\ &= -60 \div (5 \cdot 6) \\ &= -60 \div (30) \\ &= -2 \end{aligned}$$

43.
$$\begin{aligned} &35{,}000 + 3(-2000) + 1000 \\ &= 35{,}000 + (-6000) + 1000 \\ &= 29{,}000 + 1000 \\ &= 30{,}000 \end{aligned}$$
The current altitude is 30,000 feet.

45. $14(-3) + 9(+1) = -42 + 9 = -33$
The total charge is −33.

47. $7(+3) + 5(-1) + 4(+1) = 21 - 5 + 4 = +20$
The total charge is +20.

49.
$$\begin{aligned} &1(10) + 1(5) + 1(3) + (1(-1) + 1(-1)) + 2(-1) \\ &= 10 + 5 + 3 + (-1 - 1) - 2 \\ &= 15 + 3 - 2 - 2 \\ &= 18 - 4 \\ &= 14 \end{aligned}$$
Derek received 14 points.

Copyright © 2012 Pearson Education, Inc.

51. $1(20) + 1(5) + 2(3) + (2(-1) + 2(-1)) + 2(-1)$
$= 20 + 5 + 6 + (-2 - 2) - 2$
$= 25 + 6 - 4 - 2$
$= 31 - 6$
$= 25$
Vladimir received 25 points.

53. $\frac{[(30-15\div 3)+(-5)]}{(5-10)} = \frac{[(30-5)+(-5)]}{(5-10)}$
$= \frac{[25+(-5)]}{(5-10)}$
$= \frac{20}{(5-10)}$
$= \frac{20}{-5}$
$= -4$

55. $[(3+24)\div(-3)]\cdot[2+(-3)^2]$
$= [(3+24)\div(-3)]\cdot(2+9)$
$= [27\div(-3)]\cdot(2+9)$
$= -9\cdot(2+9)$
$= -9\cdot 11$
$= -99$

57. Simplify each side of the equation.
$3+x-2(-4) = 7-(-13)$
$x+3-(-8) = 7+13$
$x+3+8 = 20$
$x+11 = 20$
What number plus eleven is equal to twenty?
$9 + 11 = 20$
$x = 9$

Cumulative Review

59. $2(x + 3) = 2 \cdot x + 2 \cdot 3 = 2x + 6$

60. $3(a + 2) = 3 \cdot a + 3 \cdot 2 = 3a + 6$

61. $4(x - 2) = 4 \cdot x - 4 \cdot 2 = 4x - 8$

62. $7(x - 1) = 7 \cdot x - 7 \cdot 1 = 7x - 7$

Quick Quiz 2.5

1. $15-20\div 5(-2)^2+3 = 15-20\div 5(4)+3$
$= 15-4(4)+3$
$= 15-16+3$
$= 15+(-16)+3$
$= -1+3$
$= 2$

2. $\frac{(16\div 8-4)}{(3-5)} = \frac{(2-4)}{(3-5)}$
$= \frac{[2+(-4)]}{3-5}$
$= \frac{-2}{(3-5)}$
$= \frac{-2}{[3+(-5)]}$
$= \frac{-2}{-2}$
$= 1$

3. There are 12 hours between midnight and noon. Since the temperature dropped 5 degrees every hour for 5 of those 12 hours, it rose 8 degrees every hour for the remaining 7 hours.
$4 + 5(-5) + 7(8) = 4 + (-25) + 56 = 35$
At noon, the temperature was 35°F.

4. Answers may vary. One possible solution is to first evaluate terms with exponents.
$3^2+5(2-4) = 9+5(2-4)$
Next, evaluate operations inside parentheses.
$9 + 5(2 - 4) = 9 + 5(-2)$
Next, evaluate multiplication operations.
$9 + 5(-2) = 9 - 10$
Lastly, combine like terms.
$9 - 10 = -1$

2.6 Exercises

1. $-2x+5x \neq -10x^2$ because we do not multiply variables and coefficients when combining like terms, we add coefficients:
$-2x + 5x = (-2 + 5)x = 3x.$

3. $-6x + (-3\boxed{x}) = -9x$

5. $5y + \boxed{3}xy - 2y + 7xy = 3y + 10xy$

7. To simplify $9x + (-3xy)$, we write $9x \boxed{-} 3y$.

9. $-6(y - 1) = -6 \cdot \boxed{y} - (-6) \cdot \boxed{1} = -6y \boxed{+} 6$

11. $-8x + 3x = (-8 + 3)x = -5x$

13. $4x + (-3x) = [4 + (-3)]x = 1x = x$

15. $-5x - 7x = (-5 - 7)x = [-5 + (-7)]x = -12x$

17. $-7a - (-2a) = [-7 - (-2)]a = (-7 + 2)a = -5a$

19. $14y + (-7y) = [14 + (-7)]y = 7y$

Copyright © 2012 Pearson Education, Inc.

21. $-7x + (-6x) = [-7 + (-6)]x = -13x$

23. $7a + (-9b) = 7a - 9b$

25. $-5m + (-8n) = -5m - 8n$

27. $2x + (-y) = 2x - y$

29. $-2a - (-3b) = -2a + 3b$

31. a. $2 - 7 + 3 = 2 + (-7) + 3 = -5 + 3 = -2$

b. $2x - 7x + 3x = 2x + (-7x) + 3x$
$= -5x + 3x$
$= -2x$

33. a. $3 - 8 + 4 = 3 + (-8) + 4 = -5 + 4 = -1$

b. $3x - 8x + 4x = 3x + (-8x) + 4x$
$= -5x + 4x$
$= -1x$
$= -x$

35. a. $2 - 6 + 1 = 2 + (-6) + 1 = -4 + 1 = -3$

b. $2x - 6x + 1x = 2x + (-6x) + 1x$
$= -4x + 1x$
$= -3x$

37. $-8y + 4x + 2y = -8y + 2y + 4x$
$= (-8 + 2)y + 4x$
$= -6y + 4x$

39. $6x + 4y + (-8x) = 6x + (-8x) + 4y$
$= [6 + (-8)]x + 4y$
$= -2x + 4y$

41. $9x + 3y + (-5x) = 9x + (-5x) + 3y$
$= [9 + (-5)]x + 3y$
$= 4x + 3y$

43. $-8x - 4x - y = (-8 - 4)x - y$
$= [-8 + (-4)]x - y$
$= -12x - y$

45. $3x + 8y - 10x - 2y = 3x + 8y + (-10x) + (-2y)$
$= 3x + (-10x) + 8y + (-2y)$
$= [3 + (-10)]x + [8 + (-2)]y$
$= -7x + 6y$

47. $4x + 2y - 6x - 7 = 4x + 2y + (-6x) + (-7)$
$= 4x + (-6x) + 2y + (-7)$
$= [4 + (-6)]x + 2y + (-7)$
$= -2x + 2y - 7$

49. $4 + 3ab - 2 - 9ab = 4 + 3ab + (-2) + (-9ab)$
$= 4 + (-2) + 3ab + (-9ab)$
$= [4 + (-2)] + [3 + (-9)]ab$
$= 2 + (-6ab)$
$= 2 - 6ab$

51. $5x + 7xy - 9x - xy = 5x + 7xy + (-9x) + (-1xy)$
$= 5x + (-9x) + 7xy + (-1xy)$
$= [5 + (-9)]x + [7 + (-1)]xy$
$= -4x + 6xy$

53. $7a - 2ab - 2 - 7ab + 3a$
$= 7a + (-2ab) + (-2) + (-7ab) + 3a$
$= 7a + 3a + (-2ab) + (-7ab) + (-2)$
$= (7 + 3)a + [(-2) + (-7)]ab + (-2)$
$= 10a + (-9ab) + (-2)$
$= 10a - 9ab - 2$

55. $3a + 2x - 5a + 7ax - x$
$= 3a + 2x + (-5a) + 7ax + (-1x)$
$= 3a + (-5a) + 2x + (-1x) + 7ax$
$= [3 + (-5)]a + [2 + (-1)]x + 7ax$
$= -2a + 1x + 7ax$
$= -2a + x + 7ax$

57. $6a + 7b - 9a + 5ab - 11b$
$= 6a + 7b + (-9a) + 5ab + (-11b)$
$= 6a + (-9a) + 7b + (-11b) + 5ab$
$= [6 + (-9)]a + [7 + (-11)]b + 5ab$
$= -3a + (-4b) + 5ab$
$= -3a - 4b + 5ab$

59. $4x + 8y - 7x + 6xy - 10y$
$= 4x + 8y + (-7x) + 6xy + (-10y)$
$= 4x + (-7x) + 8y + (-10y) + 6xy$
$= [4 + (-7)]x + [8 + (-10)]y + 6xy$
$= -3x + (-2y) + 6xy$
$= -3x - 2y + 6xy$

61. Replace x with -3 and y with -2.
$x + 3y = (x) + 3(y)$
$= (-3) + 3(-2)$
$= -3 + (-6)$
$= -9$

Copyright © 2012 Pearson Education, Inc.

63. Replace m with 6 and n with −3.

$$\begin{aligned} m-6n &= (m)-6(n) \\ &= (6)-6(-3) \\ &= 6-(-18) \\ &= 6+18 \\ &= 24 \end{aligned}$$

65. Replace a with −1 and b with 5.

$$\begin{aligned} a\cdot b-6 &= (a)(b)-6 \\ &= (-1)(5)-6 \\ &= -5-6 \\ &= -5+(-6) \\ &= -11 \end{aligned}$$

67. Replace x with −9 and y with 4.

$$\frac{(x+y)}{5}=\frac{[(x)+(y)]}{5}=\frac{[(-9)+(4)]}{5}=\frac{-5}{5}=-1$$

69. Replace t with −3.

$$9t^2=9(t)^2=9(-3)^2=9(9)=81$$

71. Replace x with −5.

$$\begin{aligned} 8x-x^2 &= 8(x)-(x)^2 \\ &= 8(-5)-(-5)^2 \\ &= 8(-5)-(25) \\ &= -40-25 \\ &= -40+(-25) \\ &= -65 \end{aligned}$$

73. Replace x with −4.

$$\begin{aligned} \frac{(x^2-x)}{2} &= \frac{[(x)^2-(x)]}{2} \\ &= \frac{[(-4)^2-(-4)]}{2} \\ &= \frac{[16-(-4)]}{2} \\ &= \frac{(16+4)}{2} \\ &= \frac{20}{2} \\ &= 10 \end{aligned}$$

75. Replace a with 13 and b with 2.

$$\begin{aligned} \frac{(a-b^2)}{-3} &= \frac{[(a)-(b)^2]}{-3} \\ &= \frac{[(13)-(2)^2]}{-3} \\ &= \frac{(13-4)}{-3} \\ &= \frac{9}{-3} \\ &= -3 \end{aligned}$$

77. Replace m with 6 and n with −2.

$$\begin{aligned} \frac{(m^2+2n)}{-8} &= \frac{[(m)^2+2(n)]}{-8} \\ &= \frac{[(6)^2+2(-2)]}{-8} \\ &= \frac{[36+2(-2)]}{-8} \\ &= \frac{[36+(-4)]}{-8} \\ &= \frac{32}{-8} \\ &= -4 \end{aligned}$$

79. $-3(y+1)=-3y+(-3)=-3y-3$

81. $-9(y-1)=-9y-(-9)(1)=-9y-(-9)=-9y+9$

83.
$$\begin{aligned} -2(m-3) &= -2m-(-2)(3) \\ &= -2m-(-6) \\ &= -2m+6 \end{aligned}$$

85. $-1(x+5)=-1x+(-1)(5)=-x+(-5)=-x-5$

87. $6(-2+y)=6(-2)+6y=-12+6y$

89. $2(-4+a)=2(-4)+2a=-8+2a$

91. Replace v with −8 and t with 4.
$s=v-32t=-8-32(4)=-8-128=-136$
The skydiver is falling 136 feet per second.

Copyright © 2012 Pearson Education, Inc.

93. For $t = 1$:
$v = 72 - 32t = 72 - 32(1) = 72 - 32 = 40$
For $t = 2$:
$v = 72 - 32t = 72 - 32(2) = 72 - 64 = 8$
For $t = 3$: $v = 72 - 32t$
$= 72 - 32(3)$
$= \frac{(70-160)}{9}$
$= \frac{-90}{9}$
$= -24$
When $t = 3$, v is negative. The object is descending at 3 seconds.

95. $C = \frac{(5F-160)}{9}$
$= \frac{[5(14)-160]}{9}$
$= \frac{(70-160)}{9}$
$= \frac{-90}{9}$
$= -10°C$

97. $\frac{x^2}{7} = 13$
$\frac{7^2}{7} \stackrel{?}{=} 13$
$\frac{49}{7} \stackrel{?}{=} 13$
$7 \neq 13$
No, 7 is not a solution.

Cumulative Review

98. 6 ft + 3 ft + 6 ft + 3 ft = 18 ft
The perimeter is 18 feet.

99. 7 in. + 7 in. + 7 in. + 7 in. = 28 in.
The perimeter is 28 inches.

100. If light travels 5,580,000 miles in 30 seconds, divide the number of miles by 30 to find out how far light travels in 1 second.
$5,580,000 \div 30 = 186,000$
Light travels 186,000 miles in 1 second.
Since 1 min = 60 sec = 30 sec + 30 sec, add the number of miles light travels in 30 seconds to itself to find how far light travels in 1 minute.
$5,580,000 + 5,580,000 = 11,160,000$
Light travels 11,160,000 miles in 1 minute.

101. Since the heart beats 73 times per minute and there are 60 minutes in one hour, multiply 73 by 60 to find the number of times the heart beats in one hour.
$60(73) = 4380$
The heart beats 4380 times per hour. Since there are 24 hours in one day, multiply 4380 by 24 to find the number of times the heart beats in one day.
$24(4380) = 105,120$
The heart beats 105,120 times per day.

Quick Quiz 2.6

1. $-5x + 9y + 2 - 2y - 6x$
$= -5x + 9y + 2 + (-2y) + (-6x)$
$= -5x + (-6x) + 9y + (-2y) + 2$
$= [-5 + (-6)]x + [9 + (-2)]y + 2$
$= -11x + 7y + 2$

2. a. $-3(a - 1) = -3a - (-3) = -3a + 3$

b. $-8(x + 7) = -8x + (-8)(7)$
$= -8x + (-56)$
$= -8x - 56$

3. Replace t with 3 and v with -10.
$s = v - 32t = -10 - 32(3) = -10 - 96 = -106$
The skydiver is falling 106 feet per second.

4. Answers may vary. The possible alternatives to writing $-3b + 7$ are $7 - 3b$ or $7 + (-3b)$.

You Try It

1. $-4 \ ? \ -9$
-4 lies to the right of -9 on the number line.
$-4 > -9$

2. a. The opposite of -16 is 16.
$-(-16) = 16$

b. The opposite of 12 is -12.
$-(12) = -12$

3. a. The absolute value of a positive number is positive.
$|14| = 14$

b. The absolute value of a negative number is positive.
$|-9| = 9$

Copyright © 2012 Pearson Education, Inc.

4. a. The dot representing Boston is higher on the graph than Bangor or Boise, indicating that it was warmest in Boston.

b. The temperature is a negative number in Bangor.

5. The numbers have the same sign, so we keep the common sign.
$-8 + (-5) = -13$

6. a. The answer is negative since the negative number has the larger absolute value.
$-7 + 3 = -4$

b. The answer is positive since the positive number has the larger absolute value.
$9 + (-2) = 7$

7. a. $-11 - 4 = -11 + (-4) = -15$

b. $5 - (-7) = 5 + 7 = 12$

8. a.
$$\begin{aligned}(-2)(-3)(-5)(2) &= [(-2)(-3)][(-5)(2)]\\ &= 6(-10)\\ &= -60\end{aligned}$$

b. $(-6)(-2)(3) = 12(3) = 36$

9. a. $(-6)^2 = (-6)(-6) = 36$

b. $-6^2 = -(6)(6) = -36$

c. $(-2)^3 = (-2)(-2)(-2) = -8$

10. a. $(-18) \div (-2) = 9$

b. $-24 \div 6 = -4$

11.
$$\begin{aligned}12 - 20 \div 2(-3)^2 - 6 &= 12 - 20 \div 2(9) - 6\\ &= 12 - 10(9) - 6\\ &= 12 - 90 - 6\\ &= 12 + (-90) + (-6)\\ &= -78 + (-6)\\ &= -84\end{aligned}$$

12.
$$\begin{aligned}3a + 5b - 6a &= 3a + (-6a) + 5b\\ &= [3 + (-6)]a + 5b\\ &= -3a + 5b\end{aligned}$$

13. Replace x with -2 and y with 5.
$$\begin{aligned}-5 - xy^2 &= -5 - (-2)(5)^2\\ &= -5 - (-2)(25)\\ &= -5 - (-50)\\ &= -5 + 50\\ &= 45\end{aligned}$$

14. $-3(x - 2) = -3x - (-3)(2) = -3x - (-6) = -3x + 6$

Chapter 2 Review Problems

1. Negative numbers are numbers that are less than zero.

2. Opposites are numbers that are the same distance from zero but lie on opposite sides of zero.

3. Integers are whole numbers and their opposites.

4. Absolute value is the value of the distance between a number and 0 on the number line.

5. $-3 \ ? -1$
-3 lies to the left of -1 on the number line.
$-3 < -1$

6. $|5| \ ? \ |-13|$
$|5| = 5, |-13| = 13$
$5 < 13$
$|5| < |-13|$

7. $-9 \ ? -11$
-9 lies to the right of -11 on the number line.
$-9 > -11$

8. __+__ A profit of \$200

9. __−__ A drop in temperature of 18°

10. The opposite of 12 is __−12__.

11. $-(-(-6)) = -(6) = -6$

12. $-|-11| = -(11) = -11$

13. The number -23 has a larger absolute value than 12 because -23 is further from 0 on the number line.

14. a. The highest point on the graph corresponds to May. Justin made the most money in May.

Copyright © 2012 Pearson Education, Inc.

b. The lowest point on the graph corresponds to March. Justin lost the most money in March.

15. a. The points for January, February, and May are above the horizontal line indicating zero. Justin had a net gain in these three months.

b. The points for March and April are below the horizontal line indicating zero. Justin had a net loss in these three months.

16. a. $-43 + (-16) = -59$

b. $43 + 16 = 59$

17. a. $-27 + (-39) = -66$

b. $27 + 39 = 66$

18. $-\$25{,}000 + \$15{,}000 = -\$10{,}000$
The company had a net loss.

19. $-\$14 + \$25 = \$11$
Terry had a net profit.

20. a. $-10°\text{F} + 20°\text{F}$

b. Positive

c. $-10°\text{F} + 20°\text{F} = 10°\text{F}$

21. a. $2 + (-8) = -6$

b. $-2 + 8 = 6$

c. $-2 + (-8) = -10$

22. a. $27 + (-18) = 9$

b. $-27 + 18 = -9$

c. $-27 + (-18) = -45$

23.
$$\begin{aligned}3+(-5)+8+(-2) &= (3+8)+[(-5)+(-2)]\\ &= 11+(-7)\\ &= 4\end{aligned}$$

24.
$$\begin{aligned}&24+(-52)+(-12)+(-56)\\ &= 24+[(-52)+(-12)+(-56)]\\ &= 24+(-120)\\ &= -96\end{aligned}$$

25. Replace x with -1.
$x + 6 = (x) + 6 = (-1) + 6 = 5$

26. Replace x with -3 and y with -11.
$$\begin{aligned}-x+y+2 &= -(x)+(y)+2\\ &= -(-3)+(-11)+2\\ &= 3+(-11)+2\\ &= -11+3+2\\ &= -11+5\\ &= -6\end{aligned}$$

27.
$$\begin{aligned}&-240+350+400+(-800)\\ &= -240+(-800)+(350+400)\\ &= -1040+750\\ &= -290\end{aligned}$$
The plane is 290 feet below its initial position of 35,000 feet. This can be expressed as -290 feet.

28. $-7 - 5 = -7 + (-5) = -12$

29. $-9 - (-4) = -9 + 4 = -5$

30. $-4 - 4 = -4 + (-4) = -8$

31. $-6 - (-6) = -6 + 6 = 0$

32. $-6 - 9 + 4 = -6 + (-9) + 4 = -15 + 4 = -11$

33. $6 - (-4) + (-5) = 6 + 4 + (-5) = 10 + (-5) = 5$

34. $-4 - (-2) = -4 + 2 = -2$

35.
$$\begin{aligned}6-9-2-8 &= 6+(-9)+(-2)+(-8)\\ &= 6+[(-9)+(-2)+(-8)]\\ &= 6+(-19)\\ &= -13\end{aligned}$$

36.
$$\begin{aligned}-6-(-9)+(-1) &= -6+9+(-1)\\ &= -6+(-1)+9\\ &= (-7)+9\\ &= 2\end{aligned}$$

37. Replace y with -2.
$y-15 = (y)-15 = (-2)-15 = -2+(-15) = -17$

38. Replace x with -4 and y with -2.
$$\begin{aligned}-1-x+y &= -1-(x)+(y)\\ &= -1-(-4)+(-2)\\ &= -1+4+(-2)\\ &= -1+(-2)+4\\ &= -3+4\\ &= 1\end{aligned}$$

Copyright © 2012 Pearson Education, Inc.

39. 4th quarter gain: \$30,000
3rd quarter loss: –\$20,000
$30,000 - (-20,000) = 30,000 + 20,000 = 50,000$
The difference between the fourth quarter gain and the third quarter loss is \$50,000.

40. 1st quarter gain: \$10,000
2nd quarter loss: –\$30,000
$10,000 - (-30,000) = 10,000 + 30,000 = 40,000$
The difference between the first quarter gain and the second quarter loss is \$40,000.

41. $2300 - (-1312) = 2300 + 1312 = 3612$
The difference in altitude is 3612 feet.

42. **a.** $6(3) = 18$

b. $6(-3) = -18$

c. $-6(3) = -18$

d. $-6(-3) = 18$

43. **a.** $5(2) = 10$

b. $5(-2) = -10$

c. $-5(2) = -10$

d. $-5(-2) = 10$

44. $-7(-2) = 14$

45. $-2(5) = -10$

46. $3(-4) = -12$

47. $-4(-1) = 4$

48. $(-2)(-5)(-9) = 10(-9) = -90$

49. $$\begin{aligned}(-2)(-8)(-1)(-4) &= [(-2)(-8)][(-1)(-4)] \\ &= 16(4) \\ &= 64\end{aligned}$$

50. $$\begin{aligned}(-5)(1)(-2)(4)(-6) &= (-5)(-2)(4)(-6) \\ &= 10(4)(-6) \\ &= 40(-6) \\ &= -240\end{aligned}$$

51. $(-7)^2 = (-7)(-7) = 49$

52. $-9^2 = -(9)(9) = -81$

53. $(-6)^3 = (-6)(-6)(-6) = -216$

54. **a.** $49 \div 7 = 7$

b. $49 \div (-7) = -7$

55. **a.** $-30 \div 5 = -6$

b. $-30 \div (-5) = 6$

56. **a.** $-44 \div (-4) = 11$

b. $9(-5) = -45$

c. $(-11)(-3) = 33$

d. $\dfrac{25}{-5} = -5$

57. **a.** $12 \div (-4) = -3$

b. $5(-8) = -40$

c. $-12(-2) = 24$

d. $\dfrac{36}{-9} = -4$

58. Replace y with -1.
$y^4 = (y)^4 = (-1)^4 = (-1)(-1)(-1)(-1) = 1$

59. Replace x with -3.
$x^3 = (x)^3 = (-3)^3 = (-3)(-3)(-3) = -27$

60. Replace a with -20 and b with 5.
$\dfrac{-a}{b} = \dfrac{-(a)}{(b)} = \dfrac{-(-20)}{(5)} = \dfrac{20}{5} = 4$

61. Replace m with 6 and n with -2.
$\dfrac{-m}{-n} = \dfrac{-(m)}{-(n)} = \dfrac{-(6)}{-(-2)} = \dfrac{-6}{2} = -3$

62. $$\begin{aligned}4 - 1(6 - 9) &= 4 - 1[6 + (-9)] \\ &= 4 - 1(-3) \\ &= 4 - (-3) \\ &= 4 + 3 \\ &= 7\end{aligned}$$

Copyright © 2012 Pearson Education, Inc.

63. $3(-5)(2-6)+8 = 3(-5)[2+(-6)]+8$
$= 3(-5)(-4)+8$
$= -15(-4)+8$
$= 60+8$
$= 68$

64. $-2^2+3(-4) = -4+3(-4) = -4+(-12) = -16$

65. $\frac{(-32 \div 8+4)}{(7-9)} = \frac{(-4+4)}{(7-9)} = \frac{0}{(7-9)} = \frac{0}{-2} = 0$

66. $12+3(-5)+(-2) = 12+(-15)+(-2)$
$= 12+(-17)$
$= -5$
The temperature was −5°F at midnight.

67. $-4y+3x+9y = -4y+9y+3x$
$= (-4+9)y+3x$
$= 5y+3x$
$= 3x+5y$

68. $-6a-a = -6a-1a$
$= -6a+(-1a)$
$= [-6+(-1)]a$
$= -7a$

69. $7x+9y-6x-11y = 7x+9y+(-6x)+(-11y)$
$= 7x+(-6x)+9y+(-11y)$
$= [7+(-6)]x+[9+(-11)]y$
$= 1x+(-2y)$
$= x-2y$

70. $3+5z-7+2yz-8z$
$= 3+5z+(-7)+2yz+(-8z)$
$= 3+(-7)+5z+(-8z)+2yz$
$= [3+(-7)]+[5+(-8)]z+2yz$
$= -4+(-3z)+2yz$
$= -4-3z+2yz$

71. Replace a with 8 and b with −4.
$a+3b = (a)+3(b) = (8)+3(-4) = 8+(-12) = -4$

72. Replace x with −2 and y with −1.
$2x-y = 2(x)-(y)$
$= 2(-2)-(-1)$
$= -4-(-1)$
$= -4+1$
$= -3$

73. Replace x with −1 and y with −7.
$\frac{(x^2-y)}{4} = \frac{[(x)^2-(y)]}{4}$
$= \frac{[(-1)^2-(-7)]}{4}$
$= \frac{[1-(-7)]}{4}$
$= \frac{(1+7)}{4}$
$= \frac{8}{4}$
$= 2$

74. Replace a with −3 and b with 9.
$a^2-b = (a)^2-(b)$
$= (-3)^2-(9)$
$= 9-9$
$= 9+(-9)$
$= 0$

75. $C = \frac{(5F-160)}{9}$
$= \frac{[5(41)-160]}{9}$
$= \frac{(205-160)}{9}$
$= \frac{45}{9}$
$= 5$
The temperature is 5°C.

76. $C = \frac{(5F-160)}{9}$
$= \frac{[5(-4)-160]}{9}$
$= \frac{(-20-160)}{9}$
$= \frac{-180}{9}$
$= -20$
The temperature is −20°C.

77. $-6(x+1) = -6x+(-6)(1)$
$= -6x+(-6)$
$= -6x-6$

78. $-2(a-1) = -2a-(-2)(1) = -2a-(-2) = -2a+2$

79. $4(-2+x) = 4(-2)+4x = -8+4x$

Copyright © 2012 Pearson Education, Inc.

How Am I Doing? Chapter 2 Test

1. $-234\ ?\ -5$
   -234 lies to the left of -5 on the number line.
   $-234 < -5$

2. $|4|\ ?\ |-18|$
   $4\ ?\ 18$
   $4 < 18$
   $|4| < |-18|$

3. $\underline{\ -\ }$ 14 points

4. $-(-(-2)) = -(2) = -2$

5. The opposite of 10 is $\underline{\ -10\ }$.

6. **a.** $|12| = 12$

   b. $-|-3| = -(3) = -3$

7. **a.** $-10°F + 15°F$

   b. $-10°F + 15°F = 5°F$

8. **a.** $-6 + 8 = 2$

   b. $6 + (-8) = -2$

9. $-6 + (-4) = -10$

10. $$\begin{aligned} -20+5+(-1)+(-3) &= (-20)+(-1)+(-3)+5 \\ &= (-24)+5 \\ &= -19 \end{aligned}$$

11. $12 - 18 = 12 + (-18) = -6$

12. **a.** $-1 - 11 = -1 + (-11) = -12$

    b. $-1 - (-11) = -1 + 11 = 10$

13. $3 - (-10) = 3 + 10 = 13$

14. $$\begin{aligned} -14-3+(-6)-1 &= -14+(-3)+(-6)+(-1) \\ &= -24 \end{aligned}$$

15. $(7)(-3) = -21$

16. $(-8)(-4) = 32$

17. $(-5)(-2)(-1)(3) = 10(-1)(3) = -10(3) = -30$

18. **a.** $(-5)^2 = (-5)(-5) = 25$

    b. $(-5)^3 = (-5)(-5)(-5) = -125$

    c. $-5^2 = -(5)(5) = -25$

19. **a.** $-8 \div 2 = -4$

    b. $-8 \div (-2) = 4$

20. $\dfrac{-22}{11} = -22 \div 11 = -2$

21. $$\begin{aligned} 2-35\div 5(-3)^2-6 &= 2-35\div 5(9)-6 \\ &= 2-7(9)-6 \\ &= 2-63-6 \\ &= 2+(-63)+(-6) \\ &= -61+(-6) \\ &= -67 \end{aligned}$$

22. $$\begin{aligned} \frac{[-8+2(-3)]}{(14-21)} &= \frac{[-8+(-6)]}{(14-21)} \\ &= \frac{-14}{(14-21)} \\ &= \frac{-14}{-7} \\ &= 2 \end{aligned}$$

23. **a.** Replace x with -6 and y with -3.
    $$\begin{aligned} -7-x+y &= -7-(x)+(y) \\ &= -7-(-6)+(-3) \\ &= -7+6+(-3) \\ &= -7+(-3)+6 \\ &= -10+6 \\ &= -4 \end{aligned}$$

    b. Replace x with -7 and y with 6.
    $$\begin{aligned} -7-x+y &= -7-(x)+(y) \\ &= -7-(-7)+(6) \\ &= -7+7+6 \\ &= 0+6 \\ &= 6 \end{aligned}$$

Copyright © 2012 Pearson Education, Inc.

24. Replace x with -1 and y with -4.

$$\begin{aligned}\frac{(2x-y^2)}{-9} &= \frac{[2(x)-(y)^2]}{-9}\\ &= \frac{[2(-1)-(-4)^2]}{-9}\\ &= \frac{[2(-1)-16]}{-9}\\ &= \frac{(-2-16)}{-9}\\ &= \frac{[-2+(-16)]}{-9}\\ &= \frac{-18}{-9}\\ &= 2\end{aligned}$$

25. a. Replace x with -1.

$$x^4 = (x)^4 = (-1)^4 = (-1)(-1)(-1)(-1) = 1$$

b. Replace a with -2.

$$a^3 = (a)^3 = (-2)^3 = (-2)(-2)(-2) = -8$$

26. Replace x with -6 and y with -2.

$$\frac{-x}{y} = \frac{-(x)}{(y)} = \frac{-(-6)}{(-2)} = \frac{6}{(-2)} = 6 \div (-2) = -3$$

27.
$$\begin{aligned}5x+2y-8x-6y &= 5x+2y+(-8x)+(-6y)\\ &= 5x+(-8x)+2y+(-6y)\\ &= [5+(-8)]x+[2+(-6)]y\\ &= -3x+(-4y)\\ &= -3x-4y\end{aligned}$$

28.
$$\begin{aligned}&-3x+7xy+8y-12x-11y\\ &= -3x+7xy+8y+(-12x)+(-11y)\\ &= -3x+(-12x)+8y+(-11y)+7xy\\ &= [-3+(-12)]x+[8+(-11)]y+7xy\\ &= -15x+(-3y)+7xy\\ &= -15x-3y+7xy\end{aligned}$$

29.
$$\begin{aligned}-6(a+7) &= -6a+(-6)(7)\\ &= -6a+(-42)\\ &= -6a-42\end{aligned}$$

30.
$$\begin{aligned}-2(x-1) &= -2x-(-2)(1)\\ &= -2x-(-2)\\ &= -2x+2\end{aligned}$$

31. 1st quarter gain: \$20,000
2nd quarter loss: −\$5000
20,000 + (−5000) = 15,000
The company's overall profit at the end of the second quarter was \$15,000.

32. 3700 − (−529) = 3700 + 592 = 4292
The difference in altitude is 4292 feet.

33. Replace t with 5 and v with -7.

$$\begin{aligned}s &= v-32t\\ &= -7-32(5)\\ &= -7-160\\ &= -7+(-160)\\ &= -167\end{aligned}$$

The skydiver is falling 167 feet per second.

Copyright © 2012 Pearson Education, Inc.

Chapter 3

3.1 Exercises

1. The sum of two opposite numbers is equal to zero.

3. To solve $x - 6 = 3$, we add 6 to both sides of the equation.

5. $3 + \boxed{-3} = 0$

7. $-9 + \boxed{9} = 0$

9. $17 + \boxed{-17} = 0$

11. $-28 + \boxed{28} = 0$

13. $x + 5 + \boxed{-5} = x$

15. $m - 2 + \boxed{2} = m$

17.
$$\begin{array}{rcr} x+12 & = & 16 \\ +\ \boxed{-12} & & \boxed{-12} \\ \hline x+\boxed{0} & = & \boxed{4} \\ x & = & \boxed{4} \end{array}$$

19.
$$\begin{array}{rcr} y-16 & = & 32 \\ +\ \boxed{16} & & \boxed{16} \\ \hline y+\boxed{0} & = & \boxed{48} \\ y & = & \boxed{48} \end{array}$$

21. **a.**
$$\begin{aligned} x-8 &= 22 \\ x-8+8 &= 22+8 \\ x+0 &= 30 \\ x &= 30 \end{aligned}$$
Check:
$$\begin{aligned} x-8 &= 22 \\ 30-8 &\stackrel{?}{=} 22 \\ 22 &= 22 \checkmark \end{aligned}$$

b.
$$\begin{aligned} x+8 &= 22 \\ x+8+(-8) &= 22+(-8) \\ x+0 &= 14 \\ x &= 14 \end{aligned}$$
Check:
$$\begin{aligned} x+8 &= 22 \\ 14+8 &\stackrel{?}{=} 22 \\ 22 &= 22 \checkmark \end{aligned}$$

23. **a.**
$$\begin{aligned} x+2 &= -11 \\ x+2+(-2) &= -11+(-2) \\ x+0 &= -13 \\ x &= -13 \end{aligned}$$
Check:
$$\begin{aligned} x+2 &= -11 \\ -13+2 &\stackrel{?}{=} -11 \\ -11 &= -11 \checkmark \end{aligned}$$

b.
$$\begin{aligned} x-2 &= -11 \\ x-2+2 &= -11+2 \\ x+0 &= -9 \\ x &= -9 \end{aligned}$$
Check:
$$\begin{aligned} x-2 &= -11 \\ -9-2 &\stackrel{?}{=} -11 \\ -11 &= -11 \checkmark \end{aligned}$$

25. **a.**
$$\begin{aligned} -18 &= x+2 \\ -18+(-2) &= x+2+(-2) \\ -20 &= x+0 \\ -20 &= x \end{aligned}$$
Check:
$$\begin{aligned} -18 &= x+2 \\ -18 &\stackrel{?}{=} -20+2 \\ -18 &= -18 \checkmark \end{aligned}$$

b.
$$\begin{aligned} -18 &= x-2 \\ -18+2 &= x-2+2 \\ -16 &= x \end{aligned}$$
Check:
$$\begin{aligned} -18 &= x-2 \\ -18 &\stackrel{?}{=} -16-2 \\ -18 &= -18 \checkmark \end{aligned}$$

27.
$$\begin{aligned} y-10 &= 5 \\ y-10+10 &= 5+10 \\ y+0 &= 15 \\ y &= 15 \end{aligned}$$
Check:
$$\begin{aligned} y-10 &= 5 \\ 15-10 &\stackrel{?}{=} 5 \\ 5 &= 5 \checkmark \end{aligned}$$

29.
$$\begin{aligned} n-43 &= -74 \\ n-43+43 &= -74+43 \\ n+0 &= -31 \\ n &= -31 \end{aligned}$$
Check:
$$\begin{aligned} n-43 &= -74 \\ -31-43 &\stackrel{?}{=} -74 \\ -74 &= -74 \checkmark \end{aligned}$$

Copyright © 2012 Pearson Education, Inc.

31.
$$\begin{aligned} y+20&=-35\\ y+20+(-20)&=-35+(-20)\\ y+0&=-55\\ y&=-55 \end{aligned}$$
Check:
$$\begin{aligned} y+20&=-35\\ -55+20&\stackrel{?}{=}-35\\ -35&=-35\ \checkmark \end{aligned}$$

33.
$$\begin{aligned} 38+x&=4\\ 38+(-38)+x&=4+(-38)\\ 0+x&=-34\\ x&=-34 \end{aligned}$$
Check:
$$\begin{aligned} 38+x&=4\\ 38+(-34)&\stackrel{?}{=}4\\ 4&=4\ \checkmark \end{aligned}$$

35.
$$\begin{aligned} 1&=x-13\\ 1+13&=x-13+13\\ 14&=x+0\\ 14&=x \end{aligned}$$
Check:
$$\begin{aligned} 1&=x-13\\ 1&\stackrel{?}{=}14-13\\ 1&=1\ \checkmark \end{aligned}$$

37.
$$\begin{aligned} 20&=y+11\\ 20+(-11)&=y+11+(-11)\\ 9&=y+0\\ 9&=y \end{aligned}$$
Check:
$$\begin{aligned} 20&=y+11\\ 20&\stackrel{?}{=}9+11\\ 20&=20\ \checkmark \end{aligned}$$

39.
$$\begin{aligned} -13&=x+1\\ -13+(-1)&=x+1+(-1)\\ -14&=x+0\\ -14&=x \end{aligned}$$
Check:
$$\begin{aligned} -13&=x+1\\ -13&\stackrel{?}{=}-14+1\\ -13&=-13\ \checkmark \end{aligned}$$

41.
$$\begin{array}{rcr} 4x-3x-3&=&8\\ x-3&=&8\\ +\qquad 3&&3\\ \hline x+0&=&11\\ x&=&11 \end{array}$$
Check:
$$\begin{aligned} 4x-3x-3&=8\\ 4(11)-3(11)-3&\stackrel{?}{=}8\\ 44-33-3&\stackrel{?}{=}8\\ 8&=8\ \checkmark \end{aligned}$$

43.
$$\begin{array}{rcr} 5y-4y+1&=&-5\\ y+1&=&-5\\ +\qquad -1&&-1\\ \hline y+0&=&-6\\ y&=&-6 \end{array}$$
Check:
$$\begin{aligned} 5y-4y+1&=-5\\ 5(-6)-4(-6)+1&\stackrel{?}{=}-5\\ -30+24+1&\stackrel{?}{=}-5\\ -5&=-5\ \checkmark \end{aligned}$$

45.
$$\begin{array}{rcl} 5&=&2y-y+1\\ 5&=&y+1\\ +\ -1&&-1\\ \hline 4&=&y+0\\ 4&=&y \end{array}$$
Check:
$$\begin{aligned} 5&=2y-y+1\\ 5&\stackrel{?}{=}2(4)-4+1\\ 5&\stackrel{?}{=}8-4+1\\ 5&=5\ \checkmark \end{aligned}$$

47.
$$\begin{array}{rcr} -23+8+x&=&-2+13\\ -15+x&=&11\\ +\quad 15&&15\\ \hline x&=&26 \end{array}$$
Check:
$$\begin{aligned} -23+8+x&=-2+13\\ -23+8+26&\stackrel{?}{=}-2+13\\ 11&=11\ \checkmark \end{aligned}$$

49.
$$\begin{array}{rcl} 4-9&=&a-1+14\\ -5&=&a+13\\ +\ -13&&-13\\ \hline -18&=&a \end{array}$$
Check:
$$\begin{aligned} 4-9&=a-1+14\\ 4-9&\stackrel{?}{=}-18-1+14\\ -5&=-5\ \checkmark \end{aligned}$$

51.
$$\begin{array}{rcr} -45+9+m&=&-6+18\\ -36+m&=&12\\ +\quad 36&&36\\ \hline m&=&48 \end{array}$$
Check:
$$\begin{aligned} -45+9+m&=-6+18\\ -45+9+48&\stackrel{?}{=}-6+18\\ 12&=12\ \checkmark \end{aligned}$$

53.
$$\begin{array}{rcr} -1+11+x&=&-5+9\\ 10+x&=&4\\ +\ -10&&-10\\ \hline x&=&-6 \end{array}$$

Copyright © 2012 Pearson Education, Inc.

Check: $-1+11+x=-5+9$
$-1+11+(-6) \stackrel{?}{=} -5+9$
$4=4$ ✓

55. $3(7-11)=y-5$
$3(-4)=y-5$
$-12=y-5$
$+\quad 5 \quad 5$
$-7=y$

Check: $3(7-11)=y-5$
$3(7-11) \stackrel{?}{=} -7-5$
$3(-4) \stackrel{?}{=} -12$
$-12=-12$ ✓

57. Since $\angle a$ and $\angle b$ are supplementary angles, their sum is 180°.
$\angle a+\angle b = 180°$
$\angle a+86° = 180°$
$+\quad -86° = -86°$
$\angle a = 94°$
The measure of $\angle a$ is 94°.

59. Since $\angle x$ and $\angle y$ are supplementary angles, their sum is 180°.
$\angle x+\angle y = 180°$
$\angle x+112° = 180°$
$+\quad -112° \quad -112°$
$\angle x = 68°$
The measure of $\angle x$ is 68°.

61. Since $\angle x$ and $\angle y$ are supplementary angles, their sum is 180°.
$\angle x+\angle y = 180°$
$\angle x+43° = 180°$
$+\quad -43° \quad -43°$
$\angle x = 137°$
The measure of $\angle x$ is 137°.

63. **a.** Angle x measures 70° more than angle y: $\angle x = \angle y + 70°$.

b. $\angle x = \angle y+70°$
$125° = \angle y+70°$
$+\ -70° \quad -70°$
$55° = \angle y$
The measure of $\angle y$ is 55°.

65. **a.** Angle a measures 40° less than angle b: $\angle a = \angle b - 40°$.

b. $\angle a = \angle b-40°$
$50° = \angle b-40°$
$+\ 40° \quad 40°$
$90° = \angle b$
The measure of $\angle b$ is 90°.

67. Since $\angle a$ and $\angle b$ are supplementary angles, their sum is 180°.
$\angle a+\angle b = 180°$
$66°+(x+5°) = 180°$
$71°+x = 180°$
$+\quad -71° \quad -71°$
$x = 109°$

$\angle b = x + 5° = 109° + 5° = 114°$

Therefore, x = 109° and $\angle b$ = 114°.

69. Since $\angle a$ and $\angle b$ are supplementary angles, their sum is 180°.
$\angle a+\angle b = 180°$
$52°+(x+5°) = 180°$
$57°+x = 180°$
$+\quad -57° \quad -57°$
$x = 123°$

$\angle b = x + 5° = 123° + 5° = 128°$

Therefore, x = 123° and $\angle b$ = 128°.

71. $2^2+(5-9)=x+3^3$
$4+(-4)=x+27$
$0=x+27$
$0+(-27)=x+27+(-27)$
$-27=x$

73. $5x+1-2x=4x-2$
$3x+1=4x-2$
$3x+1+2=4x-2+2$
$3x+3=4x$
$3x+(-3x)+3=4x+(-3x)$
$3=x$

Copyright © 2012 Pearson Education, Inc.

75.
$$\begin{array}{rl} \angle a+\angle b+\angle c = & 180^\circ \\ 40^\circ+60^\circ+\angle c = & 180^\circ \\ 100^\circ+\angle c = & 180^\circ \\ +\ -100^\circ \qquad & -100^\circ \\ \hline \angle c = & 80^\circ \end{array}$$

$$\begin{array}{rl} \angle c+\angle d+\angle e = & 180^\circ \\ 80^\circ+\angle d+60^\circ = & 180^\circ \\ \angle d+140^\circ = & 180^\circ \\ + \qquad -140^\circ & -140^\circ \\ \hline \angle d = & 40^\circ \end{array}$$

$$\begin{array}{rl} \angle a+\angle b+\angle f = & 180^\circ \\ 40^\circ+60^\circ+\angle f = & 180^\circ \\ 100^\circ+\angle f = & 180^\circ \\ +\ -100^\circ \qquad & -100^\circ \\ \hline \angle f = & 80^\circ \end{array}$$

$\angle c = 80^\circ$, $\angle d = 40^\circ$, and $\angle f = 80^\circ$.

Cumulative Review

77. Seven times x: $7x$

78. The product of three and a number: $3y$

79. Eight times what number equals forty?
$8n = 40$

80. Double what number equals thirty?
$2n = 30$

81. $-8°F + 21°F = 13°F$
The high temperature was 13°F.

82. $|-16| \ ? \ |-8|$
$|-16| = 16$, $|-8| = 8$
$16 > 8$
$|-16| > |-8|$

83.
$$\begin{aligned} &(15\times 2)+(8\times 1)+[6\times(-1)]+(5\times 2)+(5\times 2) \\ &\quad +[2\times(-1)] \\ &= 30+8+(-6)+10+10+(-2) \\ &= 50 \end{aligned}$$
The player earned 50 points.

84.
$$\begin{aligned} &(12\times 2)+(14\times 1)+[8\times(-1)]+(3\times 2)+(7\times 2) \\ &\quad +[4\times(-1)] \\ &= 24+14+(-8)+6+14+(-4) \\ &= 46 \end{aligned}$$
The player earned 46 points.

Quick Quiz 3.1

1. a.
$$\begin{aligned} x+12 &= -15 \\ x+12+(-12) &= -15+(-12) \\ x+0 &= -27 \\ x &= -27 \end{aligned}$$
Check:
$$\begin{aligned} x+12 &= -15 \\ -27+12 &\stackrel{?}{=} -15 \\ -15 &= -15 \ \checkmark \end{aligned}$$

b.
$$\begin{aligned} y-7 &= -20 \\ y-7+7 &= -20+7 \\ y+0 &= -13 \\ y &= -13 \end{aligned}$$
Check:
$$\begin{aligned} y-7 &= -20 \\ -13-7 &\stackrel{?}{=} -20 \\ -20 &= -20 \ \checkmark \end{aligned}$$

2. a.
$$\begin{array}{rl} 6x-5x+2 = & 10-15 \\ x+2 = & -5 \\ + \qquad -2 & -2 \\ \hline x = & -7 \end{array}$$
Check:
$$\begin{aligned} 6x-5x+2 &= 10-15 \\ 6(-7)-5(-7)+2 &\stackrel{?}{=} 10-15 \\ -42+35+2 &\stackrel{?}{=} 10-15 \\ -5 &= -5 \ \checkmark \end{aligned}$$

b.
$$\begin{array}{rl} -8+10 = & 6-9+y \\ 2 = & -3+y \\ + \qquad 3 & 3 \\ \hline 5 = & y \end{array}$$
Check:
$$\begin{aligned} -8+10 &= 6-9+y \\ -8+10 &\stackrel{?}{=} 6-9+5 \\ 2 &= 2 \ \checkmark \end{aligned}$$

3. a. The measure of angle a is 35° less than the measure of angle b: $\angle a = \angle b - 35^\circ$.

b.
$$\begin{array}{rl} \angle a = & \angle b-35^\circ \\ 75^\circ = & \angle b-35^\circ \\ +\ 35^\circ \qquad & 35^\circ \\ \hline 110^\circ = & \angle b \end{array}$$
The measure of $\angle b$ is 110°.

4. Answers may vary. It is not correct. He should have added 9 to each side.

Copyright © 2012 Pearson Education, Inc.

3.2 Exercises

1. To solve the equation $-22x = 66$, we undo the multiplication by <u>dividing</u> both sides of the equation by <u>-22</u>.

3. $\frac{5x}{\boxed{5}} = x$

5. $\frac{-2x}{\boxed{-2}} = x$

7. $\frac{6 \cdot x}{\boxed{6}} = x$

9. $\frac{-1 \cdot x}{\boxed{-1}} = x$

11. $3x = 36$
$\frac{3x}{3} = \frac{36}{3}$
$x = 12$
Check: $3x = 36$
$3(12) \stackrel{?}{=} 36$
$36 = 36$ ✓

13. $10x = 40$
$\frac{10x}{10} = \frac{40}{10}$
$x = 4$
Check: $10x = 40$
$10(4) \stackrel{?}{=} 40$
$40 = 40$ ✓

15. $6y = -18$
$\frac{6y}{6} = \frac{-18}{6}$
$y = -3$
Check: $6y = -18$
$6(-3) \stackrel{?}{=} -18$
$-18 = -18$ ✓

17. $5m = -35$
$\frac{5m}{5} = \frac{-35}{5}$
$m = -7$
Check: $5m = -35$
$5(-7) \stackrel{?}{=} -35$
$-35 = -35$ ✓

19. $-3y = 15$
$\frac{-3y}{-3} = \frac{15}{-3}$
$y = -5$
Check: $-3y = 15$
$-3(-5) \stackrel{?}{=} 15$
$15 = 15$ ✓

21. $-7a = 49$
$\frac{-7a}{-7} = \frac{49}{-7}$
$a = -7$
Check: $-7a = 49$
$-7(-7) \stackrel{?}{=} 49$
$49 = 49$ ✓

23. $48 = 6x$
$\frac{48}{6} = \frac{6x}{6}$
$8 = x$
Check: $48 = 6x$
$48 \stackrel{?}{=} 6(8)$
$48 = 48$ ✓

25. $72 = 9x$
$\frac{72}{9} = \frac{9x}{9}$
$8 = x$
Check: $72 = 9x$
$72 \stackrel{?}{=} 9(8)$
$72 = 72$ ✓

27. $8x = 104$
$\frac{8x}{8} = \frac{104}{8}$
$x = 13$
Check: $8x = 104$
$8(13) \stackrel{?}{=} 104$
$104 = 104$ ✓

29. $-19x = -76$
$\frac{-19x}{-19} = \frac{-76}{-19}$
$x = 4$
Check: $-19x = -76$
$-19(4) \stackrel{?}{=} -76$
$-76 = -76$ ✓

Copyright © 2012 Pearson Education, Inc.

31. $2(3x) = 54$
$(2 \cdot 3)x = 54$
$6x = 54$
$\frac{6x}{6} = \frac{54}{6}$
$x = 9$
Check: $2(3x) = 54$
$2(3 \cdot 9) \stackrel{?}{=} 54$
$2(27) \stackrel{?}{=} 54$
$54 = 54$ ✓

33. $5(4x) = 40$
$(5 \cdot 4)x = 40$
$20x = 40$
$\frac{20x}{20} = \frac{40}{20}$
$x = 2$
Check: $5(4x) = 40$
$5(4 \cdot 2) \stackrel{?}{=} 40$
$5(8) \stackrel{?}{=} 40$
$40 = 40$ ✓

35. $5(x \cdot 2) = \frac{40}{2}$
$5(x \cdot 2) = 20$
$5(2x) = 20$
$(5 \cdot 2)x = 20$
$10x = 20$
$\frac{10x}{10} = \frac{20}{10}$
$x = 2$

Check: $5(x \cdot 2) = \frac{40}{2}$
$5(2 \cdot 2) \stackrel{?}{=} \frac{40}{2}$
$5(4) \stackrel{?}{=} 20$
$20 = 20$ ✓

37. $4(x \cdot 2) = \frac{96}{3}$
$4(x \cdot 2) = 32$
$4(2x) = 32$
$(4 \cdot 2)x = 32$
$8x = 32$
$\frac{8x}{8} = \frac{32}{8}$
$x = 4$

Check: $4(x \cdot 2) = \frac{96}{3}$
$4(4 \cdot 2) \stackrel{?}{=} \frac{96}{3}$
$4(8) \stackrel{?}{=} 32$
$32 = 32$ ✓

39. $-26 - 18 = 11a$
$-44 = 11a$
$\frac{-44}{11} = \frac{11a}{11}$
$-4 = a$
Check: $-26 - 18 = 11a$
$-26 - 18 \stackrel{?}{=} 11(-4)$
$-44 = -44$ ✓

41. $-4 - 4 = 8y$
$-8 = 8y$
$\frac{-8}{8} = \frac{8y}{8}$
$-1 = y$
Check: $-4 - 4 = 8y$
$-4 - 4 \stackrel{?}{=} 8(-1)$
$-8 = -8$ ✓

43. $5x - 2x = 24$
$3x = 24$
$\frac{3x}{3} = \frac{24}{3}$
$x = 8$
Check: $5x - 2x = 24$
$5(8) - 2(8) \stackrel{?}{=} 24$
$40 - 16 \stackrel{?}{=} 24$
$24 = 24$ ✓

45. $65 = 15x - 10x$
$65 = 5x$
$\frac{65}{5} = \frac{5x}{5}$
$13 = x$
Check: $65 = 15x - 10x$
$65 \stackrel{?}{=} 15(13) - 10(13)$
$65 \stackrel{?}{=} 195 - 130$
$65 = 65$ ✓

47. $-15y = 165$
$\frac{-15y}{-15} = \frac{165}{-15}$
$y = -11$

Copyright © 2012 Pearson Education, Inc.

Check: $-15y = 165$
$-15(-11) \stackrel{?}{=} 165$
$165 = 165$ ✓

49. $-4x = 3 - 23$
$-4x = -20$
$\frac{-4x}{-4} = \frac{-20}{-4}$
$x = 5$
Check: $-4x = 3 - 23$
$-4(5) \stackrel{?}{=} 3 - 23$
$-20 = -20$ ✓

51. $12x - 4x = 56$
$8x = 56$
$\frac{8x}{8} = \frac{56}{8}$
$x = 7$
Check: $12x - 4x = 56$
$12(7) - 4(7) \stackrel{?}{=} 56$
$84 - 28 \stackrel{?}{=} 56$
$56 = 56$ ✓

53. $55 = 5a$
$\frac{55}{5} = \frac{5a}{5}$
$11 = a$
Check: $55 = 5a$
$55 \stackrel{?}{=} 5(11)$
$55 = 55$ ✓

55. $(3x) \cdot 2 = \frac{36}{3}$
$(3x) \cdot 2 = 12$
$6x = 12$
$\frac{6x}{6} = \frac{12}{6}$
$x = 2$

Check: $(3x) \cdot 2 = \frac{36}{3}$
$(3 \cdot 2) \cdot 2 \stackrel{?}{=} \frac{36}{3}$
$6 \cdot 2 \stackrel{?}{=} 12$
$12 = 12$ ✓

57. $9x + 3x = -120$
$12x = -120$
$\frac{12x}{12} = \frac{-120}{12}$
$x = -10$
Check: $9x + 3x = -120$
$9(-10) + 3(-10) \stackrel{?}{=} -120$
$-90 + (-30) \stackrel{?}{=} -120$
$-120 = -120$ ✓

59. The length (L) of a building is three times the width (W): $L = 3W$.

61. The original price (R) of a ring is double the sale price (S): $R = 2S$.

63. a. The cost of a new race boat (R) is double the cost of an older model boat (B): $R = 2B$.

b. Replace R with 124,000.
$R = 2B$
$124,000 = 2B$
$\frac{124,000}{2} = \frac{2B}{2}$
$62,000 = B$
The cost of the older model boat is \$62,000.

65. a. There were three times as many children's tickets (C) sold as adults' tickets (A): $C = 3A$.

b. Replace C with 300.
$C = 3A$
$300 = 3A$
$\frac{300}{3} = \frac{3A}{3}$
$100 = A$
100 adults' tickets were sold.

67. a. The total number of soccer goals attempted (A) by a team is double the number of goals scored (S): $A = 2S$.

b. Replace A with 42.
$A = 2S$
$42 = 2S$
$\frac{42}{2} = \frac{2S}{2}$
$21 = S$
21 goals were scored.

Copyright © 2012 Pearson Education, Inc.

69. Let x = the number of shares of stock purchased, $30x$ = the purchase price, and $42x$ = the sale price.

$$360 = 42x - 30x$$
$$360 = 12x$$
$$\frac{360}{12} = \frac{12x}{12}$$
$$30 = x$$

Vu purchased 30 shares of stock.

71. Let x = the number of miles Leah drove on Monday and $2x$ the number of miles on Tuesday.

$$x + 2x = 360$$
$$3x = 360$$
$$\frac{3x}{3} = \frac{360}{3}$$
$$x = 120$$

Leah drove 120 miles on Monday.

73. a.
$$13x = 26$$
$$\frac{13x}{13} = \frac{26}{13}$$
$$x = 2$$

b.
$$-13x = 26$$
$$\frac{-13x}{-13} = \frac{26}{-13}$$
$$x = -2$$

c.
$$x + 13 = 26$$
$$x + 13 + (-13) = 26 + (-13)$$
$$x = 13$$

d.
$$x - 13 = 26$$
$$x - 13 + 13 = 26 + 13$$
$$x = 39$$

75. a.
$$13x = -26$$
$$\frac{13x}{13} = \frac{-26}{13}$$
$$x = -2$$

b.
$$-13x = -26$$
$$\frac{-13x}{-13} = \frac{-26}{-13}$$
$$x = 2$$

c.
$$x + 13 = -26$$
$$x + 13 + (-13) = -26 + (-13)$$
$$x = -39$$

d.
$$x - 13 = -26$$
$$x - 13 + 13 = -26 + 13$$
$$x = -13$$

77. Since the angles are supplementary, their sum is 180°.

$$x + 5x = 180°$$
$$6x = 180°$$
$$\frac{6x}{6} = \frac{180°}{6}$$
$$x = 30°$$

So, $5x = 5(30°) = 150°$. The angles measure 30° and 150°.

79. Since the angles are supplementary, their sum is 180°.

$$x + 4x = 180°$$
$$5x = 180°$$
$$\frac{5x}{5} = \frac{180°}{5}$$
$$x = 36°$$

So, $4x = 4(36°) = 144°$. The angles measure 36° and 144°.

81. Since the angles are supplementary, $\angle a$ is 32° less than 180°.

$$\angle a = 180° - 32°$$
$$4x = 180° - 32°$$
$$4x = 148°$$
$$\frac{4x}{4} = \frac{148°}{4}$$
$$x = 37°$$

Cumulative Review

83. 6 ft + 4 ft + 6 ft + 4 ft = 20 ft
The perimeter is 20 feet.

84. 3 in. + 3 in. + 3 in. + 3 in. = 12 in.
The perimeter is 12 inches.

85. Replace x with −7 and y with −9.
$2xy = 2(-7)(-9) = -14(-9) = 126$

86. Replace L with 2, W with 3, and H with 5.
$L \cdot W \cdot H = 2 \cdot 3 \cdot 5 = 6 \cdot 5 = 30$

Copyright © 2012 Pearson Education, Inc.

Quick Quiz 3.2

1. a.
$$7x-10x=18+45$$
$$7x-10x=63$$
$$-3x=63$$
$$\frac{-3x}{-3}=\frac{63}{-3}$$
$$x=-21$$
Check:
$$7x-10x=18+45$$
$$7(-21)-10(-21)\stackrel{?}{=}18+45$$
$$-147+210\stackrel{?}{=}63$$
$$63=63\ \checkmark$$

b.
$$-8+20=3(2x)$$
$$-8+20=(3\cdot 2)x$$
$$-8+20=6x$$
$$12=6x$$
$$\frac{12}{6}=\frac{6x}{6}$$
$$2=x$$
Check:
$$-8+20=3(2x)$$
$$-8+20\stackrel{?}{=}3(2\cdot 2)$$
$$12\stackrel{?}{=}3(4)$$
$$12=12\ \checkmark$$

2. Mindy (M) has five times as many quarters as Sara (S): $M = 5S$.

3. a. The number of blue marbles (B) is twice the number of red marbles (R): $B = 2R$.

b. Replace B with 10.
$$B=2R$$
$$10=2R$$
$$\frac{10}{2}=\frac{2R}{2}$$
$$5=R$$
There are 5 red marbles.

4. Answers may vary. First multiply $3(2x)$ to get $6x$. Then combine the like terms, $4x$ and $6x$, to get $10x$. Finally divide each side by 10 and simplify to get $x = -2$.

How Am I Doing? Sections 3.1–3.2

1.
$$y-15=-26$$
$$y-15+15=-26+15$$
$$y+0=-11$$
$$y=-11$$
Check:
$$y-15=-26$$
$$-11-15\stackrel{?}{=}-26$$
$$-26=-26\ \checkmark$$

2.
$$4x-3x-2=9$$
$$x-2=9$$
$$\underline{+\qquad 2\quad 2}$$
$$x=11$$
Check:
$$4x-3x-2=9$$
$$4(11)-3(11)-2\stackrel{?}{=}9$$
$$44-33-2\stackrel{?}{=}9$$
$$9=9\ \checkmark$$

3.
$$2-8=a-1+11$$
$$-6=a+10$$
$$\underline{+\ -10\qquad -10}$$
$$-16=a$$
Check:
$$2-8=a-1+11$$
$$2-8\stackrel{?}{=}-16-1+11$$
$$-6=-6\ \checkmark$$

4. Since the angles are supplementary, their sum is 180°.
$$\angle x+\angle y=180^\circ$$
$$\angle x+115^\circ=180^\circ$$
$$\underline{+\quad -115^\circ\ \ -115^\circ}$$
$$\angle x=65^\circ$$
The measure of $\angle x$ is 65°.

5. a. Angle x measures 30° more than angle y: $\angle x = \angle y + 30^\circ$.

b.
$$\angle x=\angle y+30^\circ$$
$$90^\circ=\angle y+30^\circ$$
$$\underline{+\ -30^\circ\qquad -30^\circ}$$
$$60^\circ=\angle y$$
The measure of $\angle y$ is 60°.

6.
$$-16-12=14a$$
$$-28=14a$$
$$\frac{-28}{14}=\frac{14a}{14}$$
$$-2=a$$

Copyright © 2012 Pearson Education, Inc.

Check: $-16-12=14a$
$-16-12 \stackrel{?}{=} 14(-2)$
$-28=-28$ ✓

7. $2(3x)=-18$
$6x=-18$
$\frac{6x}{6}=\frac{-18}{6}$
$x=-3$
Check: $2(3x)=-18$
$2[3(-3)] \stackrel{?}{=} -18$
$2(-9) \stackrel{?}{=} -18$
$-18=-18$ ✓

8. $-48-10=8x-6x$
$-48-10=2x$
$-58=2x$
$\frac{-58}{2}=\frac{2x}{2}$
$-29=x$
Check: $-48-10=8x-6x$
$-48-10 \stackrel{?}{=} 8(-29)-6(-29)$
$-48-10 \stackrel{?}{=} -232+174$
$-58=-58$ ✓

9. The length (L) of a building is four times the width (W): $L=4W$.

10. a. The number of children's tickets (C) sold is triple the number of adults' tickets (A) sold: $C=3A$.

b. Replace C with 150.
$C=3A$
$150=3A$
$\frac{150}{3}=\frac{3A}{3}$
$50=A$
50 adults' tickets were sold.

11. Let x = the number of shares of stock purchased, $50x$ = the purchase price, and $65x$ = the sales price.
$795=65x-50x$
$795=15x$
$\frac{795}{15}=\frac{15x}{15}$
$53=x$
Linda purchased 53 shares of stock.

3.3 Exercises

1. Volume, because we want to find the amount of space inside the pool.

3. a. Perimeter of a rectangle: $\underline{P=2L+2W}$

b. Perimeter of a square: $\underline{P=4s}$

c. Volume of a rectangular solid: $\underline{V=LWH}$

d. Area of a rectangle: $\underline{A=LW}$

e. Area of a square: $\underline{A=s^2}$

f. Area of a parallelogram: $\underline{A=bh}$

5. a. $P=2L+2W$

b. $P=2(2)+2(7)=4+14=18$ ft

7. a. $P=4s$

b. $P=4(11)=44$ ft

9. a. $P=4s$

b. $P=4(54)=216$ yd

11. a. $P=4s$

b. $40=4s$
$\frac{40}{4}=\frac{4s}{4}$
$10=s$
The length of each side is 10 feet.

13. a. $P=2L+2W$

b. $30=2(4W)+2W$
$30=8W+2W$
$30=10W$
$\frac{30}{10}=\frac{10W}{10}$
$3=W$
The width is 3 feet.

Copyright © 2012 Pearson Education, Inc.

15. $P = 2L + 2W$
$66 = 2(10W) + 2W$
$66 = 20W + 2W$
$66 = 22W$
$\frac{66}{22} = \frac{22W}{22}$
$3 = W$
$L = 10W = 10(3) = 30$
The width is 3 feet, and the length is 30 feet.

17. a. 50 squares with 1-inch sides can be placed in a space that is 50 square inches.

b. The area is 50 square inches.

19. 50 square tiles with sides 1 foot in length are needed to fill a space that has an area of 50 square feet.

21. a. $A = LW$

b. $A = 22(18) = 396 \text{ ft}^2$

23. a. $A = s^2$

b. $A = 10^2 = 100 \text{ in.}^2$

25. a. $A = bh$

b. $A = 12(9) = 108 \text{ ft}^2$

27. We use the formula for the area of a rectangle.
$A = LW$
$60 = 10x$
$\frac{60}{10} = \frac{10x}{10}$
$6 = x$
The unknown side is 6 feet.

29. We use the formula for the area of a parallelogram.
$A = bh$
$88 = 8h$
$\frac{88}{8} = \frac{8h}{8}$
$11 = h$
The height is 11 meters.

31. Divide the figure into two rectangles and then find the area of each rectangle separately. Add these areas together to find the area of the figure.

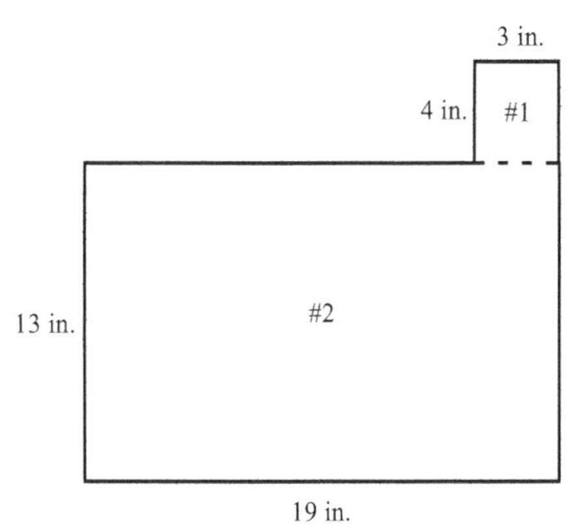

Area of rectangle #1: $A = LW = 3 \cdot 4 = 12 \text{ in.}^2$
Area of rectangle #2:
$A = LW = 13 \cdot 19 = 247 \text{ in.}^2$
Add the areas.
$12 + 247 = 259 \text{ in.}^2$
The area of the shape is 259 square inches.

33. Divide the figure into three rectangles and then find the area of each rectangle separately. Add these areas together to find the area of the figure.

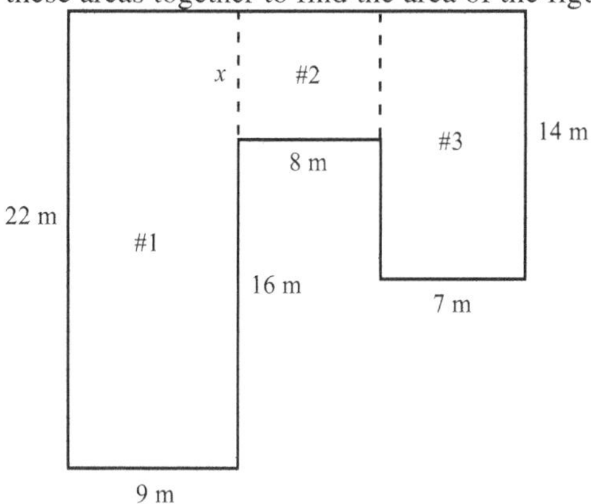

Area of rectangle #1: $A = LW = 22 \cdot 9 = 198 \text{ m}^2$
To find the area of rectangle #2, we must find the width of the rectangle, labeled x in the figure. From the figure, we see that $x + 16 = 22$. Solving this equation, we have $x = 6$, so the width of rectangle #2 is 6 meters.

Area of rectangle #2: $A = LW = 8 \cdot 6 = 48 \text{ m}^2$

Area of rectangle #3: $A = LW = 7 \cdot 14 = 98 \text{ m}^2$
Add the areas.
$198 + 48 + 98 = 344 \text{ m}^2$
The area of the shape is 344 square meters.

35. a. Consider the rectangular solid to be made up of two layers, where each layer is an array of cubes with 5 rows and 4 cubes in each

Copyright © 2012 Pearson Education, Inc.

row. There are 5(4) or 20 cubes in each layer. There are 5(4)(2) or 40 cubes in the two-layer solid.

b. The volume of the rectangular solid is 40 cubic inches.

37. We use the formula for the volume of a rectangular solid.
$V = LWH = 4 \cdot 3 \cdot 3 = 36 \text{ ft}^3$
The volume of the tank is 36 cubic feet, so 36 cubic feet of water can be placed in the tank.

39. a. $V = LWH$

b. $V = 11(5)(15) = 825 \text{ in.}^3$

41. a. $V = LWH$

b. $V = 27(10)(16) = 4320 \text{ yd}^3$

43.
$$V = LWH$$
$$300 = 10W(6)$$
$$300 = 60W$$
$$\frac{300}{60} = \frac{60W}{60}$$
$$5 = W$$
The unknown side is 5 cm.

45. $P = 4s = 4(7) = 28$ in.

47. $A = bh = 17(8) = 136 \text{ m}^2$

49. $V = LWH = 15(7)(2) = 210 \text{ yd}^3$

51. We use the formula for the area of a rectangle.
$$A = LW$$
$$30 = 6x$$
$$\frac{30}{6} = \frac{6x}{6}$$
$$5 = x$$
The unknown side is 5 feet.

53. We use the formula for the volume of a rectangular solid.
$$V = LWH$$
$$200 = L \cdot 5 \cdot 8$$
$$200 = 40L$$
$$\frac{200}{40} = \frac{40L}{40}$$
$$5 = L$$
The unknown side is 5 meters.

55. a. Multiply the number of feet by 12 to find the number of inches.
2(12) = 24
2 ft = <u>24 in.</u>
Each side of the square is 24 inches.

b. $A = s^2 = (24)^2 = 576 \text{ in.}^2$
The area of the marble is 576 square inches.

57. a. Divide each number of feet by 3 to find the number of yards.
12 ÷ 3 = 4; 12 ft = 4 yd
9 ÷ 3 = 3; 9 ft = 3 yd
The garden is 4 yards by 3 yards.

b. $A = LW = 4 \cdot 3 = 12 \text{ yd}^2$
The area of the garden is 12 square yards.

59. a. Divide each number of feet by 3 to find the number of yards.
12 ÷ 3 = 4; 12 ft = 4 yd
9 ÷ 3 = 3: 9 ft = 3 yd
The patio is 4 yards by 3 yards.
$A = LW = 4 \cdot 3 = 12 \text{ yd}^2$
The area of the patio is 12 square yards, so she must purchase 12 square yards of carpet.

b. Multiply the number of square yards by the price per square yard.
12(8) = 96
The outdoor carpet will cost $96.

61. a. Divide each number of feet by 3 to find the number of yards.
21 ÷ 3 = 7; 21 ft = 7 yd
15 ÷ 3 = 5; 15 ft = 5 yd
The room is 7 yards by 5 yards.
$A = LW = 7 \cdot 5 = 35 \text{ yd}^2$
The area of the room is 35 square yards.

b. Multiply the number of square yards by the price per square yard.
35(16) = 560
The linoleum will cost $560.

63. First change the dimensions of the garden from yards to feet.
5(3) = 15; 5 yd = 15 ft
4(3) = 12; 4 yd = 12 ft
Next find the area in square feet.
$A = LW = 15 \cdot 12 = 180 \text{ ft}^2$
The area of the garden is 180 square feet. Since each container of fertilizer covers 100 square feet, 2 containers are needed. It will cost 2($3) or $6 to fertilize the entire rose garden.

Copyright © 2012 Pearson Education, Inc.

65. To find the area that must be painted, in square feet, we find the total area of the walls and the ceiling and subtract the total area of the windows and doors.

Front wall: $A = LW = 22 \cdot 8 = 176 \text{ ft}^2$

Rear wall: $A = LW = 22 \cdot 8 = 176 \text{ ft}^2$

Side wall: $A = LW = 16 \cdot 8 = 128 \text{ ft}^2$

Side wall: $A = LW = 16 \cdot 8 = 128 \text{ ft}^2$

Ceiling: $A = LW = 22 \cdot 16 = 352 \text{ ft}^2$

Total area of walls and ceiling:

$176 + 176 + 128 + 128 + 352 = 960 \text{ ft}^2$

French door: $A = LW = 3 \cdot 7 = 21 \text{ ft}^2$

Slider: $A = LW = 6 \cdot 7 = 42 \text{ ft}^2$

First window: $A = LW = 4 \cdot 3 = 12 \text{ ft}^2$

Second window: $A = LW = 2 \cdot 4 = 8 \text{ ft}^2$

Total area of doors and windows:

$21 + 42 + 12 + 8 = 83 \text{ ft}^2$

Now subtract.

$960 - 83 = 877 \text{ ft}^2$

The area to be painted is 877 square feet. Each gallon of paint covers 400 square feet, so 2 gallons will cover 800 square feet. Each quart covers 100 square feet, so one quart will be enough for the remaining 77 square feet. Anita should purchase 2 gallons and 1 quart of paint.

67. Find the area of each rectangle. Then subtract the area of the smaller rectangle from the area of the larger rectangle.

Larger rectangle: $A = LW = 7 \cdot 5 = 35 \text{ in.}^2$

Smaller rectangle: $A = LW = 3 \cdot 2 = 6 \text{ in.}^2$

Now subtract.

$35 - 6 = 29 \text{ in.}^2$

The area of the shaded region is 29 square inches.

69. Since the last figure shown is the same as the first figure, the pattern appears to start over. Thus, the next figure in the sequence should be the same as the second figure.

Cumulative Review

71.
$$\begin{aligned} -7(x-2) &= -7x - (-7)(2) \\ &= -7x - (-14) \\ &= -7x + 14 \end{aligned}$$

72. $-7x + 3x - 2y = (-7 + 3)x - 2y = -4x - 2y$

73. $(2)(3x)(5) = 6x(5) = 5(6x) = 30x$

74.
$$\begin{aligned} -8-3-1-4 &= [-8+(-3)]+[(-1)+(-4)] \\ &= -11+(-5) \\ &= -16 \end{aligned}$$

Quick Quiz 3.3

1.
$$\begin{aligned} P &= 2L + 2W \\ 72 &= 2L + 2(2L) \\ 72 &= 2L + 4L \\ 72 &= 6L \\ \frac{72}{6} &= \frac{6L}{6} \\ 12 &= L \end{aligned}$$

The length is 12 feet.

2. Divide the figure into three rectangles and then find the area of each rectangle separately. Add these areas together to find the area of the figure.

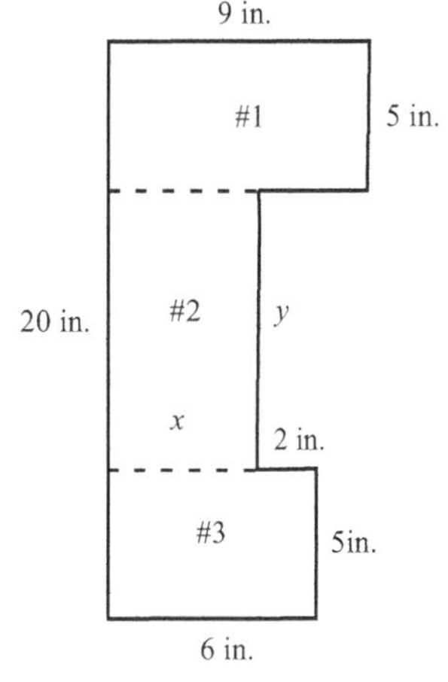

Area of rectangle #1: $A = LW = 9 \cdot 5 = 45 \text{ in}^2$

To find the area of rectangle #2, se must find the length and width of the rectangle, labeled x and y in the figure. From the figure, we see that $x + 2 = 6$. Solving this equation, we have $x = 4$, so the length of rectangle #2 is 4 inches. We also see that $5 + y + 5 = 20$. Solving this equation, we have $y = 10$, so the width of rectangle #2 is 10 inches.

Area of rectangle #2: $A = LW = 4 \cdot 10 = 40 \text{ in.}^2$

Area of rectangle #3: $A = LW = 6 \cdot 5 = 30 \text{ in.}^2$

Add the areas.

$45 + 40 + 30 = 115 \text{ in.}^2$

The area of the shape is 115 square inches.

Copyright © 2012 Pearson Education, Inc.

3. We use the formula for the volume of a rectangular solid.

$V = LWH$

$200 = 5 \cdot 5 \cdot H$

$200 = 25H$

$\frac{200}{25} = \frac{25H}{25}$

$8 = H$

The height is 8 feet.

4. Answers may vary.

a. To determine the amount of sand needed, find the volume.

b. To find the volume of a box, use the formula $V = LWH$.

c. Since the length is double the height, the length is 10 inches. The volume is given in cubic feet, and the length and height are given in inches. First change the units on these measurements so that all are in terms of the same unit. Then write an equation using the formula $V = LWH$, where V, L, and H are replaced with the values from the previous step, and solve for W.

Use Math to Save Money

1. \$3 + \$1 = \$4
It costs Mary \$4 to drive to the farther gas station.

2. \$74 – \$71 = \$3
Mary will save \$3 by buying gas at the farther station.

3. \$3 – \$4 = –\$1
The difference is –\$1.

4. It is not worth driving across town to get gas since Mary will lose money after calculating expenses.

5. \$6 + \$2 = \$8
It costs Mary \$8 to drive to the mall.

6. \$95 – \$76 = \$19
Mary will save \$19 on the sneakers by buying them on sale.

7. \$19 – \$8 = \$11
The difference is \$11.

8. It is worth driving to the mall since Mary will save \$11.

3.4 Understanding the Concept
Do I Add or Multiply Coefficients?

a. Add: $3x + 5x = 8x$.

b. Multiply: $(3x)(5x) = 15x^2$.

c. Add the opposite:
$7xy^2 - 5xy^2 = [7 + (-5)]xy^2 = 2xy^2$.

d. Multiply: $(7xy^2)(5xy^2) = 35x^2y^4$.

3.4 Exercises

1. No, $2^4 \cdot 2^2 \neq 4^6$ because we do not multiply the bases.

3. a. $2(3x^2) \neq 2 \cdot 3 + 2 \cdot x^2 = 6 + 2x^2$ because the distributive property is only used when the expression inside the parentheses is being added or subtracted.

b. $2(3x^2) = 2 \cdot 3x^2 = 6x^2$ because the parentheses means multiply.

c. $2(3 + x^2) = 2 \cdot 3 + 2x^2 = 6 + 2x^2 \neq 12x^2$ because 6 and $2x^2$ are not like terms and cannot be combined.

5. In the algebraic expression $4x^2$ the number 4 is called the coefficient or numerical coefficient.

7. a. There are five total factors of z.
$(z \cdot z \cdot z) \cdot (z \cdot z) = z^5$

b. Add the exponents.
$z^3 \cdot z^2 = z^{3+2} = z^5$

9. a. There are six total factors of x.
$(x \cdot x) \cdot (x \cdot x \cdot x \cdot x) = x^6$

b. Add the exponents.
$x^2 \cdot x^4 = x^{2+4} = x^6$

11. a. There are three factors of m.
$m \cdot m \cdot m = m^3$

 Copyright © 2012 Pearson Education, Inc.

b. Add the exponents.
$m^2 \cdot m = m^2 \cdot m^1 = m^{2+1} = m^3$

13. $2^4 \cdot 2^2 = (2 \cdot 2 \cdot 2 \cdot 2) \cdot (2 \cdot 2) = 2^6$

15. $3^5 \cdot 3^3 = (3 \cdot 3 \cdot 3 \cdot 3 \cdot 3) \cdot (3 \cdot 3 \cdot 3) = 3^8$

17. $x^5 \cdot x^2 = x^{5+2} = x^7$

19. $a \cdot a = a^1 \cdot a^1 = a^{1+1} = a^2$

21. $3^2 \cdot 3^3 = 3^{2+3} = 3^5$

23. $4 \cdot 4^5 = 4^1 \cdot 4^5 = 4^{1+5} = 4^6$

25. $8^2 \cdot 7^5 = 8^2 \cdot 7^5$

27. $x^5 \cdot y^3 = x^5 y^3$

29. $x^5 \cdot x^2 \cdot x^6 = x^{5+2+6} = x^{13}$

31. $y^3 \cdot y^6 \cdot y^5 = y^{3+6+5} = y^{14}$

33. $3^3 \cdot 3^2 \cdot 3^5 = 3^{3+2+5} = 3^{10}$

35. $2^5 \cdot 3^2 \cdot 4^7 = 2^5 \cdot 3^2 \cdot 4^7$

37. $(4y^5)(6y^7) = (4 \cdot 6)(y^5 \cdot y^7) = 24y^{12}$

39. $(6a^6)(9a^8) = (6 \cdot 9)(a^6 \cdot a^8) = 54a^{14}$

41. $(-5x)(3x^2) = (-5)(3)(x^1 \cdot x^2) = -15x^3$

43. $(-4y)(3y) = (-4)(3)(y \cdot y) = -12y^2$

45. $(5a)(7a^3)(2a^6) = (5 \cdot 7 \cdot 2)(a^1 \cdot a^3 \cdot a^6) = 70a^{10}$

47. $(x)(4x^6)(7x^5) = (1 \cdot 4 \cdot 7)(x^1 \cdot x^6 \cdot x^5) = 28x^{12}$

49. $(5x)(3y)(-2x) = (5)(3)(-2)(x \cdot y \cdot x) = -30x^2 y$

51. $(6y)(-3x)(-3y) = 6(-3)(-3)(y \cdot x \cdot y) = 54xy^2$

53. a. $5z^3 + 4$ is a binomial because there are two terms.

b. 9 is a monomial because there is one term.

c. $2x^7 - 3x^4 - 3$ is a trinomial because there are three terms.

55.
$$\begin{aligned} 2x(x^2+5) &= 2x \cdot x^2 + 2x \cdot 5 \\ &= 2x^1 \cdot 1x^2 + 2x \cdot 5 \\ &= (2 \cdot 1)(x^1 \cdot x^2) + (2 \cdot 5)x \\ &= 2x^{1+2} + 10x \\ &= 2x^3 + 10x \end{aligned}$$

57.
$$\begin{aligned} 6x^2(3x^3 - 1) &= 6x^2 \cdot 3x^3 - 6x^2 \cdot 1 \\ &= (6 \cdot 3)(x^2 \cdot x^3) - (6 \cdot 1)x^2 \\ &= 18x^{2+3} - 6x^2 \\ &= 18x^5 - 6x^2 \end{aligned}$$

59.
$$\begin{aligned} &5y^3(y^4 - 2y) + 3y^7 \\ &= 5y^3 \cdot y^4 - 5y^3 \cdot 2y + 3y^7 \\ &= 5y^3 \cdot 1y^4 - 5y^3 \cdot 2y^1 + 3y^7 \\ &= (5 \cdot 1)(y^3 \cdot y^4) - (5 \cdot 2)(y^3 \cdot y^1) + 3y^7 \\ &= 5y^{3+4} - 10y^{3+1} + 3y^7 \\ &= 5y^7 - 10y^4 + 3y^7 \\ &= 8y^7 - 10y^4 \end{aligned}$$

61.
$$\begin{aligned} &-2x^3(x^2 + 4x) + 5x^5 \\ &= -2x^3 \cdot x^2 + (-2x^3) \cdot 4x + 5x^5 \\ &= -2x^3 \cdot 1x^2 + (-2x^3) \cdot 4x^1 + 5x^5 \\ &= (-2 \cdot 1)(x^3 \cdot x^2) + (-2 \cdot 4)(x^3 \cdot x^1) + 5x^5 \\ &= -2x^{3+2} + (-8x^{3+1}) + 5x^5 \\ &= -2x^5 - 8x^4 + 5x^5 \\ &= 3x^5 - 8x^4 \end{aligned}$$

63.
$$\begin{aligned} (2y-5)(-6y^2) &= 2y(-6y^2) - 5(-6y^2) \\ &= 2(-6)(y^1 \cdot y^2) - 5(-6)y^2 \\ &= -12y^{1+2} - (-30y^2) \\ &= -12y^3 + 30y^2 \end{aligned}$$

65.
$$\begin{aligned} (2x^3 - 4x)(3x) &= 2x^3 \cdot 3x - 4x \cdot 3x \\ &= (2 \cdot 3)(x^3 \cdot x^1) - (4 \cdot 3)(x^1 \cdot x^1) \\ &= 6x^{3+1} - 12x^{1+1} \\ &= 6x^4 - 12x^2 \end{aligned}$$

Copyright © 2012 Pearson Education, Inc.

67. $A = LW$
$= (2x^4 - 5)(x^3)$
$= (2x^4)(x^3) - (5)(x^3)$
$= 2x^7 - 5x^3$

69. $A = bh$
$= (7x^3 - 4x)(x^2)$
$= (7x^3)(x^2) - (4x)(x^2)$
$= 7x^5 - 4x^3$

71. $V = LWH$
$= (7x^3 - x^2)(2x^4)(x^2)$
$= (7x^3 - x^2)(2x^6)$
$= (7x^3)(2x^6) - (x^2)(2x^6)$
$= 14x^9 - 2x^8$

73. $P = 2L + 2W = 2(4x^3) + 2(3x^2) = 8x^3 + 6x^2$

75. $P = 2L + 2W$
$= 2(5x^3 + 3x^2) + 2(2x^2 - x)$
$= 2 \cdot 5x^3 + 2 \cdot 3x^2 + 2 \cdot 2x^2 - 2 \cdot x$
$= 10x^3 + 6x^2 + 4x^2 - 2x$
$= 10x^3 + 10x^2 - 2x$

77. $A = LW$
$= 2x^2(3x^2 + 5)$
$= (2x^2)(3x^2) + (2x^2)(5)$
$= 6x^4 + 10x^2$
$P = 2L + 2W$
$= 2(3x^2 + 5) + 2(2x^2)$
$= 2(3x^2) + 2(5) + 2(2x^2)$
$= 6x^2 + 10 + 4x^2$
$= 10x^2 + 10$

79. a. $4cd + 9cd = (4 + 9)cd = 13cd$

b. $(9cd)(4cd) = (9 \cdot 4)(c \cdot c)(d \cdot d) = 36c^2d^2$

c. $4(cd + 9) = 4 \cdot cd + 4 \cdot 9 = 4cd + 36$

81. a. $9ab^5 - 7ab^5 = (9 - 7)ab^5 = 2ab^5$

b. $(9ab^5)(7ab^5) = (9 \cdot 7)(a \cdot a)(b^5 \cdot b^5)$
$= 63a^2b^{10}$

c. $9(ab^5 + 7) = 9 \cdot ab^5 + 9 \cdot 7 = 9ab^5 + 63$

83. The width of a rectangle is four more than the square of the length: $W = L^2 + 4$.

Cumulative Review

85. $\frac{-35}{7} = -35 \div 7 = -5$

86. $(-3)^3 = (-3)(-3)(-3) = -27$

87. $20{,}566 \div 312 = 65 \text{ R } 286$

$$\begin{array}{r} 65 \\ 312\overline{)20566} \\ \underline{1872} \\ 1846 \\ \underline{1560} \\ 286 \end{array}$$

88.

$$\begin{array}{r} 31{,}423 \\ \times \quad 28 \\ \hline 251\ 384 \\ 628\ 46 \\ \hline 879{,}844 \end{array}$$

89. $-1 + 2 + 0 = 1$ over par

90. $0 + 1 + (-1) + (-1) = -1$
1 under par

Quick Quiz 3.4

1. a. $x^2 \cdot x = x^2 \cdot x^1 = x^{2+1} = x^3$

b. $4^5 \cdot 4^2 = 4^{5+2} = 4^7$

c. $-3y^3(2y)(7y^4) = (-3 \cdot 2 \cdot 7)(y^3 \cdot y^1 \cdot y^4)$
$= -42y^8$

2. a. $6a(a^3 + 8a) = 6a \cdot a^3 + 6a \cdot 8a$
$= 6a^1 \cdot 1a^3 + 6a^1 \cdot 8a^1$
$= (6 \cdot 1)(a^1 \cdot a^3) + (6 \cdot 8)(a^1 \cdot a^1)$
$= 6a^{1+3} + 48a^{1+1}$
$= 6a^4 + 48a^2$

b. $(x^5 - 3)(2x^2) = (x^5)(2x^2) - 3(2x^2)$
$= (1 \cdot 2)(x^5 \cdot x^2) - (3 \cdot 2)x^2$
$= 2x^{5+2} - 6x^2$
$= 2x^7 - 6x^2$

Copyright © 2012 Pearson Education, Inc.

3. $A = LW$
$= (3x^2 + 5)(x^2)$
$= (3x^2)(x^2) + (5)(x^2)$
$= 3x^4 + 5x^2$

4. Answers may vary.

a. Multiply both terms in parentheses by $2x^4$ to obtain $2x^4 \cdot x^2 + 2x^4 \cdot y$. Then add exponents on the powers of *x* being multiplied in the first term: $2x^6 + 2x^4y$.

b. Add the exponents on the powers of *x* being multiplied: $-3x^6$.

c. To complete the problem, add the simplified forms obtained in parts (a) and (b): $(2x^6 + 2x^4y) - 3x^6$. Since $2x^6$ and $-3x^6$ are like terms, complete the simplification by combining these terms: $-x^6 + 2x^4y$.

You Try It

1.
$$\begin{aligned} 5x - 4x - 2 &= 8 \\ x - 2 &= 8 \\ +2 \quad &+2 \\ \hline x + 0 &= 10 \\ x &= 10 \end{aligned}$$
Check:
$$\begin{aligned} 5x - 4x - 2 &= 8 \\ 5(10) - 4(10) - 2 &\stackrel{?}{=} 8 \\ 50 - 40 - 2 &\stackrel{?}{=} 8 \\ 10 - 2 &\stackrel{?}{=} 8 \\ 8 &= 8 \checkmark \end{aligned}$$

2.
$$\begin{aligned} \angle a + \angle b &= 180^\circ \\ 55^\circ + \angle b &= 180^\circ \\ + -55^\circ \quad &-55^\circ \\ \hline \angle b &= 125^\circ \end{aligned}$$

3.
$$\begin{aligned} -48 &= 3(4x) \\ -48 &= 12x \\ \frac{-48}{12} &= \frac{12x}{12} \\ -4 &= x \end{aligned}$$
Check:
$$\begin{aligned} -48 &= 3(4x) \\ -48 &\stackrel{?}{=} 3(4 \cdot (-4)) \\ -48 &= -48 \checkmark \end{aligned}$$

4. a. $R = 2C$

b.
$$\begin{aligned} 12 &= 2C \\ \frac{12}{2} &= \frac{2C}{2} \\ 6 &= C \end{aligned}$$
There are 6 carnations.

5.
$$\begin{aligned} P &= 2L + 2W \\ 70 &= 2(4W) + 2W \\ 70 &= 8W + 2W \\ 70 &= 10W \\ \frac{70}{10} &= \frac{10W}{10} \\ 7 &= W \end{aligned}$$
The width is 7 feet.

6.
$$\begin{aligned} A &= LW \\ 42 &= L(6) \\ \frac{42}{6} &= \frac{6L}{6} \\ 7 &= L \end{aligned}$$
The length is 7 inches.

7.
$$\begin{aligned} A &= bh \\ 180 &= b(15) \\ \frac{180}{15} &= \frac{15b}{15} \\ 12 &= b \end{aligned}$$
The base is 12 meters.

8.
$$\begin{aligned} V &= LWH \\ 280 &= L(5)(7) \\ 280 &= 35L \\ \frac{280}{35} &= \frac{35L}{35} \\ 8 &= L \end{aligned}$$
The length is 8 centimeters.

9. a. $y^4 \cdot y^5 = y^{4+5} = y^9$

b. $6^3 \cdot 6 = 6^3 \cdot 6^1 = 6^{3+1} = 6^4$

c. $4^2 \cdot 5^3 = 4^2 \cdot 5^3$

10. $(7x^3)(-4x^2) = [7 \cdot (-4)](x^3 \cdot x^2) = -28x^5$

11. $x^5(x^3 + 5) = x^5 \cdot x^3 + x^5 \cdot 5 = x^8 + 5x^5$

Copyright © 2012 Pearson Education, Inc.

12.
$$\begin{aligned} A &= LW \\ &= (5x^3 - 3)x^4 \\ &= 5x^3 \cdot x^4 - 3 \cdot x^4 \\ &= 5x^7 - 3x^4 \end{aligned}$$

Chapter 3 Review Problems

1. Adjacent angles: two angles that share a common side
2. Supplementary angles: two angles that have a sum of 180°
3. Parallel lines: straight lines that are always the same distance apart
4. Polynomials: variable expressions that contain terms with only whole number exponents and no variables in the denominator
5. Numerical coefficient: a number that is multiplied by a variable
6. Monomial: a polynomial with one term
7. Binomial: a polynomial with two terms
8. Trinomial: a polynomial with three terms

9.
$$\begin{aligned} x - 15 &= 12 \\ x - 15 + 15 &= 12 + 15 \\ x + 0 &= 27 \\ x &= 27 \end{aligned}$$
Check: $x - 15 = 12$
$27 - 15 \stackrel{?}{=} 12$
$12 = 12$ ✓

10.
$$\begin{aligned} x + 3 &= -7 \\ x + 3 + (-3) &= -7 + (-3) \\ x + 0 &= -10 \\ x &= -10 \end{aligned}$$
Check: $x + 3 = -7$
$-10 + 3 \stackrel{?}{=} -7$
$-7 = -7$ ✓

11.
$$\begin{aligned} -8 + 1 &= y + 4 - 16 \\ -7 &= y - 12 \\ +\ 12 &\quad\ \ 12 \\ \hline 5 &= y \end{aligned}$$
Check: $-8 + 1 = y + 4 - 16$
$-8 + 1 \stackrel{?}{=} 5 + 4 - 16$
$-7 = -7$ ✓

12.
$$\begin{aligned} 6 - 13 + y &= -4 + 1 \\ -7 + y &= -3 \\ +\ 7 &\quad\ \ 7 \\ \hline y &= 4 \end{aligned}$$
Check: $6 - 13 + y = -4 + 1$
$6 - 13 + 4 \stackrel{?}{=} -4 + 1$
$-3 = -3$ ✓

13.
$$\begin{aligned} 4(2 - 6) &= x - 2 \\ 4(-4) &= x - 2 \\ -16 &= x - 2 \\ +\ 2 &\quad\ \ 2 \\ \hline -14 &= x \end{aligned}$$
Check: $4(2 - 6) = x - 2$
$4(2 - 6) \stackrel{?}{=} -14 - 2$
$4(-4) \stackrel{?}{=} -16$
$-16 = -16$ ✓

14.
$$\begin{aligned} 4x - 3x - 6 &= 6 \\ x - 6 &= 6 \\ +\ 6 &\quad\ 6 \\ \hline x &= 12 \end{aligned}$$
Check: $4x - 3x - 6 = 6$
$4(12) - 3(12) - 6 \stackrel{?}{=} 6$
$48 - 36 - 6 \stackrel{?}{=} 6$
$6 = 6$ ✓

15. Since the angles are supplementary, their sum is 180°.
$$\begin{aligned} \angle a + \angle b &= 180^\circ \\ \angle a + 81^\circ &= 180^\circ \\ +\ -81^\circ &\quad -81^\circ \\ \hline \angle a &= 99^\circ \end{aligned}$$
The measure of $\angle a$ is 99°.

16. Since the angles are supplementary, their sum is 180°.
$$\begin{aligned} \angle x + \angle y &= 180^\circ \\ 25^\circ + \angle y &= 180^\circ \\ +\ -25^\circ &\quad -25^\circ \\ \hline \angle y &= 155^\circ \end{aligned}$$
The measure of $\angle y$ is 155°.

17. a. Angle x measures 22° more than angle y: $\angle x = \angle y + 22^\circ$.

Copyright © 2012 Pearson Education, Inc.

b.
$$\angle x = \angle y + 22°$$
$$101° = \angle y + 22°$$
$$+ -22° \quad -22°$$
$$79° = \angle y$$
The measure of $\angle y$ is 79°.

18. $4y = 48$
$$\frac{4y}{4} = \frac{48}{4}$$
$$y = 12$$
Check: $4y = 48$
$$4(12) \stackrel{?}{=} 48$$
$$48 = 48 \checkmark$$

19. $7y = -63$
$$\frac{7y}{7} = \frac{-63}{7}$$
$$y = -9$$
Check: $7y = -63$
$$7(-9) \stackrel{?}{=} -63$$
$$-63 = -63 \checkmark$$

20. $6a = -42$
$$\frac{6a}{6} = \frac{-42}{6}$$
$$a = -7$$
Check: $6a = -42$
$$6(-7) \stackrel{?}{=} -42$$
$$-42 = -42 \checkmark$$

21. $7(3x) = 42$
$$21x = 42$$
$$\frac{21x}{21} = \frac{42}{21}$$
$$x = 2$$
Check: $7(3x) = 42$
$$7[3(2)] \stackrel{?}{=} 42$$
$$7(6) \stackrel{?}{=} 42$$
$$42 = 42 \checkmark$$

22. $3(y \cdot 4) = 24$
$$3(4y) = 24$$
$$12y = 24$$
$$\frac{12y}{12} = \frac{24}{12}$$
$$y = 12$$
Check: $3(y \cdot 4) = 24$
$$3(2 \cdot 4) \stackrel{?}{=} 24$$
$$3(8) \stackrel{?}{=} 24$$
$$24 = 24 \checkmark$$

23. $6(x \cdot 3) = \frac{-36}{2}$
$$6(x \cdot 3) = -18$$
$$6(3x) = -18$$
$$18x = -18$$
$$\frac{18x}{18} = \frac{-18}{18}$$
$$x = -1$$
Check: $6(x \cdot 3) = \frac{-36}{2}$
$$6[(-1) \cdot 3] \stackrel{?}{=} \frac{-36}{2}$$
$$6(-3) \stackrel{?}{=} -18$$
$$-18 = -18 \checkmark$$

24. $\frac{-48}{2} = 2(3 \cdot x)$
$$-24 = 2(3 \cdot x)$$
$$-24 = 6x$$
$$\frac{-24}{6} = \frac{6x}{6}$$
$$-4 = x$$
Check: $\frac{-48}{2} = 2(3 \cdot x)$
$$\frac{-48}{2} \stackrel{?}{=} 2[3(-4)]$$
$$-24 \stackrel{?}{=} 2(-12)$$
$$-24 = -24 \checkmark$$

25. $2(5x) = 70$
$$10x = 70$$
$$\frac{10x}{10} = \frac{70}{10}$$
$$x = 7$$
Check: $2(5x) = 70$
$$2[5(7)] \stackrel{?}{=} 70$$
$$2(35) \stackrel{?}{=} 70$$
$$70 = 70 \checkmark$$

26. $7(2y) = 28$
$$14y = 28$$
$$\frac{14y}{14} = \frac{28}{14}$$
$$y = 2$$

Copyright © 2012 Pearson Education, Inc.

Check: $7(2y) = 28$
$7[2(2)] \stackrel{?}{=} 28$
$7(4) \stackrel{?}{=} 28$
$28 = 28$ ✓

27. $6x - 4x = 20$
$2x = 20$
$\frac{2x}{2} = \frac{20}{2}$
$x = 10$
Check: $6x - 4x = 20$
$6(10) - 4(10) \stackrel{?}{=} 20$
$60 - 40 \stackrel{?}{=} 20$
$20 = 20$ ✓

28. $5x + 2x = 42$
$7x = 42$
$\frac{7x}{7} = \frac{42}{7}$
$x = 6$
Check: $5x + 2x = 42$
$5(6) + 2(6) \stackrel{?}{=} 42$
$30 + 12 \stackrel{?}{=} 42$
$42 = 42$ ✓

29. a. The length (L) of a room is three times the width (W): $L = 3W$.

b. Replace L with 36.
$L = 3W$
$36 = 3W$
$\frac{36}{3} = \frac{3W}{3}$
$12 = W$
The width of the room is 12 feet.

30. a. There are twice as many white cars (W) on the road as blue cars (B): $W = 2B$.

b. Replace W with 200.
$W = 2B$
$200 = 2B$
$\frac{200}{2} = \frac{2B}{2}$
$100 = B$
There are 100 blue cars on the road.

31. Let x = Sonia's hourly rate. Then $7x$, $8x$, and $8x$ represent her earnings for Monday, Tuesday, and Wednesday, respectively. Her total earnings for the three days were \$1150.
$7x + 8x + 8x = 1150$
$23x = 1150$
$\frac{23x}{23} = \frac{1150}{23}$
$x = 50$
Sonia charges \$50 per hour.

32. Let x = the number of miles Leo jogged on Friday and $2x$ = the number of miles he jogged Saturday.
$x + 2x = 12$
$3x = 12$
$\frac{3x}{3} = \frac{12}{3}$
$x = 4$
Leo jogged 4 miles on Friday.

33. The perimeter of a triangle is the sum of the lengths of the sides.
$x + 8 + 12 = 28$
$x + 20 = 28$
$x + 20 + (-20) = 28 + (-20)$
$x = 8$
The unknown side is 8 inches.

34. $P = 2L + 2W$
$32 = 2(3W) + 2W$
$32 = 6W + 2W$
$32 = 8W$
$\frac{32}{8} = \frac{8W}{8}$
$4 = W$
The width is 4 feet.

35. We use the formula for the area of a square.
$A = s^2 = 6^2 = 36 \text{ ft}^2$
The area is 36 square feet.

36. We use the formula for the area of a rectangle.
$A = LW = (72)(52) = 3744 \text{ in.}^2$
The area is 3744 square inches.

37. We use the formula for the area of a parallelogram.
$A = bh = 9(11) = 99 \text{ in.}^2$
The area is 99 square inches.

38. Divide the figure into three rectangles and then find the area of each rectangle separately. Add these areas together to find the area of the figure.

Copyright © 2012 Pearson Education, Inc.

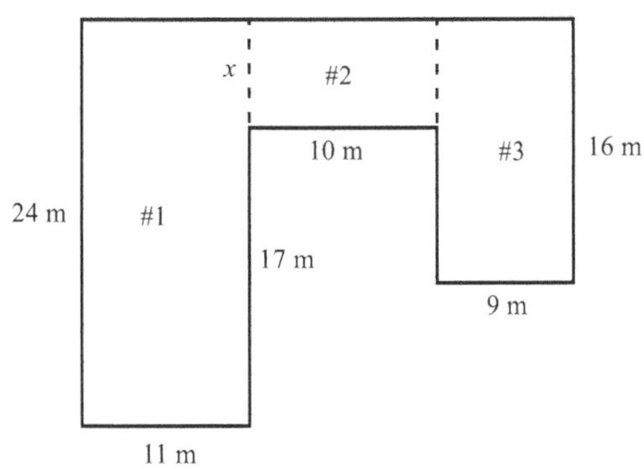

Area of rectangle #1:

$A = LW = 24 \cdot 11 = 264 \text{ m}^2$

To find the area of rectangle #2, we must find the width of the rectangle, labeled x in the figure. From the figure, we see that $x + 17 = 24$. Solving this equation, we have $x = 7$, so the width of rectangle #2 is 7 meters.

Area of rectangle #2: $A = LW = 10 \cdot 7 = 70 \text{ m}^2$

Area of rectangle #3: $A = LW = 9 \cdot 16 = 144 \text{ m}^2$

Add the areas.

$264 + 70 + 144 = 478 \text{ m}^2$

The area of the shape is 478 square meters.

39. We use the formula for the area of a rectangle.

$$\begin{aligned} A &= LW \\ 104 &= x \cdot 13 \\ 104 &= 13x \\ \frac{104}{13} &= \frac{13x}{13} \\ 8 &= x \end{aligned}$$

The unknown side is 8 centimeters.

40. We use the formula for the area of a parallelogram.

$$\begin{aligned} A &= bh \\ 96 &= x \cdot 8 \\ 96 &= 8x \\ \frac{96}{8} &= \frac{8x}{8} \\ 12 &= x \end{aligned}$$

The unknown side is 12 inches.

41. We use the formula for the volume of a rectangular solid.

$$\begin{aligned} V &= LWH \\ 108 &= 9 \cdot 6 \cdot H \\ 108 &= 54H \\ \frac{108}{54} &= \frac{54H}{54} \\ 2 &= H \end{aligned}$$

The unknown side is 2 feet.

42. We use the formula for the volume of a rectangular solid.

$$\begin{aligned} V &= LWH \\ 198 &= 6 \cdot W \cdot 11 \\ 198 &= 66W \\ \frac{198}{66} &= \frac{66W}{66} \\ 3 &= W \end{aligned}$$

The unknown side is 3 inches.

43. We use the formula for the volume of a rectangular solid.

$V = LWH = 25 \cdot 15 \cdot 2 = 750 \text{ ft}^3$

The volume of the pool is 750 cubic feet, so 750 cubic feet of water are needed to fill the pool.

44. We use the formula for the volume of a rectangular solid.

$V = LWH = 5 \cdot 5 \cdot 5 = 125 \text{ in.}^3$

The volume of the cube is 125 cubic inches, so the cube can hold 125 cubic inches of water.

45. First change the dimensions of the yard from yards to feet.

$16(3) = 48$; 16 yd = 48 ft

$12(3) = 36$; 12 yd = 36 ft

Next find the area in square feet.

$A = LW = 48 \cdot 36 = 1728 \text{ ft}^2$

The area of the yard is 1728 square feet. Since each container of weed killer covers 1000 square feet, 2 containers are needed. It will cost 2($4) or $8 to buy enough for the entire front yard.

46. a. Divide each number of feet by 3 to find the number of yards.

$18 \div 3 = 6$; 18 ft = 6 yd

$12 \div 3 = 4$; 12 ft = 4 yd

The office is 6 yards by 4 yards.

$A = LW = 6 \cdot 4 = 24 \text{ yd}^2$

The area of the office is 24 square yards, so 24 square yards of carpet must be purchased.

b. Multiply the number of square yards by the price per square yard.

$24(15) = 360$

The carpet will cost $360.

47. $2^3 \cdot 2^3 = 2^{3+3} = 2^6$

48. $7^5 \cdot 7^4 = 7^{5+4} = 7^9$

Copyright © 2012 Pearson Education, Inc.

49. $a \cdot a^6 = a^1 \cdot a^6 = a^{1+6} = a^7$

50. $4^2 \cdot 3^3 = 4^2 \cdot 3^3$

51. $(3y^4)(5y^4) = (3 \cdot 5)(y^4 \cdot y^4) = 15y^{4+4} = 15y^8$

52. $(4x^2)(3x^6) = (4 \cdot 3)(x^2 \cdot x^6) = 12x^{2+6} = 12x^8$

53. $(3a)(a^3)(7a^6) = (3 \cdot 1 \cdot 7)(a^1 \cdot a^3 \cdot a^6)$
$= 21a^{1+3+6}$
$= 21a^{10}$

54. $(3z)(y^8)(4z^3)(2y^3) = (3 \cdot 1 \cdot 4 \cdot 2)(y^8 \cdot y^3)(z^1 \cdot z^3)$
$= 24y^{8+3}z^{1+3}$
$= 24y^{11}z^4$

55. $(4x^5)(-3x^2) = (4)(-3)(x^5 \cdot x^2)$
$= -12x^{5+2}$
$= -12x^7$

56. $(-7z^7)(5z^3) = (-7)(5)(z^7 \cdot z^3)$
$= -35z^{7+3}$
$= -35z^{10}$

57. $(-3a^4)(9a^7) = (-3)(9)(a^4 \cdot a^7)$
$= -27a^{4+7}$
$= -27a^{11}$

58. $(4y^5)(-5y^9) = (4)(-5)(y^5 \cdot y^9)$
$= -20y^{5+9}$
$= -20y^{14}$

59. $3x^2 - 1$ is a binomial because there are two terms.

60. $4x^3$ is a monomial because there is one term.

61. $2xy^2 + 3x + 1$ is a trinomial because there are three terms.

62. $x(x^2 + 3) = x \cdot x^2 + x \cdot 3 = x^3 + 3x$

63. $(x^3 - 4x)6x^2 = x^3 \cdot 6x^2 - 4x \cdot 6x^2 = 6x^5 - 24x^3$

64. $A = LW$
$= (3x^2 + 6) \cdot x^3$
$= 3x^2 \cdot x^3 + 6 \cdot x^3$
$= 3x^5 + 6x^3$

65. $V = LWH$
$= 2x^4 \cdot 3x^2 \cdot 2x$
$= (2 \cdot 3 \cdot 2)(x^4 \cdot x^2 \cdot x^1)$
$= 12x^7$

How Am I Doing? Chapter 3 Test

1. $x - 2 = -8$
$x - 2 + 2 = -8 + 2$
$x + 0 = -6$
$x = -6$
Check: $x - 2 = -8$
$-6 - 2 \stackrel{?}{=} -8$
$-8 = -8$ ✓

2. $y + 3 = 72$
$y + 3 + (-3) = 72 + (-3)$
$y + 0 = 69$
$y = 69$
Check: $y + 3 = 72$
$69 + 3 \stackrel{?}{=} 72$
$72 = 72$ ✓

3. $9 - 15 = a - 3$
$-6 = a - 3$
$+ \quad 3 \quad\quad 3$
$-3 = a$
Check: $9 - 15 = a - 3$
$9 - 15 \stackrel{?}{=} -3 - 3$
$-6 = -6$ ✓

4. $3x - 2x - 7 = -1 + 6$
$x - 7 = 5$
$+ \quad 7 \quad 7$
$x = 12$
Check: $3x - 2x - 7 = -1 + 6$
$3(12) - 2(12) - 7 \stackrel{?}{=} -1 + 6$
$36 - 24 - 7 \stackrel{?}{=} -1 + 6$
$5 = 5$ ✓

Copyright © 2012 Pearson Education, Inc.

5. $12 = 5x - 2x$
$12 = 3x$
$\frac{12}{3} = \frac{3x}{3}$
$4 = x$
Check: $12 = 5x - 2x$
$12 \overset{?}{=} 5(4) - 2(4)$
$12 \overset{?}{=} 20 - 8$
$12 = 12$ ✓

6. $-3y = 42$
$\frac{-3y}{-3} = \frac{42}{-3}$
$y = -14$
Check: $-3y = 42$
$-3(-14) \overset{?}{=} 42$
$42 = 42$ ✓

7. $2(4x) = -72$
$8x = -72$
$\frac{8x}{8} = \frac{-72}{8}$
$x = -9$
Check: $2(4x) = -72$
$2[4(-9)] \overset{?}{=} -72$
$2(-36) \overset{?}{=} -72$
$-72 = -72$ ✓

8. $\frac{-16}{2} = 2 + y$
$-8 = 2 + y$
$+\ -2 \quad -2$
$-10 = y$
Check: $\frac{-16}{2} = 2 + y$
$\frac{-16}{2} \overset{?}{=} 2 + (-10)$
$-8 = -8$ ✓

9. Since the angles are supplementary, their sum is 180°.
$\angle x + \angle y = 180°$
$75° + \angle y = 180°$
$+\ -75° \qquad -75°$
$\angle y = 105°$
The measure of $\angle y$ is 105°.

10. a. Angle x measures 10° more than angle y: $\angle x = \angle y + 10°$.

b. $\angle x = \angle y + 10°$
$95° = \angle y + 10°$
$+\ -10° \qquad -10°$
$85° = \angle y$
The measure of $\angle y$ is 85°.

11. a. The number of female students (F) is triple the number of male students (M) in the class: $F = 3M$

b. Replace F with 21.
$F = 3M$
$21 = 3M$
$\frac{21}{3} = \frac{3M}{3}$
$7 = M$
There are 7 male students in the class.

12. $P = 2L + 2W$
$48 = 2(3W) + 2W$
$48 = 6W + 2W$
$48 = 8W$
$48 = \frac{8W}{8}$
$6 = W$
The width is 6 yards.

13. $A = LW = 5(3) = 15 \text{ ft}^2$
The area is 15 square feet.

14. $A = bh = 6(12) = 72 \text{ in.}^2$
The area is 72 square inches.

15. Divide the figure into two rectangles and then find the area of each rectangle separately. Add these areas together to find the area of the figure.

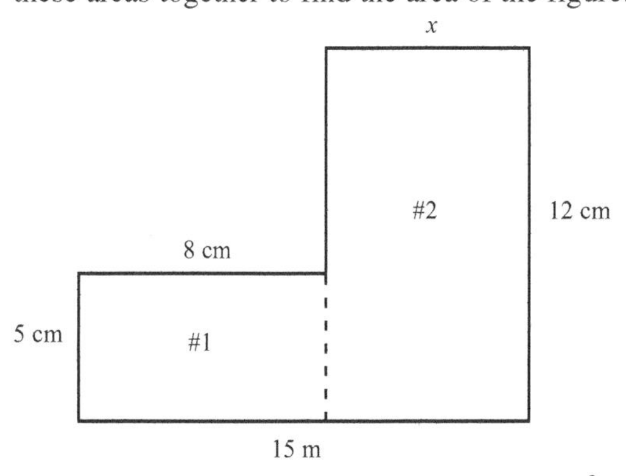

Area of rectangle #1: $A = LW = 8 \cdot 5 = 40 \text{ cm}^2$
To find the area of rectangle #2, we must find the length of the rectangle, labeled x in the figure. From the figure, we see that $x + 8 = 15$. Solving this equation, we have $x = 7$, so the

Copyright © 2012 Pearson Education, Inc.

length of rectangle #2 is 7 centimeters.
Area of rectangle #2: $A = LW = 7 \cdot 12 = 84 \text{ cm}^2$
Add the areas.
$40 + 84 = 124 \text{ cm}^2$
The area of the shape is 124 square centimeters.

16. We use the formula for the volume of a rectangular solid.
$$V = LWH$$
$$288 = 12 \cdot W \cdot 8$$
$$288 = 96W$$
$$\frac{288}{96} = \frac{96W}{96}$$
$$3 = W$$
The unknown side is 3 inches.

17. $A = LW$
$= (4x^3 + x^2)(x^2)$
$= (4x^3)(x^2) + (x^2)(x^2)$
$= 4x^5 + x^4$

18. Divide the number of feet by 3 to find the number of yards.
$9 \div 3 = 3$; 9 ft = 3 yd
$6 \div 3 = 2$; 6 ft = 2 yd
The entryway is 3 yards by 2 yards.
$A = LW = 3 \cdot 2 = 6 \text{ yd}^2$
The area of the entryway is 6 square yards, so 6 square yards of tile must be purchased.

19. We use the formula for the volume of a rectangular solid.
$V = LWH = 30(40)(10) = 12{,}000 \text{ ft}^3$
The volume of the hole is 12,000 cubic feet, so they will need to haul away 12,000 cubic feet of dirt.

20. $y^3 \cdot y^2 = y^{3+2} = y^5$

21. $z \cdot z^3 = z^1 \cdot z^3 = z^{1+3} = z^4$

22. $2^3 \cdot 3^2 = 2^3 \cdot 3^2$

23. $(5x)(x^3)(x^4) = (5 \cdot 1 \cdot 1)(x^1 \cdot x^3 \cdot x^4)$
$= 5x^{1+3+4}$
$= 5x^8$

24. $(-8x^2)(-9x^4) = (-8)(-9)(x^2 \cdot x^4)$
$= 72x^{2+4}$
$= 72x^6$

25. $5y(y^4 + 8) = 5y \cdot y^4 + 5y \cdot 8 = 5y^5 + 40y$

26. $(3x^2 - 5x)6x^3 = 3x^2 \cdot 6x^3 - 5x \cdot 6x^3$
$= 18x^5 - 30x^4$

Cumulative Test for Chapters 1–3

1. 5280
The number to the right of the hundreds place is 5 or more. We add 1 in the hundreds place and replace the digits to the right with zeros.
5300

2.
$$\begin{array}{r} 18{,}700 \\ -\ \ 896 \\ \hline 17{,}804 \end{array}$$

3. a. Subtract the expenses from the income.
$$\begin{array}{r} 167{,}350 \\ -\ 86{,}000 \\ \hline 81{,}350 \end{array}$$
The profit was $81,350.

b. Divide the profit by 2.
$$\begin{array}{r} 40675 \\ 2\overline{)81350} \\ \underline{8} \\ 013 \\ \underline{12} \\ 15 \\ \underline{14} \\ 10 \\ \underline{10} \\ 0 \end{array}$$
Each owner received $40,675.

4. $6(7x)(2) = (6 \cdot 7 \cdot 2)x = 84x$

5. $3(y \cdot 8) = 3(8y) = (3 \cdot 8)y = 24y$

6.
$$\begin{array}{r} 209 \\ \times\ \ 67 \\ \hline 1\,463 \\ 12\,54\ \\ \hline 14{,}003 \end{array}$$

Copyright © 2012 Pearson Education, Inc.

7. $2844 \div 14 = 203 \text{ R2}$

$$\begin{array}{r} 203 \\ 14\overline{)2844} \\ \underline{28} \\ 04 \\ \underline{00} \\ 44 \\ \underline{42} \\ 2 \end{array}$$

8. a. Double some number equals 28: $2x = 28$

b. The sum of a number and 9 equals 16: $x + 9 = 16$.

9.
$$\begin{aligned} |-17| &\ ?\ |-2| \\ 17 &\ ?\ 2 \\ 17 &> 2 \\ |-17| &> |-2| \end{aligned}$$

10. $5 + (-6) = -1$

11. $-10 - 8 = -10 + (-8) = -18$

12. $-7 - 6 - 4 - 8 = -7 + (-6) + (-4) + (-8) = -25$

13. $(-18) \div (-9) = 2$

14. $(-2)^5 = (-2)(-2)(-2)(-2)(-2) = -32$

15.
$$\begin{aligned} (-2)(-1)(4)(3)(-2) &= [(-2)(-1)][(4)(3)](-2) \\ &= 2(12)(-2) \\ &= 24(-2) \\ &= -48 \end{aligned}$$

16.
$$\begin{aligned} -4 + 15 \div 5(-3)^2 - 1 &= -4 + 15 \div 5(9) - 1 \\ &= -4 + 3(9) - 1 \\ &= -4 + 27 - 1 \\ &= -4 + 27 + (-1) \\ &= -4 + (-1) + 27 \\ &= -5 + 27 \\ &= 22 \end{aligned}$$

17. a.
$$\begin{aligned} 3mn - 7mn + 4m &= 3mn + (-7mn) + 4m \\ &= [3 + (-7)]mn + 4m \\ &= -4mn + 4m \end{aligned}$$

b.
$$\begin{aligned} 5 + 9y + 3 + y &= 5 + 3 + 9y + y \\ &= (5 + 3) + (9 + 1)y \\ &= 8 + 10y \end{aligned}$$

18. Replace x with -1 and y with -3.
$$\begin{aligned} x - 2y^3 + 1 &= (x) - 2(y)^3 + 1 \\ &= (-1) - 2(-3)^3 + 1 \\ &= -1 - 2(-27) + 1 \\ &= -1 - (-54) + 1 \\ &= -1 + 54 + 1 \\ &= -1 + 55 \\ &= 54 \end{aligned}$$

19. $-2(x - 3) = -2 \cdot x - (-2) \cdot 3 = -2x + 6$

20.
$$\begin{array}{rcr} -36 + y &=& -2 \\ +\ 36 & & 36 \\ \hline y &=& 34 \end{array}$$

21.
$$\begin{array}{rcr} 2x - x + 5 &=& -6 \\ x + 5 &=& -6 \\ +\quad -5 & & -5 \\ \hline x &=& -11 \end{array}$$

22.
$$\begin{aligned} \frac{120}{10} &= -3y \\ 12 &= -3y \\ \frac{12}{-3} &= \frac{-3y}{-3} \\ -4 &= y \end{aligned}$$

23. $P = 2L + 2W = 2(3) + 2(2) = 10$ in.
The perimeter is 10 inches.

24. We use the formula for the volume of a rectangular solid.
$$\begin{aligned} V &= LWH \\ 30 &= L \cdot 2 \cdot 3 \\ 30 &= 6L \\ \frac{30}{6} &= \frac{6L}{6} \\ 5 &= L \end{aligned}$$
The unknown side is 5 meters.

25. Since the angles are supplementary, their sum is 180°.
$$\begin{array}{rcr} \angle a + \angle b &=& 180^\circ \\ \angle a + 105^\circ &=& 180^\circ \\ +\quad -105^\circ & & -105^\circ \\ \hline \angle a &=& 75^\circ \end{array}$$
The measure of $\angle a$ is 75°.

26. a. The number of children (C) is double the number of adults (A): $C = 2A$.

Copyright © 2012 Pearson Education, Inc.

b. Replace C with 140.

$$C = 2A$$
$$140 = 2A$$
$$\frac{140}{2} = \frac{2A}{2}$$
$$70 = A$$

There are 70 adults at the amusement park.

27. $(-3x^2)(4x^6) = (-3)(4)(x^2 \cdot x^6)$
$= -12x^{2+6}$
$= -12x^8$

28. $x^3(2x^2 + 5) = x^3 \cdot 2x^2 + x^3 \cdot 5 = 2x^5 + 5x^3$

Copyright © 2012 Pearson Education, Inc.

Chapter 4

4.1 Exercises

1. A number is divisible by 2 if it is even.

3. A number is divisible by 5 if the last digit is 5 or 0.

5. With this method, the divisors and quotient are the prime factors. Therefore, the divisor (number you divide by) must be prime for your factors to be prime.

7. 165 is not divisible by 2 because it is not even.

9. 232 is not divisible by 3 because the sum of the digits, $2 + 3 + 2 = 7$, is not divisible by 3.

11. 102 is even and therefore divisible by 2. The sum of the digits, $1 + 0 + 2 = 3$, is divisible by 3 and therefore 102 is divisible by 3. The last digit is not a 0 or 5, therefore 102 is not divisible by 5.

13. 705 is not even and therefore not divisible by 2. The sum of the digits, $7 + 0 + 5 = 12$, is divisible by 3 and therefore 705 is divisible by 3. The last digit is 5, therefore 7 is divisible by 5.

15. 330 is even and therefore divisible by 2. The sum of the digits, $3 + 3 + 0 = 6$, is divisible by 3 and therefore 330 is divisible by 3. The last digit is 0, therefore 330 is divisible by 5.

17. 22,971 is not even and therefore is not divisible by 2. The sum of the digits, $2 + 2 + 9 + 7 + 1 = 21$, is divisible by 3 and therefore 22,971 is divisible by 3. The last digit is not a 0 or 5, therefore 22,971 is not divisible by 5.

19. 0, 1: neither; 17: prime; 9, 40, 8, 15, 22: composite

21. a. $8 = 2 \cdot 4$ or $8 \cdot 1$

b. $8 = 2 \cdot 2 \cdot 2$ or 2^3

23. $28 = 2 \cdot \boxed{2} \cdot 7 = 2^{\boxed{2}} \cdot 7$

25. $75 = 2 \cdot \boxed{5} \cdot 5 = 3 \cdot 5^{\boxed{2}}$

27. a.

```
     [5]
  2) 10
[3]) 30
  5)150
```

b. $150 = 5 \cdot 3 \cdot 2 \cdot 5 = 2 \cdot 3 \cdot 5^2$

29. a.

```
    220
    / \
  22  [10]
  /\   /\
 2 [11] [2] 5
```

b. $220 = 2 \cdot 11 \cdot 2 \cdot 5 = 2^2 \cdot 5 \cdot 11$

31.

```
   5
3)15
```

$15 = 3 \cdot 5$

33.

```
  2
2)4
5)20
```

$20 = 5 \cdot 2 \cdot 2 = 2 \cdot 2 \cdot 5$ or $2^2 \cdot 5$

35.

```
    24
   /  \
  4 · 6
 /\   /\
2 · 2 2 · 3
```

$24 = 2 \cdot 2 \cdot 2 \cdot 3$ or $2^3 \cdot 3$

37.

```
   2
5)10
7)70
```

$70 = 7 \cdot 5 \cdot 2 = 2 \cdot 5 \cdot 7$

39.

```
     64
    /  \
   8  ·  8
  /\     /\
 2 · 4  2 · 4
    /\     /\
   2 · 2  2 · 2
```

$64 = 2 \cdot 2 \cdot 2 \cdot 2 \cdot 2 \cdot 2$ or 2^6

Copyright © 2012 Pearson Education, Inc.

41.
```
    80
   /  \
  8  · 10
 / \   / \
2 · 4  2 · 5
   / \
  2 · 2
```
$80 = 2 \cdot 2 \cdot 2 \cdot 2 \cdot 5$ or $2^4 \cdot 5$

43.
```
   5
5)25
3)75
```
$75 = 3 \cdot 5 \cdot 5$ or $3 \cdot 5^2$

45.
```
   3
3) 9
5)45
```
$45 = 5 \cdot 3 \cdot 3 = 3 \cdot 3 \cdot 5$ or $3^2 \cdot 5$

47.
```
    3
 3) 9
11)99
```
$99 = 11 \cdot 3 \cdot 3 = 3 \cdot 3 \cdot 11$ or $3^2 \cdot 11$

49.
```
300
 / \
3 · 100
    /  \
  10 · 10
  / \  / \
 2 · 5 2 · 5
```
$300 = 3 \cdot 2 \cdot 5 \cdot 2 \cdot 5 = 2 \cdot 2 \cdot 3 \cdot 5 \cdot 5$ or $2^2 \cdot 3 \cdot 5^2$

51.
```
  110
  / \
 10 · 11
 /\
2 · 5
```
$110 = 2 \cdot 5 \cdot 11$

53.
```
   17
2) 34
2) 68
2)136
```
$136 = 2 \cdot 2 \cdot 2 \cdot 17$ or $2^3 \cdot 17$

55.
```
    90
   /  \
  9  · 10
 / \   / \
3 · 3  2 · 5
```
$90 = 3 \cdot 3 \cdot 2 \cdot 5$ or $2 \cdot 3^2 \cdot 5$

57.
```
   225
   /  \
  9  · 25
 / \   / \
3 · 3  5 · 5
```
$225 = 3 \cdot 3 \cdot 5 \cdot 5 = 3^2 \cdot 5^2$

59. Using a calculator:
$1309 \div 7 = 187$
$187 \div 11 = 17$
$1309 = 7 \cdot 11 \cdot 17$

61. Using a calculator:
$2737 \div 7 = 391$
$391 \div 17 = 23$
$2737 = 7 \cdot 17 \cdot 23$

63. Answers may vary, but the number must end in the digit 5 or 0 and the sum of the digits must be divisible by 3.

65. a. The first six consecutive positive odd numbers are 1, 3, 5, 7, 9, and 11.

b. The completed first portion of the table is shown below.

1					
1	3				
1	3	5			
1	3	5	7		
1	3	5	7	9	
1	3	5	7	9	11

c. Write the sum of each row in the seventh column.
Row 3: $1 + 3 + 5 = 9$
Row 4: $1 + 3 + 5 + 7 = 16$
Row 5: $1 + 3 + 5 + 7 + 9 = 25$
Row 6: $1 + 3 + 5 + 7 + 9 + 11 = 36$

Copyright © 2012 Pearson Education, Inc.

1						1
1	3					4
1	3	5				9
1	3	5	7			16
1	3	5	7	9		25
1	3	5	7	9	11	36

d. In the eighth column, write the exponent form of each number in the seventh column.
$9 = 3 \cdot 3 = 3^2$; $16 = 4 \cdot 4 = 4^2$; $25 = 5 \cdot 5 = 5^2$; $36 = 6 \cdot 6 = 6^2$

1						1	1^2
1	3					4	2^2
1	3	5				9	3^2
1	3	5	7			16	4^2
1	3	5	7	9		25	5^2
1	3	5	7	9	11	36	6^2

e. The list of numbers is the set of the first 6 perfect square whole numbers.

f. The list of numbers is the set of the first 6 whole numbers squared.

g. Based on observation, the next two numbers in the seventh column should be the next two perfect square whole numbers, which are 49 and 64. The next two numbers in the eight column should be the next two whole numbers squared, 7^2 and 8^2.

Copyright © 2012 Pearson Education, Inc.

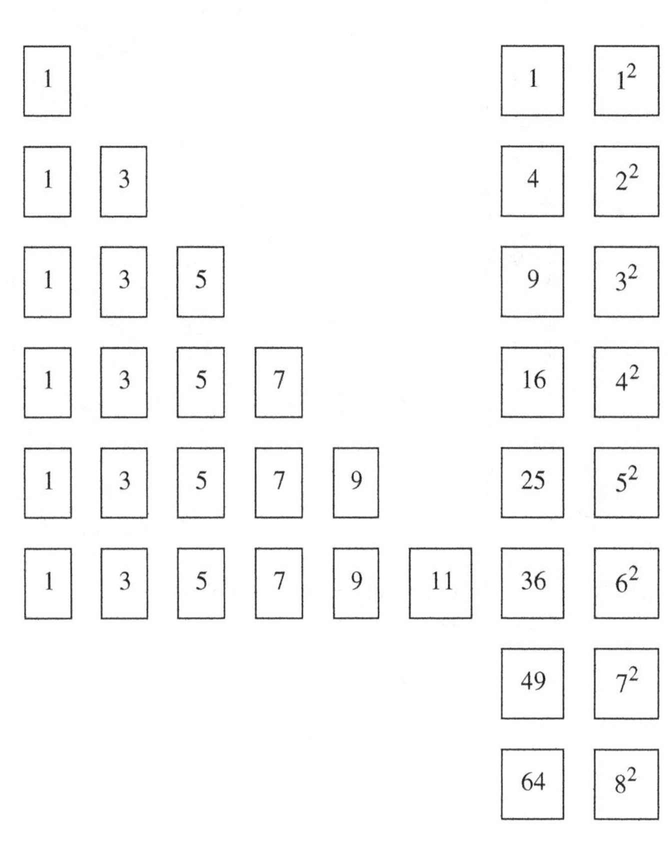

h. We use the following pattern: The sum of the first n consecutive odd numbers equals n^2.
The sum of the first 12 consecutive odd numbers equals 12^2.
The sum of the first 20 consecutive odd numbers equals 20^2.

Cumulative Review

67. $(2x^2)(5x^3y) = (2 \cdot 5) \cdot (x^2 \cdot x^3) \cdot y = 10x^5y$

68. $7y^2 + 2y^2 = (7+2)y^2 = 9y^2$

69. $5x + 3x + 2 = (5+3)x + 2 = 8x + 2$

70 $(5x)(3x)(2) = (5 \cdot 3 \cdot 2) \cdot (x \cdot x) = 30x^2$

71. Since the angles are supplementary, their sum is 180°.

$$\begin{aligned} \angle a + \angle b &= 180° \\ \angle a + 81° &= 180° \\ + \quad -81° & \quad -81° \\ \hline \angle a &= 99° \end{aligned}$$

The measure of $\angle a$ is 99°.

 Copyright © 2012 Pearson Education, Inc.

72. Since the angles are supplementary, their sum is 180°.

$\angle x + \angle y = 180°$

$\angle x + 62° = 180°$

$+ \quad -62° \quad -62°$

$\angle x = 118°$

The measure of $\angle x$ is 118°.

Quick Quiz 4.1

1. 504 is even and therefore divisible by 2. The sum of the digits, 5 + 0 + 4 = 9, is divisible by 3 and therefore 504 is divisible by 3. The last digit is not a 0 or 5, so 504 is not divisible by 5.

2. The number 57 is composite since it is divisible by the numbers other than 1 and itself, such as 3.

3.
```
315
/ \
3 · 105
    / \
   3 · 35
      / \
     5 · 7
```
$315 = 3 \cdot 3 \cdot 5 \cdot 7$ or $3^2 \cdot 5 \cdot 7$

4. a. Answers may vary. One possible solution is by noticing that the number ends in an even number.

b. If any whole number ends with an even number, the number has a factor of 2.

4.2 Exercises

1. Fractions are a set of numbers used to describe <u>part</u> of a whole quantity. In the fraction $\frac{2}{3}$, the 2 is called the <u>numerator</u> and the 3 is called the <u>denominator</u>.

3. To change the mixed number $6\frac{2}{3}$ to an improper fraction, multiply $\underline{3} \times \underline{6}$, and then add $\underline{2}$ to the product. The improper fraction is written as $\frac{\boxed{20}}{\boxed{3}}$.

5. Three out of five parts are shaded, or $\frac{3}{5}$.

7. Three out of four parts are shaded, or $\frac{3}{4}$.

9. $\frac{7}{0}$; division by 0 is undefined.

11. $\frac{0}{z} = 0$; any fraction with 0 in the numerator and a nonzero denominator equals 0.

13. $\frac{44}{44} = 1$; any nonzero number divided by itself is 1.

15. $\frac{0}{9} = 0$; any fraction with 0 in the numerator and a nonzero denominator equals 0.

17. $\frac{8}{0}$; division by 0 is undefined.

19. $\frac{y}{y} = 1$; any nonzero number divided by itself is 1.

21. The baseball player had 5 base hits out of 12 times; the fraction is $\frac{5}{12}$.

23. First we must find the total number of customers.
57 + 37 = 94 total customers
Since 37 of the 94 customers are men, the fraction is $\frac{37}{94}$.

25. First we find the number of dancers who are not juniors.
26 − 9 = 17 non-juniors
Since 17 out of 26 dancers are not juniors, the fraction is $\frac{17}{26}$.

27. First we must find the number of games the team lost (did not win).
32 − 21 = 11 games lost
Since the team lost 11 out of 32 games the fraction is $\frac{11}{32}$.

Copyright © 2012 Pearson Education, Inc.

29. First we must find the total amount of deductions from Arnold's paycheck.
$24 + $22 + $58 + $31 = $135
Now we find the total for state and federal income tax.
$24 + $58 = $82
The fractional part for either state or federal income tax is $\frac{82}{135}$.

31. First we must find the total amount of deductions from Arnold's paycheck.
$24 + $22 + $58 + $31 = $135
Now we find the part that is not for insurance.
$135 − $31 = $104
The fractional part that is not for insurance is $\frac{104}{135}$.

33. $\frac{11}{9}$ is an improper fraction because the numerator is larger than the denominator.

35. $\frac{8}{8}$ is an improper fraction because the numerator is equal to the denominator.

37. $\frac{5}{6}$ is a proper fraction because the numerator is less than the denominator.

39. $7\frac{1}{9}$ is a mixed number because a whole number is added to a fraction.

41. $8\overline{)15}$ = 1 R 7; $\begin{array}{r}8\\ \hline 7\end{array}$ $\quad \frac{15}{8} = 1\frac{7}{8}$

43. $5\overline{)48}$ = 9 R 3; $\begin{array}{r}45\\ \hline 3\end{array}$ $\quad \frac{48}{5} = 9\frac{3}{5}$

45. $2\overline{)41}$ = 20 R 1; $\begin{array}{r}4\\ \hline 1\end{array}$ $\quad \frac{41}{2} = 20\frac{1}{2}$

47. $5\overline{)32}$ = 6 R 2; $\begin{array}{r}30\\ \hline 2\end{array}$ $\quad \frac{32}{5} = 6\frac{2}{5}$

49. $5\overline{)47}$ = 9 R 2; $\begin{array}{r}45\\ \hline 2\end{array}$ $\quad \frac{47}{5} = 9\frac{2}{5}$

51. $\frac{33}{33} = 1$; any nonzero number divided by itself is 1.

53. $8\frac{3}{7} = \frac{(7\cdot 8)+3}{7} = \frac{56+3}{7} = \frac{59}{7}$

55. $24\frac{1}{4} = \frac{(4\cdot 24)+1}{4} = \frac{96+1}{4} = \frac{97}{4}$

57. $15\frac{2}{3} = \frac{(3\cdot 15)+2}{3} = \frac{45+2}{3} = \frac{47}{3}$

59. $33\frac{1}{3} = \frac{(3\cdot 33)+1}{3} = \frac{99+1}{3} = \frac{100}{3}$

61. $8\frac{9}{10} = \frac{(10\cdot 8)+9}{10} = \frac{80+9}{10} = \frac{89}{10}$

63. $8\frac{7}{15} = \frac{(15\cdot 8)+7}{15} = \frac{120+7}{15} = \frac{127}{15}$

Cumulative Review

65. $(-5)(-8) = 40$

66. $(-7)(9) = -63$

67.
```
   63
   / \
  9 · 7
 /\
3 · 3
```
$63 = 3\cdot 3\cdot 7$ or $3^2\cdot 7$

68.
```
    54
   /  \
  6  ·  9
 / \   / \
2 · 3 3 · 3
```
$54 = 2\cdot 3\cdot 3\cdot 3$ or $2\cdot 3^3$

Copyright © 2012 Pearson Education, Inc.

Quick Quiz 4.2

1. $\frac{0}{9} = 0$; any fraction with 0 in the numerator and a nonzero denominator equals 0.

2. $4\frac{3}{5} = \frac{(5 \cdot 4)+3}{5} = \frac{20+3}{5} = \frac{23}{5}$

3. First find the total number of pets.
   $8 + 5 + 2 = 15$ pets
   Since 8 of the 15 pets are cats, the fraction is $\frac{8}{15}$.

4. Answers may vary. One possible solution is to multiply the whole number portion by the denominator. This product is then added to the numerator.

4.3 Exercises

1. To build an equivalent fraction we <u>multiply</u> the numerator and denominator by the <u>same</u> number.

3. $\frac{3 \cdot \boxed{2}}{7 \cdot 2} = \frac{\boxed{6}}{14}$

5. $\frac{7 \cdot \boxed{3}}{5x \cdot 3} = \frac{\boxed{21}}{15x}$

7. $\frac{7 \cdot \boxed{y}}{8 \cdot \boxed{y}} = \frac{7y}{8y}$

9. $\frac{2 \cdot \boxed{y}}{9 \cdot \boxed{y}} = \frac{\boxed{2y}}{9y}$

11. a. $\frac{7}{9} = \frac{7 \cdot 4}{9 \cdot 4} = \frac{28}{36}$

    b. $\frac{7}{9} = \frac{7 \cdot 5x}{9 \cdot 5x} = \frac{35x}{45x}$

13. a. $\frac{4}{11} = \frac{4 \cdot 4}{11 \cdot 4} = \frac{16}{44}$

    b. $\frac{4}{11} = \frac{4 \cdot 5x}{11 \cdot 5x} = \frac{20x}{55x}$

15. $\frac{3}{8} = \frac{3 \cdot 4}{8 \cdot 4} = \frac{12}{32}$

17. $\frac{5}{6} = \frac{5 \cdot 5}{6 \cdot 5} = \frac{25}{30}$

19. $\frac{9}{13} = \frac{9 \cdot 3}{13 \cdot 3} = \frac{27}{39}$

21. $\frac{35}{40} = \frac{35 \cdot 2}{40 \cdot 2} = \frac{70}{80}$

23. $\frac{8}{9} = \frac{8 \cdot y}{9 \cdot y} = \frac{8y}{9y}$

25. $\frac{3}{7} = \frac{3 \cdot 4y}{7 \cdot 4y} = \frac{12y}{28y}$

27. $\frac{3}{6} = \frac{3 \cdot 3a}{6 \cdot 3a} = \frac{9a}{18a}$

29. $\frac{5}{7} = \frac{5 \cdot 3x}{7 \cdot 3x} = \frac{15x}{21x}$

31. $\frac{20}{25} = \frac{2 \cdot 2 \cdot \cancel{5}}{5 \cdot \cancel{5}} = \frac{2 \cdot 2}{5} = \frac{4}{5}$

33. $\frac{12}{16} = \frac{\cancel{2} \cdot \cancel{2} \cdot 3}{\cancel{2} \cdot \cancel{2} \cdot 2 \cdot 2} = \frac{3}{2 \cdot 2} = \frac{3}{4}$

35. $\frac{30}{36} = \frac{\cancel{2} \cdot \cancel{3} \cdot 5}{\cancel{2} \cdot 2 \cdot \cancel{3} \cdot 3} = \frac{5}{2 \cdot 3} = \frac{5}{6}$

37. $\frac{16}{28} = \frac{\cancel{2} \cdot \cancel{2} \cdot 2 \cdot 2}{\cancel{2} \cdot \cancel{2} \cdot 7} = \frac{2 \cdot 2}{7} = \frac{4}{7}$

39. $\frac{24}{36} = \frac{\cancel{2} \cdot \cancel{2} \cdot 2 \cdot \cancel{3}}{\cancel{2} \cdot \cancel{2} \cdot \cancel{3} \cdot 3} = \frac{2}{3}$

41. $\frac{30}{85} = \frac{2 \cdot 3 \cdot \cancel{5}}{\cancel{5} \cdot 17} = \frac{2 \cdot 3}{17} = \frac{6}{17}$

43. $\frac{48}{56} = \frac{\cancel{2} \cdot \cancel{2} \cdot \cancel{2} \cdot 2 \cdot 3}{\cancel{2} \cdot \cancel{2} \cdot \cancel{2} \cdot 7} = \frac{6}{7}$

45. $\frac{36}{72} = \frac{\cancel{2} \cdot \cancel{2} \cdot \cancel{3} \cdot \cancel{3}}{\cancel{2} \cdot \cancel{2} \cdot 2 \cdot \cancel{3} \cdot \cancel{3}} = \frac{1}{2}$

Copyright © 2012 Pearson Education, Inc.

47. $\frac{49}{35}=\frac{7\cdot\cancel{7}}{5\cdot\cancel{7}}=\frac{7}{5}$ or $1\frac{2}{5}$

49. $\frac{75}{60}=\frac{\cancel{3}\cdot\cancel{5}\cdot 5}{2\cdot 2\cdot\cancel{3}\cdot\cancel{5}}=\frac{5}{2\cdot 2}=\frac{5}{4}$ or $1\frac{1}{4}$

51. a. $\frac{-12}{18}=-\frac{12}{18}=-\frac{\cancel{2}\cdot 2\cdot\cancel{3}}{\cancel{2}\cdot\cancel{3}\cdot 3}=-\frac{2}{3}$

b. $\frac{12}{-18}=-\frac{12}{18}=-\frac{\cancel{2}\cdot 2\cdot\cancel{3}}{\cancel{2}\cdot\cancel{3}\cdot 3}=-\frac{2}{3}$

c. $-\frac{12}{18}=-\frac{\cancel{2}\cdot 2\cdot\cancel{3}}{\cancel{2}\cdot\cancel{3}\cdot 3}=-\frac{2}{3}$

53. $\frac{-15}{30}=-\frac{15}{30}=-\frac{\cancel{15}\cdot 1}{\cancel{15}\cdot 2}=-\frac{1}{2}$

55. $-\frac{42}{48}=-\frac{42}{48}=-\frac{\cancel{6}\cdot 7}{\cancel{6}\cdot 8}=-\frac{7}{8}$

57. $\frac{30}{-42}=-\frac{30}{42}=-\frac{\cancel{6}\cdot 5}{\cancel{6}\cdot 7}=-\frac{5}{7}$

59. $-\frac{16}{18}=-\frac{\cancel{2}\cdot 8}{\cancel{2}\cdot 9}=-\frac{8}{9}$

61. $\frac{21a}{24a}=\frac{\cancel{3}\cdot 7\cdot\cancel{a}}{2\cdot 2\cdot 2\cdot\cancel{3}\cdot\cancel{a}}=\frac{7}{2\cdot 2\cdot 2}=\frac{7}{8}$

63. $\frac{20y}{35y}=\frac{2\cdot 2\cdot\cancel{5}\cdot\cancel{y}}{\cancel{5}\cdot 7\cdot\cancel{y}}=\frac{2\cdot 2}{7}=\frac{4}{7}$

65. $\frac{24xy}{42x}=\frac{\cancel{6}\cdot 2\cdot 2\cdot\cancel{x}\cdot y}{\cancel{6}\cdot 7\cdot\cancel{x}}=\frac{2\cdot 2\cdot y}{7}=\frac{4y}{7}$

67. $\frac{14y}{28xy}=\frac{\cancel{14}\cdot 1\cdot\cancel{y}}{\cancel{14}\cdot 2\cdot x\cdot\cancel{y}}=\frac{1}{2x}$

69. $\frac{27x^2}{45x}=\frac{\cancel{9}\cdot 3\cdot\cancel{x}\cdot x}{\cancel{9}\cdot 5\cdot\cancel{x}}=\frac{3x}{5}$

71. $\frac{20y}{24y^2}=\frac{\cancel{4}\cdot 5\cdot\cancel{y}}{\cancel{4}\cdot 2\cdot 3\cdot\cancel{y}\cdot y}=\frac{5}{2\cdot 3\cdot y}=\frac{5}{6y}$

73. $\frac{36n^2}{-42n}=-\frac{36n^2}{42n}=-\frac{\cancel{6}\cdot 6\cdot\cancel{n}\cdot n}{\cancel{6}\cdot 7\cdot\cancel{n}}=-\frac{6\cdot n}{7}=-\frac{6n}{7}$

75. $\frac{-35x}{45x^2}=-\frac{35x}{45x^2}$

$=-\frac{\cancel{5}\cdot 7\cdot\cancel{x}}{\cancel{5}\cdot 3\cdot 3\cdot\cancel{x}\cdot x}$

$=-\frac{7}{3\cdot 3\cdot x}$

$=-\frac{7}{9x}$

77. First find the number of questions that Le Mar answered incorrectly.
$36-22=14$ questions

The fraction is $\frac{14}{36}$. Now simplify.

$\frac{14}{36}=\frac{7\cdot\cancel{2}}{18\cdot\cancel{2}}=\frac{7}{18}$

Le Mar answered $\frac{7}{18}$ of the questions incorrectly.

79. First find the total number of questions.
$36+20+25=81$ total questions
Now find the number answered correctly.
$22+15+20=57$ questions

The fraction is $\frac{57}{81}$. Now simplify.

$\frac{57}{81}=\frac{19\cdot\cancel{3}}{27\cdot\cancel{3}}=\frac{19}{27}$

Alexsandra answered $\frac{19}{27}$ of the questions correctly.

81. First find the total number of shark attacks.
$83+68+62+57+65+61+62+71+60+61$
$=650$ total shark attacks
Since 62 of these shark attacks happened in 2006, the fraction is $\frac{62}{650}$. Now simplify.

$\frac{62}{650}=\frac{31\cdot\cancel{2}}{325\cdot\cancel{2}}=\frac{31}{325}$

$\frac{31}{325}$ of the shark attacks happened in 2006.

Copyright © 2012 Pearson Education, Inc.

83. First find the total number of shark attacks for all years in the bar graph.
83 + 68 + 62 + 57 + 65 + 61 + 62 + 71 + 60 + 61 = 650 total shark attacks
Now find the total numbers of shark attacks for just the years 2008 and 2009.
60 + 61 = 121 shark attacks
The fraction is $\frac{121}{650}$, which does not simplify.
$\frac{121}{650}$ of the shark attacks happened in 2008 or 2009.

85. $$\frac{40a^2b^2c^4}{88ab} = \frac{\cancel{8}\cdot 5\cdot \cancel{a}\cdot a\cdot \cancel{b}\cdot b\cdot c\cdot c\cdot c\cdot c}{\cancel{8}\cdot 11\cdot \cancel{a}\cdot \cancel{b}} = \frac{5\cdot a\cdot b\cdot c\cdot c\cdot c\cdot c}{11} = \frac{5abc^4}{11}$$

87. $$\frac{256xy^3}{300yz^5} = \frac{\cancel{4}\cdot 2\cdot 2\cdot 2\cdot 2\cdot 2\cdot 2\cdot x\cdot \cancel{y}\cdot y\cdot y}{\cancel{4}\cdot 3\cdot 5\cdot 5\cdot \cancel{y}\cdot z\cdot z\cdot z\cdot z\cdot z} = \frac{2\cdot 2\cdot 2\cdot 2\cdot 2\cdot 2\cdot x\cdot y\cdot y}{3\cdot 5\cdot 5\cdot z\cdot z\cdot z\cdot z\cdot z} = \frac{64xy^2}{75z^5}$$

89. First find the area of the garden.
$A = L\cdot W = 90\cdot 80 = 7200\text{ in}^2$
Now assuming the length is 120 inches and the area is 7200 square inches, find the width.
$$A = L\cdot W$$
$$7200 = 120W$$
$$\frac{7200}{120} = \frac{120W}{120}$$
$$60 = W$$
The width will be 60 inches.

Cumulative Review

90. $x^4\cdot x^3\cdot x^3 = x^{4+3+3} = x^{10}$

91. $2^4\cdot 2^6 = 2^{4+6} = 2^{10}$

92. $(-3a)(2a^4) = (-3)(2)(a^1\cdot a^4) = -6a^{1+4} = -6a^5$

93. a. The number of peanuts, P, in a mix is double the number of cashews, C, $P = 2C$.

b. Replace P with 34 and solve the equation.
$$P = 2C$$
$$34 = 2C$$
$$\frac{34}{2} = \frac{2C}{2}$$
$$17 = C$$
There are 17 cashews in the mix.

94. Find the areas of both rooms.
Rectangle: $A = L\cdot W = 13\cdot 16 = 208\text{ ft}^2$
Square: $A = s^2 = (15)^2 = 225\text{ ft}^2$
The square room has the greater area. Find the perimeters of both rooms.
Rectangle:
$P = 2L + 2W = 2(13) + 2(16) = 26 + 32 = 58\text{ ft}$
Square: $P = 4s = 4(15) = 60\text{ ft}$
The square room has the greater perimeter.

Quick Quiz 4.3

1. $\frac{3}{7} = \frac{3\cdot 8}{7\cdot 8} = \frac{24}{56}$

2. $\frac{-24}{36} = -\frac{24}{36} = -\frac{\cancel{2}\cdot \cancel{2}\cdot 2\cdot \cancel{3}}{\cancel{2}\cdot \cancel{2}\cdot \cancel{3}\cdot 3} = -\frac{2}{3}$

3. $\frac{33a^2}{44a} = \frac{\cancel{11}\cdot 3\cdot \cancel{a}\cdot a}{\cancel{11}\cdot 4\cdot \cancel{a}} = \frac{3\cdot a}{4} = \frac{3a}{4}$

4. Answers may vary. One quick way to determine that the fraction may be reduced is to see that both the numerator and the denominator are divisible by three. The common factor of three will cancel.

Use Math to Save Money

1. The total of the original investments is $2500 + $2500 + $2500 + $2500 = $10,000.

Large-cap: $\frac{\$2500}{\$10,000} = \frac{2500\cdot 1}{2500\cdot 4} = \frac{1}{4}$

Small-cap: $\frac{\$2500}{\$10,000} = \frac{2500\cdot 1}{2500\cdot 4} = \frac{1}{4}$

Bonds: $\frac{\$2500}{\$10,000} = \frac{2500\cdot 1}{2500\cdot 4} = \frac{1}{4}$

International: $\frac{\$2500}{\$10,000} = \frac{2500\cdot 1}{2500\cdot 4} = \frac{1}{4}$

$\frac{1}{4}$ of the original total was invested in each category.

Copyright © 2012 Pearson Education, Inc.

2. The total of the current investments is $6000 + $5000 + $4000 + $5000 = $20,000.

Large-cap: $\frac{\$6000}{\$20,000} = \frac{2000 \cdot 3}{2000 \cdot 10} = \frac{3}{10}$

Small-cap: $\frac{\$5000}{\$20,000} = \frac{5000 \cdot 1}{5000 \cdot 4} = \frac{1}{4}$

Bonds: $\frac{\$4000}{\$20,000} = \frac{4000 \cdot 1}{4000 \cdot 5} = \frac{1}{5}$

International: $\frac{\$5000}{\$20,000} = \frac{5000 \cdot 1}{5000 \cdot 4} = \frac{1}{4}$

$\frac{1}{4}$ of the current total is invested in each of small-cap and international mutual funds, $\frac{3}{10}$ in large-cap mutual funds, and $\frac{1}{5}$ in bonds.

3. Large-cap and bonds have changed, since neither accounts for exactly $\frac{1}{4}$ of the total amount.

4. To maintain the original fractions, Jason should have $\frac{\$20,000}{4} = \$5000$ in each category.

5. Since Jason has more than $5000 in large-cap mutual funds, he should sell $6000 − $5000 = $1000 worth of large-cap funds. Since Jason has less than $5000 in bonds, he should buy $5000 − $4000 = $1000 worth of bonds.

How Am I Doing? Sections 4.1–4.3

1. 312 is even and therefore divisible by 2. The sum of the digits, 3 + 1 + 2 = 6, is divisible by 3 and therefore 312 is divisible by 3. The last digit is not a 0 or 5, so 312 is not divisible by 5.

2.

```
     120
     / \
    4 · 30
   /\   / \
  2·2  2 · 15
           /\
          3 · 5
```

$120 = 2 \cdot 2 \cdot 2 \cdot 3 \cdot 5$ or $2^3 \cdot 3 \cdot 5$

3. a. $\frac{0}{3} = 0$; any fraction with 0 in the numerator and a nonzero denominator equals 0.

b. $\frac{3}{0}$; division by 0 is undefined.

c. $\frac{a}{a} = 1$; any nonzero number divided by itself is 1.

d. $\frac{22}{22} = 1$; any nonzero number divided by itself is 1.

4. $7\overline{)51}$ = 7 R 2; $\;51 - 49 = 2$ $\qquad \frac{51}{7} = 7\frac{2}{7}$

5. $5\frac{3}{4} = \frac{(4 \cdot 5) + 3}{4} = \frac{20 + 3}{4} = \frac{23}{4}$

6. Find the total number of people in the class.
17 + 19 = 36 people
Since 19 of 36 people were men, the fraction is $\frac{19}{36}$.

7. $\frac{3}{7} = \frac{3 \cdot 5}{7 \cdot 5} = \frac{15}{35}$

8. $\frac{2}{9} = \frac{2 \cdot 3y}{9 \cdot 3y} = \frac{6y}{27y}$

9. $\frac{20}{-42} = -\frac{20}{42} = -\frac{\cancel{2} \cdot 2 \cdot 5}{\cancel{2} \cdot 3 \cdot 7} = -\frac{2 \cdot 5}{3 \cdot 7} = -\frac{10}{21}$

10. $\frac{25y}{45y^2} = \frac{\cancel{5} \cdot 5 \cdot \cancel{y}}{\cancel{5} \cdot 9 \cdot y \cdot \cancel{y}} = \frac{5}{9 \cdot y} = \frac{5}{9y}$

11. First find the number of windows that were not shattered.
42 − 7 = 35
Since 35 of the 42 windows were not shattered, the fraction is $\frac{35}{42}$. Now simplify.

$\frac{35}{42} = \frac{\cancel{7} \cdot 5}{\cancel{7} \cdot 6} = \frac{5}{6}$

Thus, $\frac{5}{6}$ of the windows were not shattered by the hurricane.

 Copyright © 2012 Pearson Education, Inc.

4.4 Exercises

1. a. We add the coefficients and the variables stay the same.
$15x^3 + 5x^3 = (15+5)x^3 = 20x^3$

b. We multiply coefficients and then add exponents of like bases.
$(15x^3)(5x^3) = (15 \cdot 5)x^{3+3} = 75x^6$

c. We simplify coefficients and then subtract exponents of like bases.
$\frac{15x^3}{5x} = \frac{3 \cdot \cancel{5} \cdot x^{3-1}}{\cancel{5}} = 3x^2$

d. We multiply exponents.
$(x^3)^2 = x^{3\cdot 2} = x^6$

3. $\frac{7^4}{7^3} = 7^{4-3} = 7^1 = 7$

5. $\frac{a^8}{a^3} = a^{8-3} = a^5$

7. $\frac{5^8}{5^9} = \frac{1}{5^{9-8}} = \frac{1}{5^1} = \frac{1}{5}$

9. $\frac{3}{3^6} = \frac{3^1}{3^6} = \frac{1}{3^{6-1}} = \frac{1}{3^5}$

11. $\frac{z^4}{y^8} = \frac{z^4}{y^8}$; we cannot divide; the bases are not the same.

13. $\frac{9^3}{8^8} = \frac{9^3}{8^8}$; we cannot divide; the bases are not the same.

15. $\frac{z^8}{z^8} = 1$; any nonzero number divided by itself equals 1.

17. $\frac{y^3z^4}{y^5z^7} = \frac{1}{y^{5-3}z^{7-4}} = \frac{1}{y^2z^3}$

19. $\frac{m^9 3^6}{m^7 3^7} = \frac{m^{9-7}}{3^{7-6}} = \frac{m^2}{3^1} = \frac{m^2}{3}$

21. $\frac{a^5 7^4}{a^3 7^7} = \frac{a^{5-3}}{7^{7-4}} = \frac{a^2}{7^3}$

23. $\frac{b^9 9^9}{b^7 9^{11} 3^0} = \frac{b^9 9^9}{b^7 9^{11} \cdot 1} = \frac{b^{9-7}}{9^{11-9}} = \frac{b^2}{9^2}$

25. $\frac{4^6 ab^0}{4^9 a^4 b} = \frac{4^6 a \cdot 1}{4^9 a^4 b} = \frac{1}{4^{9-6} a^{4-1} b} = \frac{1}{4^3 a^3 b}$

27. $\frac{20y^4}{35y} = \frac{\cancel{5} \cdot 2 \cdot 2 \cdot y^4}{\cancel{5} \cdot 7 \cdot y^1} = \frac{2 \cdot 2 \cdot y^{4-1}}{7} = \frac{4y^3}{7}$

29. $\frac{9a^4}{27a^3} = \frac{\cancel{9} \cdot a^4}{\cancel{9} \cdot 3 \cdot a^3} = \frac{a^{4-3}}{3} = \frac{a^1}{3} = \frac{a}{3}$

31. $\frac{56x^9y^0}{64x^3} = \frac{56x^9 \cdot 1}{64x^3}$
$= \frac{\cancel{8} \cdot 7 \cdot x^9}{\cancel{8} \cdot 2 \cdot 2 \cdot 2 \cdot x^3}$
$= \frac{7 \cdot x^{9-3}}{2 \cdot 2 \cdot 2}$
$= \frac{7x^6}{8}$

33. $\frac{12x^4y^2}{15xy^3} = \frac{\cancel{3} \cdot 2 \cdot 2 \cdot x^4 \cdot y^2}{\cancel{3} \cdot 5 \cdot x^1 \cdot y^3}$
$= \frac{2 \cdot 2 \cdot x^{4-1}}{5 \cdot y^{3-2}}$
$= \frac{4x^3}{5y^1}$
$= \frac{4x^3}{5y}$

35. a. Add the exponents.
$z^2 \cdot z^2 \cdot z^2 = z^{2+2+2} = z^6$

b. Multiply the exponents.
$(z^2)^3 = z^{2\cdot 3} = z^6$

37. a. Add the exponents.
$x^4 \cdot x^4 = x^{4+4} = x^8$

b. Multiply the exponents.
$(x^4)^2 = x^{4\cdot 2} = x^8$

Copyright © 2012 Pearson Education, Inc.

39. $(z^2)^4 = z^{2\cdot4} = z^8$

41. $(3^5)^4 = 3^{5\cdot4} = 3^{20}$

43. $(b^1)^6 = b^{1\cdot6} = b^6$

45. $(x^0)^4 = x^{0\cdot4} = x^0 = 1$

47. $(y^3)^3 = y^{3\cdot3} = y^9$

49. $(2^4)^5 = 2^{4\cdot5} = 2^{20}$

51. $(x^2)^0 = x^{2\cdot0} = x^0 = 1$

53. $(6^3)^9 = 6^{3\cdot9} = 6^{27}$

55. $(x^2)^2 = x^{2\cdot2} = x^4$

57. $(3y^2)^6 = (3^1\cdot y^2)^6 = 3^{1\cdot6}\cdot y^{2\cdot6} = 3^6y^{12}$

59. $(4x^2)^3 = (4^1\cdot x^2)^3 = 4^{1\cdot3}\cdot x^{2\cdot3} = 4^3x^6$

61. $(3a^4)^8 = (3^1\cdot a^4)^8 = 3^{1\cdot8}\cdot a^{4\cdot8} = 3^8a^{32}$

63. $(2^2x^5)^3 = 2^{2\cdot3}\cdot x^{5\cdot3} = 2^6x^{15}$

65. $(8^3n^4)^6 = 8^{3\cdot6}\cdot n^{4\cdot6} = 8^{18}n^{24}$

67. $\left(\frac{4}{x}\right)^2 = \left(\frac{4^1}{x^1}\right)^2 = \frac{4^{(1)(2)}}{x^{(1)(2)}} = \frac{4^2}{x^2}$ or $\frac{16}{x^2}$

69. $\left(\frac{a}{b}\right)^7 = \left(\frac{a^1}{b^1}\right)^7 = \frac{a^{(1)(7)}}{b^{(1)(7)}} = \frac{a^7}{b^7}$

71. $\left(\frac{3}{x}\right)^3 = \left(\frac{3^1}{x^1}\right)^3 = \frac{3^{(1)(3)}}{x^{(1)(3)}} = \frac{3^3}{x^3}$ or $\frac{27}{x^3}$

73. $\left(\frac{m}{n}\right)^4 = \left(\frac{m^1}{n^1}\right)^4 = \frac{m^{(1)(4)}}{n^{(1)(4)}} = \frac{m^4}{n^4}$

75. $\left(\frac{x}{6}\right)^2 = \left(\frac{x^1}{6^1}\right)^2 = \frac{x^{(1)(2)}}{6^{(1)(2)}} = \frac{x^2}{6^2}$ or $\frac{x^2}{36}$

77. $\left(\frac{3}{7}\right)^2 = \left(\frac{3^1}{7^1}\right)^2 = \frac{3^{(1)(2)}}{7^{(1)(2)}} = \frac{3^2}{7^2}$ or $\frac{9}{49}$

79.
$$\begin{aligned}\frac{25x^2y^3z^4}{135x^7y} &= \frac{\cancel{5}\cdot5\cdot x^2\cdot y^3\cdot z^4}{\cancel{5}\cdot3\cdot3\cdot3\cdot x^7\cdot y^1}\\ &= \frac{5\cdot y^{3-1}\cdot z^4}{3\cdot3\cdot3\cdot x^{7-2}}\\ &= \frac{5y^2z^4}{27x^5}\end{aligned}$$

81.
$$\begin{aligned}\frac{156a^0b^8}{144b^6c^9} &= \frac{\cancel{12}\cdot13\cdot1\cdot b^8}{\cancel{12}\cdot2\cdot2\cdot3\cdot b^6\cdot c^9}\\ &= \frac{13\cdot b^{8-6}}{2\cdot2\cdot3\cdot c^9}\\ &= \frac{13b^2}{12c^9}\end{aligned}$$

83. $\left(\frac{2y}{5x}\right)^2 = \frac{(2y)^2}{(5x)^2} = \frac{2^2\cdot y^2}{5^2\cdot x^2} = \frac{4y^2}{25x^2}$

85.
$$\begin{aligned}\left(\frac{3a^2}{2b^3}\right)^3 &= \frac{(3a^2)^3}{(2b^3)^3}\\ &= \frac{3^3\cdot(a^2)^3}{2^3\cdot(b^3)^3}\\ &= \frac{27\cdot a^{2\cdot3}}{8\cdot b^{3\cdot3}}\\ &= \frac{27a^6}{8b^9}\end{aligned}$$

87. **a.** $15x^3 + 5x^3 = (15+5)x^3 = 20x^3$

b. $(15x^3)(5x^3) = (15\cdot5)x^{3+3} = 75x^6$

c. $(x^3)^3 = x^{3\cdot3} = x^9$

d. $\frac{15x^3}{5x^5} = \frac{\cancel{5}\cdot3\cdot x^3}{\cancel{5}\cdot x^5} = \frac{3}{x^{5-3}} = \frac{3}{x^2}$

89. **a.** $3x^3 + 9x^3 = (3+9)x^3 = 12x^3$

b. $(3x^3)(9x^3) = (3\cdot9)x^{3+3} = 27x^6$

Copyright © 2012 Pearson Education, Inc.

c. $(3x^3)^3 = 3^3 x^{3\cdot 3} = 3^3 x^9$ or $27x^9$

d. $\frac{3x^3}{9x^4} = \frac{\cancel{3}\cdot x^3}{\cancel{3}\cdot 3\cdot x^4} = \frac{1}{3\cdot x^{4-3}} = \frac{1}{3\cdot x^1} = \frac{1}{3x}$

91. a. $12x^4 + 3x^4 = (12+3)x^4 = 15x^4$

b. $(12x^4)(3x^4) = (12\cdot 3)x^{4+4} = 36x^8$

c. $(2x^4)^4 = 2^4 x^{4\cdot 4} = 2^4 x^{16}$ or $16x^{16}$

d. $\frac{2x^4}{3x} = \frac{2x^{4-1}}{3} = \frac{2x^3}{3}$

93. a. $5y^2 + 15y^2 = (5+15)y^2 = 20y^2$

b. $(5y^2)(15y^2) = (5\cdot 15)y^{2+2} = 75y^4$

c. $(5y^2)^2 = 5^2 y^{2\cdot 2} = 5^2 y^4$ or $25y^4$

d. $\frac{5y^2}{15y^7} = \frac{\cancel{5}\cdot y^2}{\cancel{5}\cdot 3\cdot y^7} = \frac{1}{3\cdot y^{7-2}} = \frac{1}{3y^5}$

95. The first and last number in each row is 1, so the first and last number in next row should also be 1.

1						1

Along with the second diagonal from both ends, the numbers are the consecutive whole numbers 1, 2, 3, 4, 5, so the next numbers in these diagonals should be 6.

1	6				6	1

Furthermore, each number in the interior of the triangle is the sum of the two numbers diagonally above it. So the remaining numbers in the next row are 5 + 10 = 15, 10 + 10 = 20, and 10 + 5 = 15.

1	6	15	20	15	6	1

There are other patterns that you may find that will produce the same numbers.

Cumulative Review

96. $3x = 42$
$\frac{3x}{3} = \frac{42}{3}$
$x = 14$

97. $48 = 16x$
$\frac{48}{16} = \frac{16x}{16}$
$3 = x$ or $x = 3$

98. $(18-4) = 7x$
$14 = 7x$
$\frac{14}{7} = \frac{7x}{7}$
$2 = x$ or $x = 2$

99. $13x = 130$
$\frac{13x}{13} = \frac{130}{13}$
$x = 10$

100. $A = LW = (3x+3)(2x^2)$
$= (3x)(2x^2) + (3)(2x^2)$
$= 6x^3 + 6x^2$

101. For the first two years, Kristina will pay $620 per month for 9 months of room and board.
2(9)($620) = $11,160
For the last two years, Kristina will pay $350 rent and $250 food expenses per month for 9 months.
2(9)($350 + $250) = 2(9)($600) = $10,800
Tuition and books will be $7500 per year for 4 years.
4($7500) = $30,000
Add these expenses.
$11,160 + $10,800 + $30,000 = $51,960
Subtract the amount in the trust fund from this amount.
$51,960 − $45,000 = $6960
Kristina should plan to borrow $6960.

Quick Quiz 4.4

1. $\frac{a^5 b}{a^3 b^2} = \frac{a^5\cdot b^1}{a^3\cdot b^2} = \frac{a^{5-3}}{b^{2-1}} = \frac{a^2}{b^1} = \frac{a^2}{b}$

2. $(7x^3)^2 = (7^1\cdot x^3)^2 = 7^{1\cdot 2}\cdot x^{3\cdot 2} = 7^2 x^6$ or $49x^6$

Copyright © 2012 Pearson Education, Inc.

3. $\left(\frac{y}{2}\right)^4 = \left(\frac{y^1}{2^1}\right)^4 = \frac{y^{(1)(4)}}{2^{(1)(4)}} = \frac{y^4}{2^4} = \frac{y^4}{16}$

4. Answers may vary. One possible approach is to simplify the rational expression inside the parentheses first. The result is a quantity $2x^2$, raised to the third power. Distribution of the exponents to the quantity $2x^2$ yields $8x^6$.

4.5 Exercises

1. A ratio compares amounts with the same units and a rate compares amounts with different units.

3. $25 \text{ to } 45 = \frac{25}{45} = \frac{\cancel{5}\cdot 5}{\cancel{5}\cdot 9} = \frac{5}{9}$

5. $35:10 = \frac{35}{10} = \frac{\cancel{5}\cdot 7}{\cancel{5}\cdot 2} = \frac{7}{2}$

7. $54:70 = \frac{54}{70} = \frac{\cancel{2}\cdot 27}{\cancel{2}\cdot 35} = \frac{27}{35}$

9. $34 \text{ min to } 12 \text{ min} = \frac{34 \cancel{\text{min}}}{12 \cancel{\text{min}}} = \frac{\cancel{2}\cdot 17}{\cancel{2}\cdot 6} = \frac{17}{6}$

11. $14 \text{ gal to } 35 \text{ gal} = \frac{14 \cancel{\text{gal}}}{35 \cancel{\text{gal}}} = \frac{2\cdot \cancel{7}}{5\cdot \cancel{7}} = \frac{2}{5}$

13. $17 \text{ hr to } 41 \text{ hr} = \frac{17 \cancel{\text{hr}}}{41 \cancel{\text{hr}}} = \frac{17}{41}$

15. $\$121 \text{ to } \$432 = \frac{\cancel{\$}121}{\cancel{\$}423} = \frac{121}{423}$

17. a. $\frac{\text{chlorine}}{\text{water}} \Rightarrow \frac{15 \cancel{\text{mL}}}{35 \cancel{\text{mL}}} = \frac{\cancel{5}\cdot 3}{\cancel{5}\cdot 7} = \frac{3}{7}$

b. $\frac{\text{water}}{\text{chlorine}} \Rightarrow \frac{35 \cancel{\text{mL}}}{15 \cancel{\text{mL}}} = \frac{\cancel{5}\cdot 7}{\cancel{5}\cdot 3} = \frac{7}{3}$

19. a. $\frac{\text{wins}}{\text{losses}} \Rightarrow \frac{29}{13}$

b. $\frac{\text{losses}}{\text{wins}} \Rightarrow \frac{13}{29}$

21. $\frac{\text{corn flakes}}{\text{margarine}} \Rightarrow \frac{8 \cancel{\text{g}}}{20 \cancel{\text{g}}} = \frac{\cancel{4}\cdot 2}{\cancel{4}\cdot 5} = \frac{2}{5}$

23. $\frac{\text{shortest}}{\text{longest}} \Rightarrow \frac{3970 \cancel{\text{ft}}}{6532 \cancel{\text{ft}}} = \frac{\cancel{2}\cdot 1985}{\cancel{2}\cdot 3266} = \frac{1985}{3266}$

25. We divide $410 \div 19$ to find the unit rate.

$$\begin{array}{r} 21 \text{ R } 11 \\ 19\overline{)410} \\ \underline{38} \\ 30 \\ \underline{19} \\ 11 \end{array}$$

$\frac{410 \text{ cal}}{19 \text{ g of fat}} = 21\frac{11}{19}$ cal per g of fat

27. We divide $300 \div 20$ to find the unit rate.

$\frac{300 \text{ mi}}{15 \text{ gal}} = 20$ mi/gal

29. We divide $304 \div 38$ to find the unit rate.

$\frac{\$304}{38 \text{ hr}} = \$8$ per hour

31. We divide $320 \div 6$ to find the unit rate.

$$\begin{array}{r} 53 \text{ R } 2 \\ 6\overline{)320} \\ \underline{30} \\ 20 \\ \underline{18} \\ 2 \end{array}$$

$53\frac{2}{6} = 53\frac{1}{3}$

$\frac{320 \text{ mi}}{6 \text{ hr}} = 53\frac{1}{3}$ mph

33. We divide $616 \div 28$ to find the unit rate.

$\frac{616 \text{ mi}}{28 \text{ gal}} = 22$ mi/gal

The car can be driven 22 miles on one gallon of gas.

35. We divide $108 \div 9$ to find the unit rate.

$\frac{\$108}{9 \text{ books}} = \$12$ per book

The cost is \$12 per book.

Copyright © 2012 Pearson Education, Inc.

37. a. Students per instructor:

$$\frac{\text{students}}{\text{instructor}} \Rightarrow \frac{90 \text{ students}}{5 \text{ instructors}} = \frac{18 \text{ students}}{1 \text{ instructor}} = 18 \text{ students per instructor}$$

b. Students per tutor:

$$\frac{\text{students}}{\text{tutor}} \Rightarrow \frac{30 \text{ students}}{2 \text{ tutors}} = \frac{15 \text{ students}}{1 \text{ tutor}} = 15 \text{ students per tutor}$$

c. Since every 15 students require 1 tutor, we divide 90 by 15 to find how many tutors are needed for 90 students.
$90 \div 15 = 6$ tutors for 90 students.

39. a. Box of 8: $\frac{\$96}{8} = \$12$ per glass

Box of 6: $\frac{\$78}{6} = \$13$ per glass

b. The box of 8 glasses is the better buy.

41. a. 4 used CDs: $\frac{\$32}{4} = \$8$ per CD

6 used CDs: $\frac{\$48}{6} = \$8$ per CD

b. Both offer the same unit price.

43. $\frac{3161 \text{ sales in August}}{21 \text{ days}}$

We divide to find the unit rate.

$$\begin{array}{r} 101\frac{30}{31} \\ 31\overline{)3161} \\ \underline{31} \\ 61 \\ \underline{31} \\ 30 \end{array}$$

We round to 102 because $\frac{30}{31}$ is close to 1.

The store had an average of 102 sales per day in August.

Cumulative Review

45.
$$\begin{aligned} x-12 &= 25 \\ x-12+12 &= 25+12 \\ x &= 37 \end{aligned}$$
Check:
$$\begin{aligned} x-12 &= 25 \\ 37-12 &\stackrel{?}{=} 25 \\ 25 &= 25 \checkmark \end{aligned}$$

46.
$$\begin{aligned} x+15 &= 40 \\ x+15+(-15) &= 40+(-15) \\ x &= 25 \end{aligned}$$
Check:
$$\begin{aligned} x+15 &= 40 \\ 25+15 &\stackrel{?}{=} 40 \\ 40 &= 40 \checkmark \end{aligned}$$

47.
$$\begin{aligned} 5x-4x+6 &= 14 \\ x+6 &= 14 \\ x+6-6 &= 14-6 \\ x &= 8 \end{aligned}$$
Check:
$$\begin{aligned} 5x-4x+6 &= 14 \\ 5(8)-4(8)+6 &\stackrel{?}{=} 14 \\ 40-32+6 &\stackrel{?}{=} 14 \\ 14 &= 14 \checkmark \end{aligned}$$

48.
$$\begin{aligned} -2-5+a &= 15 \\ -7+a &= 15 \\ -7+7+a &= 15+7 \\ a &= 22 \end{aligned}$$
Check:
$$\begin{aligned} -2-5+a &\stackrel{?}{=} 15 \\ -2-5+22 &\stackrel{?}{=} 15 \\ 15 &= 15 \checkmark \end{aligned}$$

49. 101°F – (–8F) = 109°F

50. First play: –4 yd
Second play: 8 yd
Third play: –5 yd
–4 yd + 8 yd + (–5 yd) = –1 yd
There was a net loss of 1 yard after the third play.

Quick Quiz 4.5

1. $\frac{\text{votes for Mark}}{\text{votes for Sara}} = \frac{45}{75}$ or $\frac{3}{5}$

Copyright © 2012 Pearson Education, Inc.

2. We divide $30 \div 12$ to find the unit rate.

$$12\overline{)30} = 2 \text{ R } 6, \quad 30 - 24 = 6 \qquad 2\frac{6}{12} = 2\frac{1}{2}$$

$$\frac{30 \text{ hot dogs}}{12 \text{ min}} = 2\frac{1}{2} \text{ hot dogs per minute}$$

Reza cooked $2\frac{1}{2}$ hot dogs per minute.

3. 36 bars: $\frac{\$72}{36} = \$2$ per bar

20 bars: $\frac{\$60}{20} = \$3$ per bar

\$72 for 36 bars is the better buy.

4. Answers may vary. One possible solution is to write a ratio.

$$\frac{8 \text{ sales people}}{160 \text{ customers}} = \frac{1 \text{ sales people}}{20 \text{ customers}}$$

The answer is that there are 0.05 sales people per customer, or 1 sales person per 20 customers.

4.6 Understanding the Concept Reducing a Proportion

1. A ratio is a fraction and reducing does not change the value of the fraction.

4.6 Exercises

1. When the numerator and denominator of a fraction are equal, the fraction is equal to 1. Thus both fractions are equal to 1.

3. $\frac{2}{7} = \frac{24}{84}$

5. $\frac{12 \text{ goals}}{7 \text{ games}} = \frac{24 \text{ goals}}{14 \text{ games}}$

7. $\frac{3}{8} = \frac{18}{48}$

9. Restate as follows: 14 crackers is to 6 grams of fat as 70 crackers is to 30 grams of fat.

$$\frac{14 \text{ crackers}}{6 \text{ grams of fat}} = \frac{70 \text{ crackers}}{30 \text{ grams of fat}}$$

11. $\frac{3\frac{1}{2} \text{ rotations}}{2 \text{ min}} = \frac{14 \text{ rotations}}{8 \text{ min}}$

13. Restate as follows: 4 made is to 7 attempts as 12 made is to 21 attempts.

$$\frac{4 \text{ made}}{7 \text{ attempts}} = \frac{12 \text{ made}}{21 \text{ attempts}}$$

15. $\frac{5}{8} \stackrel{?}{=} \frac{30}{45}$

$45 \cdot 5 = 225$

$8 \cdot 30 = 240$

Since $225 \neq 240$, we know that $\frac{5}{8} \neq \frac{30}{45}$.

17. $\frac{6}{11} \stackrel{?}{=} \frac{42}{77}$

$77 \cdot 6 = 462$

$11 \cdot 42 = 462$

Since $462 = 462$, we know that $\frac{6}{11} = \frac{42}{77}$.

19. $\frac{2}{7} \stackrel{?}{=} \frac{8}{28}$

$28 \cdot 2 = 56$

$7 \cdot 8 = 56$

The cross products are equal. Thus $\frac{2}{7} = \frac{8}{28}$. This is a proportion.

21. $\frac{14}{19} \stackrel{?}{=} \frac{26}{29}$

$29 \cdot 14 = 406$

$19 \cdot 26 = 494$

The cross products are not equal. Thus $\frac{14}{19} \neq \frac{26}{29}$. This is not a proportion.

23. $\frac{2 \text{ American dollars}}{11 \text{ French francs}} \stackrel{?}{=} \frac{65 \text{ American dollars}}{135 \text{ French francs}}$

$135 \cdot 2 = 270$

$11 \cdot 65 = 715$

The cross products are not equal. This is not a proportion.

25.

$$\frac{x}{8} = \frac{5}{2}$$
$$2 \cdot x = 8 \cdot 5$$
$$2x = 40$$
$$\frac{2x}{2} = \frac{40}{2}$$
$$x = 20$$

Copyright © 2012 Pearson Education, Inc.

Check: $\frac{20}{3} \stackrel{?}{=} \frac{5}{2}$
$2 \cdot 20 \stackrel{?}{=} 8 \cdot 5$
$40 = 40$ ✓

27. $\frac{12}{x} = \frac{2}{5}$
$5 \cdot 12 = x \cdot 2$
$60 = 2x$
$\frac{60}{2} = \frac{2x}{2}$
$30 = x$

Check: $\frac{12}{30} \stackrel{?}{=} \frac{2}{5}$
$5 \cdot 12 \stackrel{?}{=} 30 \cdot 2$
$60 = 60$ ✓

29. $\frac{12}{18} = \frac{x}{21}$
$21 \cdot 12 = 18 \cdot x$
$252 = 18x$
$\frac{252}{18} = \frac{18x}{18}$
$14 = x$

Check: $\frac{12}{18} \stackrel{?}{=} \frac{14}{21}$
$21 \cdot 12 \stackrel{?}{=} 18 \cdot 14$
$252 = 252$ ✓

31. $\frac{15}{6} = \frac{10}{x}$
$x \cdot 15 = 6 \cdot 10$
$15x = 60$
$\frac{15x}{15} = \frac{60}{15}$
$x = 4$

Check: $\frac{15}{6} \stackrel{?}{=} \frac{10}{4}$
$4 \cdot 15 \stackrel{?}{=} 6 \cdot 10$
$60 = 60$ ✓

33. $\frac{80 \text{ gal}}{24 \text{ acres}} = \frac{20 \text{ gal}}{n \text{ acres}}$
$\frac{80}{24} = \frac{20}{n}$
$n \cdot 80 = 24 \cdot 20$
$80n = 480$
$\frac{80n}{80} = \frac{480}{80}$
$n = 6$

35. $\frac{n \text{ gram}}{15 \text{ liters}} = \frac{12 \text{ gram}}{45 \text{ liters}}$
$\frac{n}{15} = \frac{12}{45}$
$45 \cdot n = 15 \cdot 12$
$45n = 180$
$\frac{45n}{45} = \frac{180}{45}$
$n = 4$

37. $\frac{400 \text{ words}}{5 \text{ min}} \stackrel{?}{=} \frac{675 \text{ words}}{9 \text{ min}}$
$9 \cdot 400 = 3600$
$5 \cdot 675 = 3375$
$3600 \neq 3375$
This is not a proportion.
No, Mark and John do not type at the same rate.

39. $\frac{18}{30} \stackrel{?}{=} \frac{15}{25}$
$25 \cdot 18 = 450$
$30 \cdot 15 = 450$
$450 = 450$
This is a proportion.
Yes, 18 out of 30 slices of cake is the same amount as 15 out of 25 slices of cake.

41. $\frac{3}{4} \stackrel{?}{=} \frac{4}{5}$
$5 \cdot 3 = 15$
$4 \cdot 4 = 16$
$15 \neq 16$
This is not a proportion.
No, the two parcels do not yield the same amount of land.

43. Let n represent the number of calories in 5 servings of cereal. 2 servings is to 126 calories as 5 servings is to n calories.
$\frac{2 \text{ servings}}{126 \text{ calories}} = \frac{5 \text{ servings}}{n \text{ calories}}$
$\frac{2}{126} = \frac{5}{n}$
$\frac{1}{63} = \frac{5}{n}$
$n \cdot 1 = 63 \cdot 5$
$n = 315$
There are 315 calories in 5 servings.

Copyright © 2012 Pearson Education, Inc.

45. Let n represent the number of calories in a 40-ounce jar. 1 ounce is to 170 calories as 40 ounces is to n calories.

$$\frac{1 \text{ ounce}}{170 \text{ calories}} = \frac{40 \text{ ounces}}{n \text{ calories}}$$
$$\frac{1}{170} = \frac{40}{n}$$
$$n \cdot 1 = 170 \cdot 40$$
$$n = 6800$$

There are 6800 calories in a 40-ounce jar of Deluxe Mixed Nuts.

47. Let n represent the number of millions of Snickers made in 7 days. 1 day is to 16 million as 7 days is to n million.

$$\frac{1 \text{ day}}{16 \text{ million}} = \frac{7 \text{ days}}{n \text{ million}}$$
$$\frac{1}{16} = \frac{7}{n}$$
$$n \cdot 1 = 16 \cdot 7$$
$$n = 112$$

At the given rate, 112 million Snickers bars are made in 7 days.

49. Let x represent the height of the second building.

$$\frac{210 \text{ ft}}{3 \text{ in.}} = \frac{x \text{ ft}}{5 \text{ in.}}$$
$$\frac{210}{3} = \frac{x}{5}$$
$$\frac{70}{1} = \frac{x}{5}$$
$$5 \cdot 70 = 1 \cdot x$$
$$350 = x$$

The second building is 350 feet tall.

51. Let x represent the weight of the 150-pound person on Pluto.

$$\frac{120 \text{ lb}}{8 \text{ lb}} = \frac{150 \text{ lb}}{x \text{ lb}}$$
$$\frac{120}{8} = \frac{150}{x}$$
$$\frac{15}{1} = \frac{150}{x}$$
$$x \cdot 15 = 1 \cdot 150$$
$$15x = 150$$
$$\frac{15x}{15} = \frac{150}{15}$$
$$x = 10$$

The 150-pound person weighs approximately 10 pounds on Pluto.

53. Let n represent the number of shares received by the person holding 850 shares.

$$\frac{8 \text{ shares received}}{5 \text{ shares held}} = \frac{n \text{ shares received}}{850 \text{ shares held}}$$
$$\frac{8}{5} = \frac{n}{850}$$
$$850 \cdot 8 = 5 \cdot n$$
$$6800 = 5n$$
$$\frac{6800}{5} = \frac{5n}{5}$$
$$1360 = n$$

The person holding 850 shares will receive 1360 shares.

55. Let x represent the thickness of the second wire.

$$\frac{100 \text{ watt}}{30 \text{ mm}} = \frac{140 \text{ watt}}{x \text{ mm}}$$
$$\frac{100}{30} = \frac{140}{x}$$
$$x \cdot 10 = 3 \cdot 140$$
$$10x = 420$$
$$\frac{10x}{10} = \frac{420}{10}$$
$$x = 42$$

The speaker wire would need to be 42 mm thick.

57. Let x represent the width of the enclosed area.

$$\frac{12 \text{ ft}}{18 \text{ ft}} = \frac{x \text{ ft}}{30 \text{ ft}}$$
$$\frac{12}{18} = \frac{x}{30}$$
$$\frac{2}{3} = \frac{x}{30}$$
$$30 \cdot 2 = 3 \cdot x$$
$$60 = 3x$$
$$\frac{60}{3} = \frac{3x}{3}$$
$$20 = x$$

The enclosed area should be 20 feet wide.

59. Let x represent the amount Kelvey invested.

$$\frac{\$6}{\$7} = \frac{\$2400}{\$x}$$
$$\frac{6}{7} = \frac{2400}{x}$$
$$x \cdot 6 = 7 \cdot 2400$$
$$6x = 16,800$$
$$\frac{6x}{6} = \frac{16,800}{6}$$
$$x = 2800$$

Kelvey invested \$2800.

Copyright © 2012 Pearson Education, Inc.

61. $\frac{1}{3}$ is to $\frac{1}{8}$ as $\frac{1}{4}$ is to $\frac{3}{32}$.

$$\frac{\frac{1}{3}}{\frac{1}{8}} = \frac{\frac{1}{4}}{\frac{3}{32}}$$

63. Let x represent the amount of federal withholding.

$$\frac{\$325}{\$1950} = \frac{\$40}{\$x}$$
$$\frac{325}{1950} = \frac{40}{x}$$
$$\frac{1}{6} = \frac{40}{x}$$
$$x \cdot 1 = 6 \cdot 40$$
$$x = 240$$

The federal withholding is $240.

65. Let x represent the amount of the retirement deduction.

$$\frac{\$325}{\$1950} = \frac{\$32}{\$x}$$
$$\frac{325}{1950} = \frac{32}{x}$$
$$\frac{1}{6} = \frac{32}{x}$$
$$x \cdot 1 = 6 \cdot 32$$
$$x = 192$$

The retirement deduction is $192.

67. Find the sum of the deductions.
$240 + $132 + $192 + $96 = $660
Subtract the total deductions from the monthly salary.
$1950 − $660 = $1290
Renee takes home $1290.

69. Let W represent the new width.

$$\frac{\text{old length}}{\text{old width}} = \frac{\text{new length}}{\text{new width}}$$
$$\frac{2 \text{ in.}}{3 \text{ in.}} = \frac{6 \text{ in.}}{W \text{ in.}}$$
$$\frac{2}{3} = \frac{6}{W}$$
$$W \cdot 2 = 3 \cdot 6$$
$$2W = 18$$
$$\frac{2W}{2} = \frac{18}{2}$$
$$W = 9$$

The width should be 9 inches. Let H represent the new width.

$$\frac{\text{old length}}{\text{old height}} = \frac{\text{new length}}{\text{new height}}$$
$$\frac{2 \text{ in.}}{5 \text{ in.}} = \frac{6 \text{ in.}}{H \text{ in.}}$$
$$\frac{2}{5} = \frac{6}{H}$$
$$H \cdot 2 = 5 \cdot 6$$
$$2H = 30$$
$$\frac{2H}{2} = \frac{30}{2}$$
$$H = 15$$

The height should be 15 inches.

71. a.
$$\frac{24}{4} \stackrel{?}{=} \frac{36}{6}$$
$$6 \cdot 24 = 144$$
$$4 \cdot 36 = 144$$
$$144 = 144$$

Yes, the ratios form a proportion: $\frac{24}{4} = \frac{36}{6}$.

b.
$$\frac{24}{3} \stackrel{?}{=} \frac{36}{5}$$
$$5 \cdot 24 = 120$$
$$3 \cdot 36 = 108$$
$$120 \neq 108$$

No, the ratios do not form a proportion: $\frac{24}{3} \neq \frac{36}{5}$.

73. a. 1 + 1 = 2; 1 + 2 =3 ; 2 + 3 = 5; 3 + 5 = 8; 5 + 8 = 13; 8 + 13 = 21; 13 + 21 = 34
The numbers are filled into the boxes below.
[2] [3] [5] [8] [13] [21] [34]

b. To find the next number in the sequence, we add the two preceding numbers.

c. 21 + 34 = 55; 34 + 55 = 89; 55 + 89 = 144
The next three numbers are 55, 89, and 144.

Cumulative Review

74. Two times a number added to six is equal to twenty-eight: $6 + 2x = 28$.

75. Twenty divided by a number is equal to five: $20 \div n = 5$.

76. Eight subtracted from some number is equal to nine: $x - 8 = 9$.

Copyright © 2012 Pearson Education, Inc.

77. Four plus three times a number is equal to nineteen: $4 + 3x = 19$.

78. $5(-1)+50(+5)+10(-3) = -5+250+(-30) = 215$
Victor earned 215 points on the final exam.

79. $0(-1)+50(+5)+15(-3) = 0+250+(-45) = 205$
Victor would have earned 205 points on the final exam.

Quick Quiz 4.6

1. $\frac{18}{30} \stackrel{?}{=} \frac{34}{56}$
$56 \cdot 18 = 1008$
$30 \cdot 34 = 1020$
Since $1008 \neq 1020$, $\frac{18}{30} \neq \frac{34}{56}$. No, this is not a proportion.

2. $\frac{12}{15} = \frac{n}{90}$
$\frac{4}{5} = \frac{n}{90}$
$90 \cdot 4 = 5 \cdot n$
$360 = 5n$
$\frac{360}{5} = \frac{5n}{5}$
$72 = n$

3. Let x represent the number of miles.
$\frac{2 \text{ in.}}{72 \text{ mi}} = \frac{9 \text{ in.}}{x \text{ mi}}$
$\frac{2}{72} = \frac{9}{x}$
$\frac{1}{36} = \frac{9}{x}$
$x \cdot 1 = 36 \cdot 9$
$x = 324$
9 inches represents 324 miles.

4. Answers may vary. One possible method is to write a proportion. Let x = amount to Justin.
$\frac{5}{7} = \frac{x}{840}$
$5(840) = 7x$
$4200 = 7x$
$600 = x$
Justin will receive \$600 when Sara receives \$840.

You Try It

1. 4731 is not even and therefore not divisible by 2. The sum of the digits, 4 + 7 + 3 + 1 = 15, is divisible by 3 and therefore 4731 is divisible by 3. The last digit is not 0 or 5, so 4731 is not divisible by 5.

2.
```
98
/ \
2 · 49
    / \
    7 · 7
```
$98 = 2 \cdot 7 \cdot 7$ or $2 \cdot 7^2$

3. $7\overline{)45}$ with quotient 6; 42; remainder 3

$\frac{45}{7} = 6\frac{3}{7}$

4. $5\frac{1}{5} = \frac{(5 \cdot 5)+1}{5} = \frac{25+1}{5} = \frac{26}{5}$

5. $\frac{7}{8} = \frac{7 \cdot 3x}{8 \cdot 3x} = \frac{21x}{24x}$

6. $\frac{27xy}{36y} = \frac{3 \cdot \cancel{9} \cdot x \cdot \cancel{y}}{4 \cdot \cancel{9} \cdot \cancel{y}} = \frac{3x}{4}$

7. a. $\frac{x^7}{x^2} = x^{7-2} = x^5$

b. $\frac{4^3}{4^6} = \frac{1}{4^{6-3}} = \frac{1}{4^3} = \frac{1}{64}$

8. a. $(3x^5)^6 = 3^6 x^{5 \cdot 6} = 3^6 x^{30}$

b. $\left(\frac{x}{3}\right)^2 = \frac{x^2}{3^2} = \frac{x^2}{9}$

9. "15 dollars to 26 dollars" can be written:
the ratio of 15 to 26
15 : 26
$\frac{15}{26}$

10. $\frac{16 \text{ staff}}{120 \text{ children}} = \frac{\cancel{8} \cdot 2 \text{ staff}}{\cancel{8} \cdot 15 \text{ children}} = \frac{2 \text{ staff}}{15 \text{ children}}$

Copyright © 2012 Pearson Education, Inc.

11. $\frac{427 \text{ boxes}}{7 \text{ hours}} = \frac{61 \cdot \cancel{7} \text{ boxes}}{\cancel{7} \text{ hours}} = 61$ boxes per hour

12. $\frac{35}{55} = \frac{14}{22}$

13. $\frac{14}{9} \stackrel{?}{=} \frac{42}{27}$
$27 \cdot 14 = 378$
$9 \cdot 42 = 378$
$378 = 378$; therefore $\frac{14}{9} = \frac{42}{27}$.

14. $\frac{6}{17} \stackrel{?}{=} \frac{13}{27}$
$27 \times 6 \stackrel{?}{=} 17 \times 13$
$162 \neq 221$
This is not a proportion.

15. $\frac{7}{n} = \frac{35}{115}$
$115 \cdot 7 = n \cdot 35$
$805 = n \cdot 35$
$\frac{805}{35} = n$
$23 = n$

16. $\frac{4 \text{ products}}{15 \text{ calls}} = \frac{n \text{ products}}{60 \text{ calls}}$
$60 \times 4 = 15 \times n$
$240 = 15 \times n$
$\frac{240}{15} = n$
$16 = n$
The solicitor should sell 16 products after 60 calls.

Chapter 4 Review Problems

1. Prime number: a whole number greater than 1 that is divisible only by itself and 1
2. Composite number: a whole number greater than 1 that is divisible by a whole number other than itself and 1
3. Proper fraction: a fraction that describes a quantity less than 1
4. Improper fraction: a fraction that describes a quantity greater than or equal to one
5. Mixed number: the sum of a whole number greater than zero and a proper fraction
6. Equivalent fractions: fractions that look different but have the same value
7. Ratio: a comparison of two quantities that have the same units
8. Rate: a ratio that compares different units.
9. Proportion: two rates or ratios that are equal
10. 588,640 is even and therefore divisible by 2. The sum of the digits, 5 + 8 + 8 + 6 + 4 + 0 = 31, is not divisible by 3 and therefore 588,640 is not divisible by 3. The last digit is 0, therefore 588,640 is divisible by 5.
11. 41,595 is not even and therefore not divisible by 2. The sum of the digits, 4 + 1 + 5 + 9 + 5 = 24, is divisible by 3 and therefore 41,595 is divisible by 3. The last digit is 5, therefore 41,595 is divisible by 5.
12. 0: neither; 7, 11: prime;
21, 50, 25, 51: composite
13. 1: neither; 7, 13, 41: prime;
32, 12, 50, 6: composite

14.
```
      36
     /  \
    4  ·  9
   /\    /\
  2 · 2  3 · 3
```
$36 = 2 \cdot 2 \cdot 3 \cdot 3$ or $2^2 \cdot 3^2$

15.
```
56
/\
7 · 8
   / \
   2 · 4
      / \
      2 · 2
```
$56 = 2 \cdot 2 \cdot 2 \cdot 7$ or $2^3 \cdot 7$

16.
```
    17
5) 85
5)425
```
$425 = 5 \cdot 5 \cdot 17 = 5^2 \cdot 17$

17.

```
      90
     /  \
    9 · 10
   / \   / \
  3 · 3 2 · 5
```

$90 = 2 \cdot 3 \cdot 3 \cdot 5 = 2 \cdot 3^2 \cdot 5$

18. $\frac{3}{0}$; division by 0 is undefined.

19. $\frac{0}{3} = 0$; any fraction with 0 in the numerator and a nonzero denominator equals 0.

20. $\frac{y}{y} = 1$; any nonzero number divided by itself is 1.

21. 20 of the 69 students are men; the fraction is $\frac{20}{69}$.

22. 8 of the 25 dishes are desserts, so 25 − 8 = 17 are not desserts; the fraction is $\frac{17}{25}$.

23.

```
   8 R 3
5)43
  40
   3
```

$\frac{43}{5} = 8\frac{3}{5}$

24.

```
   9 R 1
6)55
  54
   1
```

$\frac{55}{6} = 9\frac{1}{6}$

25.

```
   7
8)56
  56
   0
```

$\frac{56}{8} = 7$

26. $2\frac{1}{3} = \frac{(3 \cdot 2) + 1}{3} = \frac{6 + 1}{3} = \frac{7}{3}$

27. $6\frac{3}{5} = \frac{(5 \cdot 6) + 3}{5} = \frac{30 + 3}{5} = \frac{33}{5}$

28. $10\frac{2}{5} = \frac{(5 \cdot 10) + 2}{5} = \frac{50 + 2}{5} = \frac{52}{5}$

29. $\frac{2}{9} = \frac{2 \cdot 2}{9 \cdot 2} = \frac{4}{18}$

30. $\frac{3}{4} = \frac{3 \cdot 9}{4 \cdot 9} = \frac{27}{36}$

31. $\frac{4}{5} = \frac{4 \cdot 7x}{5 \cdot 7x} = \frac{28x}{35x}$

32. $\frac{6}{11} = \frac{6 \cdot 3y}{11 \cdot 3y} = \frac{18y}{33y}$

33. $\frac{55}{75} = \frac{\cancel{5} \cdot 11}{\cancel{5} \cdot 3 \cdot 5} = \frac{11}{3 \cdot 5} = \frac{11}{15}$

34. $\frac{48}{54} = \frac{\cancel{2} \cdot 2 \cdot 2 \cdot 2 \cdot \cancel{3}}{\cancel{2} \cdot \cancel{3} \cdot 3 \cdot 3} = \frac{2 \cdot 2 \cdot 2}{3 \cdot 3} = \frac{8}{9}$

35. $\frac{108}{36} = \frac{\cancel{36} \cdot 3}{\cancel{36} \cdot 1} = \frac{3}{1} = 3$

36. $\frac{175}{75} = \frac{\cancel{25} \cdot 7}{\cancel{25} \cdot 3} = \frac{7}{3}$ or $2\frac{1}{3}$

37. $\frac{25x}{60x} = \frac{\cancel{5} \cdot 5 \cdot \cancel{x}}{2 \cdot 2 \cdot 3 \cdot \cancel{5} \cdot \cancel{x}} = \frac{5}{2 \cdot 2 \cdot 3} = \frac{5}{12}$

38. $\frac{84x}{105xy} = \frac{2 \cdot 2 \cdot \cancel{3} \cdot \cancel{7} \cdot \cancel{x}}{\cancel{3} \cdot 5 \cdot \cancel{7} \cdot \cancel{x} \cdot y} = \frac{2 \cdot 2}{5 \cdot y} = \frac{4}{5y}$

39. $\frac{-16}{18} = -\frac{16}{18} = \frac{\cancel{2} \cdot 2 \cdot 2 \cdot 2}{\cancel{2} \cdot 3 \cdot 3} = -\frac{2 \cdot 2 \cdot 2}{3 \cdot 3} = -\frac{8}{9}$

40. $\frac{24}{-36} = -\frac{24}{36} = -\frac{\cancel{12} \cdot 2}{\cancel{12} \cdot 3} = -\frac{2}{3}$

41. $\frac{y^5}{y^3} = y^{5-3} = y^2$

42. $\frac{3^2}{3^3} = \frac{1}{3^{3-2}} = \frac{1}{3^1} = \frac{1}{3}$

43. $\frac{8^2}{3^4} = \frac{8^2}{3^4}$

Copyright © 2012 Pearson Education, Inc.

44. $\frac{x^5y^3}{x^2y^9}=\frac{x^{5-2}}{y^{9-3}}=\frac{x^3}{y^6}$

45. $\frac{2^3x^0}{2^6x^9}=\frac{2^3\cdot 1}{2^6x^9}=\frac{1}{2^{6-3}x^9}=\frac{1}{2^3x^9}$ or $\frac{1}{8x^9}$

46. $\frac{3^2y^0}{3^3y^6}=\frac{3^2\cdot 1}{3^3y^6}=\frac{1}{3^{3-2}y^6}=\frac{1}{3^1y^6}=\frac{1}{3y^6}$

47. $\frac{20x^5}{35x^9}=\frac{\not{5}\cdot 4\cdot x^5}{\not{5}\cdot x^9}=\frac{4}{7\cdot x^{9-5}}=\frac{4}{7x^4}$

48. $\frac{18y^6}{6y^4}=\frac{\not{6}\cdot 3y^6}{\not{6}\cdot y^4}=3y^{6-4}=3y^2$

49. $(3y^2)^3=(3^1\cdot y^2)^3=3^{1\cdot 3}y^{2\cdot 3}=3^3y^6$ or $27y^6$

50. $(2^4x)^2=(2^4\cdot x^1)^2=2^{4\cdot 2}x^{1\cdot 2}=2^8x^2$ or $256x^2$

51. $\left(\frac{3}{y}\right)^2=\left(\frac{3^1}{y^1}\right)=\frac{3^{(1)(2)}}{y^{(1)(2)}}=\frac{3^2}{y^2}$ or $\frac{9}{y^2}$

52. $\left(\frac{x}{2}\right)^3=\left(\frac{x^1}{2^1}\right)^3=\frac{x^{(1)(3)}}{2^{(1)(3)}}=\frac{x^3}{2^3}$ or $\frac{x^3}{8}$

53. $30 \text{ to } 46=\frac{30}{46}=\frac{\not{2}\cdot 3\cdot 5}{\not{2}\cdot 23}=\frac{3\cdot 5}{23}=\frac{15}{23}$

54. $15:35=\frac{15}{35}=\frac{3\cdot\not{5}}{\not{5}\cdot 7}=\frac{3}{7}$

55. a. $\frac{20 \text{ mL}}{55 \text{ mL}}=\frac{4\cdot\not{5}}{\not{5}\cdot 11}=\frac{4}{11}$

b. $\frac{55 \text{ mL}}{20 \text{ mL}}=\frac{\not{5}\cdot 11}{4\cdot\not{5}}=\frac{11}{4}$

56. We divide $35 \div 7$ to find the unit rate.
$\frac{\$35}{7 \text{ washcloths}}=\$5$ per washcloth

57. We divide $112 \div 28$ to find the unit rate.
$\frac{112 \text{ in.}}{28 \text{ hr}}=4$ in. per hr

58. We divide $286 \div 11$ to find the unit rate.
$\frac{286 \text{ mi}}{11 \text{ gal}}=26$ mi/gal

59. We divide $3090 \div 30$ to find the unit rate.
$\frac{3090 \text{ words}}{30 \text{ min}}=103$ words per min
Tan can type 103 words per minute.

60. a. Secretaries per lawyer:
$\frac{\text{secretaries}}{\text{lawyer}} \Rightarrow \frac{32 \text{ secretaries}}{16 \text{ lawyers}}$
$=\frac{2 \text{ secretaries}}{1 \text{ lawyer}}$
$=2$ secretaries per lawyer

b. Paralegals per lawyer:
$\frac{\text{paralegals}}{\text{lawyer}} \Rightarrow \frac{12 \text{ paralegals}}{4 \text{ lawyers}}$
$=\frac{3 \text{ paralegals}}{1 \text{ lawyer}}$
$=3$ paralegals per lawyer

c. Since every lawyer requires 3 paralegals, we multiply 60(3) to find how many paralegals are needed for 60 lawyers.
60(3) = 180 paralegals for 60 lawyers

61. a. 6CDs: $\frac{\$72}{6}=\$12$ per CD

8 CDs: $\frac{\$96}{8}=\$12$ per CD

b. Both have the same unit price.

62. a. $\frac{\text{April profit}}{\text{May profit}} \Rightarrow \frac{\$24,000}{\$31,000}=\frac{\not{1000}\cdot 24}{\not{1000}\cdot 31}=\frac{24}{31}$

b. $\frac{\$45,000 \text{ profit in June}}{30 \text{ days}}$

We divide $45,000 \div 30$ to find the unit rate.
$45,000 \div 30 = 1500$
The store had an average profit of $1500 per day in June.

63. $\frac{3}{7}=\frac{21}{49}$

64. $\frac{2 \text{ teachers}}{50 \text{ students}}=\frac{6 \text{ teachers}}{150 \text{ students}}$

Copyright © 2012 Pearson Education, Inc.

65. 2 inches is to 190 miles as 6 inches is to 570 miles.
$$\frac{2 \text{ in.}}{190 \text{ mi}} = \frac{6 \text{ in.}}{570 \text{ mi}}$$

66. 234 miles is to 9 gallons as 468 miles is to 18 gallons.
$$\frac{234 \text{ mi}}{9 \text{ gal}} = \frac{468 \text{ mi}}{18 \text{ gal}}$$

67. $\frac{3}{4} \stackrel{?}{=} \frac{50}{70}$

$70 \cdot 3 = 210$
$4 \cdot 50 = 200$

Since $210 \neq 200$, we know that $\frac{3}{4} \neq \frac{50}{70}$.

68. $\frac{13}{91} \stackrel{?}{=} \frac{12}{84}$

$84 \cdot 13 = 1092$
$91 \cdot 12 = 1092$

Since $1092 = 1092$, we know that $\frac{13}{91} = \frac{12}{84}$.

69. $\frac{3}{7} \stackrel{?}{=} \frac{12}{28}$

$28 \cdot 3 = 84$
$7 \cdot 12 = 84$

The cross products are equal. Thus $\frac{3}{7} = \frac{12}{28}$. This is a proportion.

70. $\frac{4 \text{ goals}}{7 \text{ attempts}} \stackrel{?}{=} \frac{20 \text{ goals}}{46 \text{ attempts}}$

$46 \cdot 4 = 184$
$7 \cdot 20 = 140$
The cross products are not equal. This is not a proportion.

71. $\frac{x}{30} = \frac{2}{15}$

$15 \cdot x = 30 \cdot 2$
$15x = 60$
$\frac{15x}{15} = \frac{60}{15}$
$x = 4$

Check: $\frac{4}{30} \stackrel{?}{=} \frac{2}{15}$
$15 \cdot 4 \stackrel{?}{=} 30 \cdot 2$
$60 = 60$ ✓

72. $\frac{6}{5} = \frac{54}{x}$

$x \cdot 6 = 5 \cdot 54$
$6x = 270$
$\frac{6x}{6} = \frac{270}{6}$
$x = 45$

Check: $\frac{6}{5} \stackrel{?}{=} \frac{54}{45}$
$45 \cdot 6 \stackrel{?}{=} 5 \cdot 54$
$270 = 270$ ✓

73. $\frac{17 \text{ qt}}{47 \text{ ft}^2} = \frac{n \text{ qt}}{94 \text{ ft}^2}$

$\frac{17}{47} = \frac{n}{94}$
$94 \cdot 17 = 47 \cdot n$
$1598 = 47n$
$\frac{1598}{47} = \frac{47n}{47}$
$34 = n$

74. Let n represent the number of miles.
1 inch is to 120 miles as 3 inches is to n miles.
$$\frac{1 \text{ in.}}{120 \text{ mi}} = \frac{3 \text{ in.}}{n \text{ mi}}$$
$\frac{1}{120} = \frac{3}{n}$
$n \cdot 1 = 120 \cdot 3$
$n = 360$
3 inches represents 360 miles.

75. Let n represent the speed in miles per hour.
75 revolutions per minute is to 12 miles per hour as 100 revolutions per minute is to n miles per hour.
$$\frac{75 \text{ rpm}}{12 \text{ mph}} = \frac{100 \text{ rpm}}{n \text{ mph}}$$
$\frac{75}{12} = \frac{100}{n}$
$n \cdot 75 = 12 \cdot 100$
$75n = 1200$
$\frac{75n}{75} = \frac{1200}{75}$
$n = 16$
At 100 revolutions per minute, Dale goes 16 miles per hour.

 Copyright © 2012 Pearson Education, Inc.

76. Let x represent the new width.
4 feet is to 7 feet as x feet is to 14 feet.

$$\frac{4 \text{ ft}}{7 \text{ ft}} = \frac{x \text{ ft}}{14 \text{ ft}}$$
$$\frac{4}{7} = \frac{x}{14}$$
$$14 \cdot 4 = 7 \cdot x$$
$$56 = 7x$$
$$\frac{56}{7} = \frac{7x}{7}$$
$$8 = x$$

The patio cover should be 8 feet wide.

77. a. Let x represent the amount of Gloria's federal withholding.

$$\frac{\$60}{\$400} = \frac{x}{\$340}$$
$$\frac{60}{400} = \frac{x}{340}$$
$$\frac{3}{20} = \frac{x}{340}$$
$$340 \cdot 3 = 20 \cdot x$$
$$1020 = 20x$$
$$\frac{1020}{20} = \frac{20x}{20}$$
$$51 = x$$

Gloria's federal withholding is $51.

b. Let x represent the amount of Gloria's state withholding.

$$\frac{\$20}{\$400} = \frac{x}{\$340}$$
$$\frac{20}{400} = \frac{x}{340}$$
$$\frac{1}{20} = \frac{x}{340}$$
$$340 \cdot 1 = 20 \cdot x$$
$$340 = 20x$$
$$\frac{340}{20} = \frac{20x}{20}$$
$$17 = x$$

Gloria's state withholding is $17.

How Am I Doing? Chapter 4 Test

1. 230 is even and therefore divisible by 2. 230 ends in 0 and is therefore divisible by 5. Since the sum of the digits, 2 + 3 + 0 = 5, is not divisible by 3, 230 is not divisible by 3.

2. a. $27 = 3^3$: composite

b. 1: neither

c. 19: prime

3.
```
  84
  /\
 7 · 12
     /\
    3 · 4
       / \
      2 · 2
```
$84 = 2 \cdot 2 \cdot 3 \cdot 7$ or $2^2 \cdot 3 \cdot 7$

4.
```
     120
    /   \
  10 ·   12
  / \    / \
 2 · 5  3 · 4
           /\
          2 · 2
```
$120 = 2 \cdot 2 \cdot 2 \cdot 3 \cdot 5$ or $2^3 \cdot 3 \cdot 5$

5. a. $\frac{0}{4} = 0$; zero divided by any nonzero number is equal to 0.

b. $\frac{t}{t} = 1$; any nonzero number divided by itself is equal to 1.

c. $\frac{12}{0}$; division by 0 is undefined.

6. 17 out of 36 students are men. The fraction is $\frac{17}{36}$.

7. a. First we must find the total number of classes.
12 + 7 + 16 = 35 classes
16 out of 35 classes are for adults. The fraction is $\frac{16}{35}$.

b. Subtract the number of classes for children from the total number of classes to find the number of classes not for children.
35 − 12 = 23 classes
23 out of 35 classes are not for children. The fraction is $\frac{23}{35}$.

Copyright © 2012 Pearson Education, Inc.

8. $5\overline{)8}$ = 1 R 3; $\frac{5}{3}$ (8 − 5 = 3)

$\frac{8}{5} = 1\frac{3}{5}$

9. $7\frac{1}{6} = \frac{(6\cdot 7)+1}{6} = \frac{42+1}{6} = \frac{43}{6}$

10. $\frac{7}{8} = \frac{7\cdot 5}{8\cdot 5} = \frac{35}{40}$

11. $\frac{4}{9} = \frac{4\cdot 3y}{9\cdot 3y} = \frac{12y}{27y}$

12. $\frac{-18}{56} = -\frac{18}{56} = -\frac{\not{2}\cdot 3\cdot 3}{\not{2}\cdot 2\cdot 2\cdot 7} = -\frac{3\cdot 3}{2\cdot 2\cdot 7} = -\frac{9}{28}$

13. $\frac{16x}{32x^2y} = \frac{\not{2}\cdot\not{2}\cdot\not{2}\cdot\not{2}\cdot\not{x}}{\not{2}\cdot\not{2}\cdot\not{2}\cdot\not{2}\cdot 2\cdot\not{x}\cdot x\cdot y} = \frac{1}{2xy}$

14. $\frac{y^3z^4}{y^7z} = \frac{z^{4-1}}{y^{7-3}} = \frac{z^3}{y^4}$

15. $\frac{8^2}{7^3} = \frac{8^2}{7^3}$; the bases are different, so this cannot be simplified.

16. $\frac{42x^7y^6}{36x^0y^9} = \frac{\not{6}\cdot 7\cdot x^7\cdot y^6}{\not{6}\cdot 2\cdot 3\cdot 1\cdot y^9} = \frac{7\cdot x^7}{2\cdot 3\cdot y^{9-6}} = \frac{7x^7}{6y^3}$

17.
$$\begin{aligned}(2y^4)^3 &= (2^1\cdot y^4)^3\\ &= 2^{(1)(3)}\cdot y^{(4)(3)}\\ &= 2^3y^{12} \text{ or } 8y^{12}\end{aligned}$$

18. $(x^3)^5 = x^{3\cdot 5} = x^{15}$

19. $\left(\frac{x}{3}\right)^2 = \left(\frac{x^1}{3^1}\right)^2 = \frac{x^{(1)(2)}}{3^{(1)(2)}} = \frac{x^2}{3^2} = \frac{x^2}{9}$

20. a. $\frac{18}{4} = \frac{\not{2}\cdot 9}{\not{2}\cdot 2} = \frac{9}{2}$

The ratio of wins to losses is $\frac{9}{2}$.

b. $\frac{4}{18} = \frac{\not{2}\cdot 2}{\not{2}\cdot 9} = \frac{2}{9}$

The ratio of losses to wins is $\frac{2}{9}$.

21. We divide 150 ÷ 20 to find the unit rate.

$20\overline{)150}$ = 7 R 10; 150 − 140 = 10

$7\frac{10}{20} = 7\frac{1}{2}$

$\frac{150 \text{ calories}}{20 \text{ min}} = 7\frac{1}{2}$ cal per min

22. We divide 525 ÷ 25 to find the unit rate.

$25\overline{)525}$ = 21; 52 − 50 = 2, 25 − 25 = 0

The car can be driven 21 miles on one gallon of gas.

23. a. 12 reams: $\frac{\$36}{12} = \$3$ per ream

48 reams: $\frac{\$96}{48} = \$2$ per ream

b. 48 reams for $96 is the better deal.

24. $\frac{2 \text{ inches}}{225 \text{ miles}} = \frac{6 \text{ inches}}{675 \text{ miles}}$

25. $\frac{20}{52} \stackrel{?}{=} \frac{5}{13}$

$13\cdot 20 = 260$

$52\cdot 5 = 260$

The cross products are equal. Thus $\frac{20}{52} = \frac{5}{13}$.

Yes, this is a proportion.

26.
$$\begin{aligned}\frac{4}{6} &= \frac{20}{x}\\ \frac{2}{3} &= \frac{20}{x}\\ x\cdot 2 &= 3\cdot 20\\ 2x &= 60\\ \frac{2x}{2} &= \frac{60}{2}\\ x &= 30\end{aligned}$$

Copyright © 2012 Pearson Education, Inc.

27. $\frac{4}{16} \stackrel{?}{=} \frac{6}{24}$

$24 \cdot 4 = 96$
$16 \cdot 6 = 96$

The cross products are equal, so $\frac{4}{16} = \frac{6}{24}$. Yes, 4 out of 16 slices is the same as 6 out of 24 slices.

28. Let x represent the amount of fertilizer needed for 1600 square feet. 2 tablespoons is to 400 square feet as x tablespoons is to 1600 square feet.

$$\frac{2 \text{ Tb}}{400 \text{ ft}^2} = \frac{x \text{ Tb}}{1600 \text{ ft}^2}$$
$$\frac{2}{400} = \frac{x}{1600}$$
$$\frac{1}{200} = \frac{x}{1600}$$
$$1600 \cdot 1 = 200 \cdot x$$
$$1600 = 200x$$
$$\frac{1600}{200} = \frac{200x}{200}$$
$$8 = x$$

To fertilize a lawn measuring 1600 square feet, 8 tablespoons of fertilizer are needed.

Copyright © 2012 Pearson Education, Inc.

Chapter 5

5.1 Exercises

1. To multiply fractions, multiply numerator times numerator and denominator times denominator.

3. To split $\frac{1}{4}$ into 6 equal parts, we divide because we want to split an amount into equal parts.

5. Answers may vary but any word problem that requires taking $\frac{1}{3}$ of 90 or uses repeated addition of $\frac{1}{3}$ is correct.

7. $\frac{1}{2}\cdot\frac{3}{7}=\frac{3}{\boxed{14}}$

9. $\frac{1}{4}\div\frac{3}{7}=\frac{1}{4}\cdot\frac{\boxed{7}}{\boxed{3}}=\frac{7}{12}$

11. $\frac{1}{2}\cdot\frac{\boxed{5}}{\boxed{6}}=\frac{5}{12}$

13. $\frac{5}{7}\div\frac{4}{3}=\frac{5}{7}\boxed{\cdot}\frac{\boxed{3}}{\boxed{4}}=\frac{15}{28}$

15. $\frac{1}{4}$ of $\frac{1}{3}=\frac{1}{4}\cdot\frac{1}{3}=\frac{1\cdot1}{4\cdot3}=\frac{1}{12}$

17. $\frac{5}{21}$ of $\frac{7}{8}=\frac{5}{21}\cdot\frac{7}{8}=\frac{5\cdot7}{21\cdot8}=\frac{5\cdot\cancel{7}}{3\cdot\cancel{7}\cdot8}=\frac{5}{24}$

19. $\frac{7}{12}\cdot\frac{8}{28}=\frac{7\cdot8}{12\cdot28}=\frac{\cancel{7}\cdot\cancel{2}\cdot\cancel{2}\cdot\cancel{2}}{\cancel{2}\cdot\cancel{2}\cdot3\cdot\cancel{2}\cdot2\cdot\cancel{7}}=\frac{1}{6}$

21. $\frac{3}{20}\cdot\frac{8}{9}=\frac{3\cdot8}{20\cdot9}=\frac{\cancel{3}\cdot\cancel{2}\cdot\cancel{2}\cdot2}{\cancel{2}\cdot\cancel{2}\cdot5\cdot\cancel{3}\cdot3}=\frac{2}{15}$

23. The product is positive since there are two negative signs and 2 is even.
$\frac{-3}{8}\cdot\left(\frac{14}{-6}\right)=\frac{3\cdot14}{8\cdot6}=\frac{\cancel{3}\cdot\cancel{2}\cdot7}{\cancel{2}\cdot2\cdot2\cdot2\cdot\cancel{3}}=\frac{7}{8}$

25. The product is negative since there is one negative sign and 1 is odd.
$\frac{16}{11}\cdot\left(\frac{-18}{36}\right)=-\frac{16\cdot18}{11\cdot36}=\frac{2\cdot2\cdot2\cdot\cancel{2}\cdot\cancel{18}}{11\cdot\cancel{2}\cdot\cancel{18}}=-\frac{8}{11}$

27. The product is positive since there are two negative signs and 2 is even.
$\frac{-2}{21}\cdot\left(\frac{-14}{18}\right)=\frac{2\cdot14}{21\cdot18}=\frac{2\cdot\cancel{2}\cdot\cancel{7}}{3\cdot\cancel{7}\cdot\cancel{2}\cdot3\cdot3}=\frac{2}{27}$

29. The product is negative since there is one negative sign and 1 is odd.
$-14\cdot\frac{1}{28}=\frac{-14}{1}\cdot\frac{1}{28}=-\frac{14\cdot1}{1\cdot28}=-\frac{\cancel{14}\cdot1}{1\cdot2\cdot\cancel{14}}=-\frac{1}{2}$

31. $\frac{6}{35}\cdot5=\frac{6}{35}\cdot\frac{5}{1}=\frac{6\cdot5}{35\cdot1}=\frac{2\cdot3\cdot\cancel{5}}{\cancel{5}\cdot7\cdot1}=\frac{6}{7}$

33. $\frac{2x}{3}\cdot\frac{3x}{5}=\frac{2\cdot x\cdot\cancel{3}\cdot x}{\cancel{3}\cdot5}=\frac{2\cdot x^{1+1}}{5}=\frac{2x^2}{5}$

35. $$\begin{aligned}\frac{6x^4}{7}\cdot28x&=\frac{6x^4}{7}\cdot\frac{28x}{1}\\&=\frac{2\cdot3\cdot x^4\cdot2\cdot2\cdot\cancel{7}\cdot x}{\cancel{7}\cdot1}\\&=\frac{2\cdot3\cdot2\cdot2\cdot x^{4+1}}{1}\\&=24x^5\end{aligned}$$

37. $$\begin{aligned}8x^2\cdot\frac{3x^3}{2}&=\frac{8x^2}{1}\cdot\frac{3x^3}{2}\\&=\frac{2\cdot2\cdot\cancel{2}\cdot x^2\cdot3\cdot x^3}{1\cdot\cancel{2}}\\&=\frac{2\cdot2\cdot3\cdot x^{2+3}}{1}\\&=12x^5\end{aligned}$$

39. $\frac{2}{10}\cdot\frac{6}{8}=\frac{2\cdot6}{10\cdot8}=\frac{\cancel{2}\cdot\cancel{2}\cdot3}{\cancel{2}\cdot5\cdot\cancel{2}\cdot2\cdot2}=\frac{3}{20}$

41. $\frac{6x}{25}\cdot\frac{15}{12x^2}=\frac{\cancel{6}\cdot x\cdot3\cdot\cancel{5}}{5\cdot\cancel{5}\cdot\cancel{6}\cdot2\cdot x^2}=\frac{3}{5\cdot2\cdot x^{2-1}}=\frac{3}{10x}$

Copyright © 2012 Pearson Education, Inc.

43. $\frac{-3y^3}{20}\cdot\frac{12}{21y^2}=-\frac{3\cdot y^3\cdot \cancel{3}\cdot \cancel{4}}{\cancel{4}\cdot 5\cdot \cancel{3}\cdot 7\cdot y^2}$
$=-\frac{3\cdot y^{3-2}}{5\cdot 7}$
$=-\frac{3y}{35}$

45. $\frac{3x^2}{15}\cdot\frac{18x^3}{20}=\frac{\cancel{3}\cdot x^2\cdot \cancel{2}\cdot 3\cdot 3\cdot x^3}{\cancel{3}\cdot 5\cdot \cancel{2}\cdot 2\cdot 5}$
$=\frac{3\cdot 3\cdot x^{2+3}}{5\cdot 2\cdot 5}$
$=\frac{9x^5}{50}$

47. $A=\frac{1}{2}bh$
$=\frac{1}{2}\cdot 12\text{ m}\cdot 8\text{ m}$
$=\frac{1\cdot 12\text{ m}\cdot 8\text{ m}}{2\cdot 1\cdot 1}$
$=\frac{1\cdot 12\cdot 8\cdot \text{m}\cdot \text{m}}{2}$
$=\frac{96\text{ m}^2}{2}$
$=48\text{ m}^2$

49. $A=\frac{1}{2}bh$
$=\frac{1}{2}\cdot 21\text{ in.}\cdot 40\text{ in.}$
$=\frac{1\cdot 21\text{ in.}\cdot 40\text{ in.}}{2\cdot 1\cdot 1}$
$=\frac{1\cdot 21\cdot 40\cdot \text{ in.}\cdot \text{ in.}}{2}$
$=\frac{840\text{ in.}^2}{2}$
$=420\text{ in.}^2$

51. To find the reciprocal, we invert the fraction.
$\frac{1}{3}\to\frac{3}{1}=3$

53. To find the reciprocal, we write 5 as a fraction and then invert the fraction.
$5=\frac{5}{1}\to\frac{1}{5}$

55. To find the reciprocal, we invert the fraction.
$\frac{2}{-5}\to\frac{-5}{2}=-\frac{5}{2}$

57. To find the reciprocal, we invert the fraction.
$\frac{-x}{y}\to\frac{y}{-x}=-\frac{y}{x}$

59. $\frac{6}{14}\div\frac{3}{8}=\frac{6}{14}\cdot\frac{8}{3}=\frac{\cancel{2}\cdot \cancel{3}\cdot 2\cdot 2\cdot 2}{\cancel{2}\cdot 7\cdot \cancel{3}}=\frac{8}{7}$

61. $\frac{7}{24}\div\frac{9}{16}=\frac{7}{24}\cdot\frac{16}{9}=\frac{7\cdot \cancel{8}\cdot 2}{3\cdot \cancel{8}\cdot 3\cdot 3}=\frac{14}{27}$

63. There is one negative sign. The number 1 is odd, so the answer is negative.
$\frac{-1}{12}\div\frac{3}{4}=\frac{-1}{12}\cdot\frac{4}{3}=-\frac{1\cdot \cancel{4}}{\cancel{4}\cdot 3\cdot 3}=-\frac{1}{9}$

65. There are two negative signs. The number 2 is even, so the answer is positive.
$\frac{-7}{24}\div\left(\frac{7}{-8}\right)=\frac{-7}{24}\cdot\left(\frac{-8}{7}\right)=\frac{\cancel{7}\cdot \cancel{8}}{3\cdot \cancel{8}\cdot \cancel{7}}=\frac{1}{3}$

67. $\frac{8x^6}{15}\div\frac{16x^2}{5}=\frac{8x^6}{15}\cdot\frac{5}{16x^2}$
$=\frac{\cancel{8}\cdot x^6\cdot \cancel{5}}{3\cdot \cancel{5}\cdot 2\cdot \cancel{8}\cdot x^2}$
$=\frac{x^{6-2}}{3\cdot 2}$
$=\frac{x^4}{6}$

69. $\frac{7x^4}{12}\div\frac{-28}{36x^2}=\frac{7x^4}{12}\cdot\frac{36x^2}{-28}$
$=-\frac{\cancel{7}\cdot x^4\cdot \cancel{12}\cdot 3\cdot x^2}{\cancel{12}\cdot 2\cdot 2\cdot \cancel{7}}$
$=-\frac{3\cdot x^{4+2}}{2\cdot 2}$
$=-\frac{3x^6}{4}$

71. $14\div\frac{2}{7}=\frac{14}{1}\cdot\frac{7}{2}=\frac{\cancel{2}\cdot 7\cdot 7}{1\cdot \cancel{2}}=\frac{49}{1}=49$

73. $\frac{7}{22}\div 14=\frac{7}{22}\cdot\frac{1}{14}=\frac{\cancel{7}\cdot 1}{2\cdot 11\cdot 2\cdot \cancel{7}}=\frac{1}{44}$

Copyright © 2012 Pearson Education, Inc.

75. $21x^4 \div \frac{7x}{3} = \frac{21x^4}{1} \cdot \frac{3}{7x}$
$= \frac{3 \cdot \cancel{7} \cdot x^4 \cdot 3}{1 \cdot \cancel{7} \cdot x}$
$= \frac{3 \cdot 3 \cdot x^{4-1}}{1}$
$= 9x^3$

77. $22x^3 \div \frac{11}{6x^5} = \frac{22x^3}{1} \cdot \frac{6x^5}{11}$
$= \frac{2 \cdot \cancel{11} \cdot x^3 \cdot 2 \cdot 3 \cdot x^5}{1 \cdot \cancel{11}}$
$= \frac{2 \cdot 2 \cdot 3 \cdot x^{3+5}}{1}$
$= 12x^8$

79. a. $\frac{1}{15} \cdot \frac{25}{21} = \frac{\cancel{5} \cdot 5}{\cancel{5} \cdot 3 \cdot 3 \cdot 7} = \frac{5}{63}$

b. $\frac{1}{15} \div \frac{25}{21} = \frac{1}{15} \cdot \frac{21}{25} = \frac{1 \cdot \cancel{3} \cdot 7}{\cancel{3} \cdot 5 \cdot 5} = \frac{7}{125}$

81. a. $\frac{2x^2}{3} \div \frac{12}{21x^5} = \frac{2x^2}{3} \cdot \frac{21x^5}{12}$
$= \frac{\cancel{2}x^2 \cdot \cancel{3} \cdot 7x^5}{\cancel{3} \cdot \cancel{2} \cdot 2 \cdot 3}$
$= \frac{7x^{2+5}}{2 \cdot 3}$
$= \frac{7x^7}{6}$

b. $\frac{2x^2}{3} \cdot \frac{12}{21x^5} = \frac{2x^3 \cdot \cancel{3} \cdot 2 \cdot 2}{\cancel{3} \cdot 3 \cdot 7}$
$= \frac{2 \cdot 2 \cdot 2}{3 \cdot 7 \cdot x^{5-2}}$
$= \frac{8}{21x^3}$

83. $\frac{5x^7}{-27} \cdot \frac{-9}{20x^4} = \frac{\cancel{5}x^7 \cdot \cancel{9}}{3 \cdot \cancel{9} \cdot 5 \cdot 2 \cdot 2x^4} = \frac{x^{7-4}}{3 \cdot 2 \cdot 2} = \frac{x^3}{12}$

85. $\frac{12x^6}{35} \div \frac{-16}{25x^2} = \frac{12x^6}{35} \cdot \frac{25x^2}{-16}$
$= -\frac{\cancel{4} \cdot 3 \cdot x^6 \cdot \cancel{5} \cdot 5x^2}{\cancel{5} \cdot 7 \cdot 2 \cdot 2 \cdot \cancel{4}}$
$= -\frac{3 \cdot 5 \cdot x^{6+2}}{7 \cdot 2 \cdot 2}$
$= -\frac{15x^8}{28}$

87. The word *of* indicates multiplication.
$\frac{2}{15} \cdot \$1350 = \frac{2}{15} \cdot \frac{\$1350}{1} = \$180$
$180 is withheld each week.

89. Since we need to find out how many $\frac{3}{4}$-foot lengths are in 12 feet, we divide.
$12 \div \frac{3}{4} = 12 \cdot \frac{4}{3} = \frac{12 \cdot 4}{3} = \frac{\cancel{3} \cdot 4 \cdot 4}{\cancel{3}} = 16$
Babette can make 16 pipes.

91. We multiply to find the total number of miles.
$32 \cdot \frac{1}{4} = \frac{32}{1} \cdot \frac{1}{4} = \frac{\cancel{4} \cdot 8 \cdot 1}{1 \cdot \cancel{4}} = 8$
Julie runs 8 miles.

93. Since we need to find out how many $\frac{3}{4}$-quart bottles are in 120 quarts, we divide.
$120 \div \frac{3}{4} = 120 \cdot \frac{4}{3} = \frac{120 \cdot 4}{3} = \frac{\cancel{3} \cdot 40 \cdot 4}{\cancel{3}} = 160$
160 bottles can be made from each vat.

95. $\frac{5}{14} \div \frac{2}{21} \div \left(\frac{15}{-3}\right) = \frac{5}{14} \div \frac{2}{21} \div (-5)$
$= \frac{5}{14} \cdot \frac{21}{2} \div (-5)$
$= \frac{5 \cdot 3 \cdot \cancel{7}}{2 \cdot \cancel{7} \cdot 2} \div (-5)$
$= \frac{15}{4} \div (-5)$
$= \frac{15}{4} \cdot \left(\frac{1}{-5}\right)$
$= -\frac{3 \cdot \cancel{5} \cdot 1}{4 \cdot \cancel{5}}$
$= -\frac{3}{4}$

Copyright © 2012 Pearson Education, Inc.

97. We divide to find the number of years.

$$4 \div \frac{2}{3} = \frac{4}{1} \cdot \frac{3}{2} = \frac{\cancel{2} \cdot 2 \cdot 3}{1 \cdot \cancel{2}} = \frac{6}{1} = 6$$

The house will sink 4 inches in 6 years.

99. We multiply to find the number of pizzas.

$$17 \cdot \frac{3}{8} = \frac{17}{1} \cdot \frac{3}{8} = \frac{51}{8} = 6\frac{3}{8}$$

James should order 7 pizzas.

101. $\frac{3}{4} \cdot \frac{x}{27} = \frac{4}{9}$

Simplify the left side of the equation.

$$\frac{\cancel{3} \cdot x}{4 \cdot \cancel{3} \cdot 9} = \frac{4}{9}$$

$$\frac{x}{36} = \frac{4}{9}$$

The equation is now a proportion. Find the cross products.

$$9 \cdot x = 36 \cdot 4$$

$$9x = 144$$

$$\frac{9x}{9} = \frac{144}{9}$$

$$x = 16$$

103. $\frac{2}{7}, \frac{4}{7}, \frac{x}{7}, \frac{8}{7}, \frac{10}{7}, \frac{y}{7}, \ldots$

The numerators of the fractions are consecutive even numbers. The even number following 4 is 6, so $x = 6$. The even number following 10 is 12, so $y = 12$.

105. $\frac{1}{2}, \frac{2}{6}, \frac{4}{18}, \frac{x}{54}, \frac{16}{y}, \frac{32}{486}, \ldots$

Each numerator is twice the preceding numerator. Twice 4 is 8, so $x = 8$. Each denominator is three times the preceding denominator. Since $3 \cdot 54$ is 162, $y = 162$.

Cumulative Review

107. $\frac{2}{3} = \frac{2 \cdot 5}{3 \cdot 5} = \frac{10}{15}$

108. $\frac{3}{4} = \frac{3 \cdot 5}{4 \cdot 5} = \frac{15}{20}$

109.

120
/ \
10 · 12
/ \ / \
2 · 5 2 · 6
/ \
2 · 3

$$120 = 2 \cdot 2 \cdot 2 \cdot 3 \cdot 5 = 2^3 \cdot 3 \cdot 5$$

110.

145
/ \
5 · 29

$$145 = 5 \cdot 29$$

Quick Quiz 5.1

1. $\frac{3}{14}$ of $\frac{14}{27} = \frac{3}{\cancel{14}} \cdot \frac{\cancel{14}}{27} = \frac{3 \cdot 14}{14 \cdot 27} = \frac{\cancel{3} \cdot \cancel{14}}{\cancel{14} \cdot \cancel{3} \cdot 3 \cdot 3} = \frac{1}{9}$

2. The product is negative since there is one negative sign and 1 is odd.

$$\frac{-2x^2}{5} \cdot \frac{15x^4}{8} = \frac{\cancel{2} \cdot x^2 \cdot 3 \cdot \cancel{5} \cdot x^4}{\cancel{5} \cdot \cancel{2} \cdot 2 \cdot 2}$$

$$= -\frac{3 \cdot x^{2+4}}{2 \cdot 2}$$

$$= -\frac{3x^6}{4}$$

3. $\frac{5x^5}{8} \div \frac{x^3}{20} = \frac{5x^5}{8} \cdot \frac{20}{x^3}$

$$= \frac{5 \cdot x^5 \cdot \cancel{4} \cdot 5}{\cancel{4} \cdot 2 \cdot x^3}$$

$$= \frac{5 \cdot 5 \cdot x^{5-3}}{2}$$

$$= \frac{25x^2}{2}$$

4. Answers may vary. One possible solution is to first write both terms with a denominator.

$$\frac{-16x^2}{3} \div \frac{8x}{1}$$

Next rewrite the expression as a multiplication problem by multiplying the first term by the reciprocal of the second term.

$$\frac{-16x^2}{3} \cdot \frac{1}{8x}$$

Multiply, and simplify.

$$-\frac{2x}{3}$$

Copyright © 2012 Pearson Education, Inc.

5.2 Exercises

1. We know that 12 is a multiple of 3 because the sum of its digits is divisible by 3, and therefore a multiple of 3. We know that 12 is not a multiple of 5 because the ones place value is neither 0 nor 5.

3. a. The first four multiples of 6 are 6, 12, 18, and 24; the first four multiples of 8 are 8, 16, 24, and 32.

b. The multiple 24 is common to both 6 and 8.

5. a. The first five multiples of 2 are 2, 4, 6, 8, and 10; the first five multiples of 5 are 5, 10, 15, 20, and 25.

b. The multiple 10 is common to both 2 and 5.

7. a. The first four multiples of $12x$ are $12x$, $24x$, $36x$, and $48x$; the first four multiples of $18x$ are $18x$, $36x$, $54x$, and $72x$.

b. The multiple $36x$ is common to both $12x$ and $18x$.

9. The LCM needs two factors of 2 since 28 has two factors of 2. Thus, we need one more factor of 2.
The LCM needs two factors of 3 since 90 has two factors of 3. Thus, we need one more factor of 3.

11. The LCM needs two factors of 7 since $49x$ has two factors of 7. Thus, we need one more factor of 7.

The LCM needs two factors of x since $10x^2$ has two factors of x. Thus, we need one more factor of x.

13. Some multiples of 5: 5, 10, 15, 20
Some multiples of 15: 15, 30, 45, 60
The smallest common multiple is 15.
LCM = 15

15. $8 = 2 \cdot 2 \cdot 2$
$28 = 2 \cdot 2 \cdot 7$
$\text{LCM} = 2 \cdot 2 \cdot 2 \cdot 7 = 56$

17. Some multiples of 15: 15, 30, 45, 60
Some multiples of 20: 20, 40, 60, 80
The smallest common multiple is 60.
LCM = 60

19. Some multiples of 40: 40, 80, 120, 160
Some multiples of 60: 60, 120, 180, 240
The smallest common multiple is 120.
LCM = 120

21. $5 = 5$
$8 = 2 \cdot 2 \cdot 2$
$12 = 2 \cdot 2 \cdot 3$
$\text{LCM} = 2 \cdot 2 \cdot 2 \cdot 3 \cdot 5 = 120$

23. $7 = 7$
$14 = 2 \cdot 7$
$20 = 2 \cdot 2 \cdot 5$
$\text{LCM} = 2 \cdot 2 \cdot 5 \cdot 7 = 140$

25. $4x = 2 \cdot 2 \cdot x$
$18x = 2 \cdot 3 \cdot 3 \cdot x$
$\text{LCM} = 2 \cdot 2 \cdot 3 \cdot 3 \cdot x = 36x$

27. $21a = 3 \cdot 7 \cdot a$
$81a = 3 \cdot 3 \cdot 3 \cdot 3 \cdot a$
$\text{LCM} = 3 \cdot 3 \cdot 3 \cdot 3 \cdot 7 \cdot a = 567a$

29. $18x = 2 \cdot 3 \cdot 3 \cdot x$
$45x^2 = 3 \cdot 3 \cdot 5 \cdot x \cdot x$
$\text{LCM} = 2 \cdot 3 \cdot 3 \cdot 5 \cdot x \cdot x = 90x^2$

31. $22x^2 = 2 \cdot 11 \cdot x \cdot x$
$4x^3 = 2 \cdot 2 \cdot x \cdot x \cdot x$
$\text{LCM} = 2 \cdot 2 \cdot 11 \cdot x \cdot x \cdot x = 44x^3$

33. $12x^2 = 2 \cdot 2 \cdot 3 \cdot x \cdot x$
$5x = 5 \cdot x$
$3x^3 = 3 \cdot x \cdot x \cdot x$
$\text{LCM} = 2 \cdot 2 \cdot 3 \cdot 5 \cdot x \cdot x \cdot x = 60x^3$

35. $12x = 2 \cdot 2 \cdot 3 \cdot x$
$14 = 2 \cdot 7$
$4x^2 = 2 \cdot 2 \cdot x \cdot x$
$\text{LCM} = 2 \cdot 2 \cdot 3 \cdot 7 \cdot x \cdot x = 84x^2$

37. We must find the LCM of 4 and 6.
$4 = 2 \cdot 2$
$6 = 2 \cdot 3$
$\text{LCM} = 2 \cdot 2 \cdot 3 = 12$
Jessica and Luis will meet to begin their next lap together in 12 minutes.

Copyright © 2012 Pearson Education, Inc.

39. We must find the LCM of 6 and 8.
$6 = 2 \cdot 3$
$8 = 2 \cdot 2 \cdot 2$
$\text{LCM} = 2 \cdot 2 \cdot 2 \cdot 3 = 24$
Each machine will begin labeling a carton of juice at the same time again in 24 minutes.

41. When we include the break times, the time between the beginning of events on the first field is 20 + 15 or 35 minutes and the time between the beginning of events on the second field is 30 + 15 or 45 minutes.
We find the LCM of 35 and 45.
$35 = 5 \cdot 7$
$45 = 3 \cdot 3 \cdot 5$
$\text{LCM} = 3 \cdot 3 \cdot 5 \cdot 7 = 315$
Events will start at the same time again after 315 minutes, which is 5 hours 15 minutes. Since the events start at 8 A.M. this will occur at 1:15 P.M.

43. $2x^3 = 2 \cdot x \cdot x \cdot x$
$8x^2 = 2 \cdot 2 \cdot 2 \cdot x \cdot y \cdot y$
$10x^2y = 2 \cdot 5 \cdot x \cdot x \cdot y$
$\text{LCM} = 2 \cdot 2 \cdot 2 \cdot 5 \cdot x \cdot x \cdot x \cdot y \cdot y = 40x^3y^2$

45. $2z^2 = 2 \cdot z \cdot z$
$5xyz = 5 \cdot x \cdot y \cdot z$
$15xy = 3 \cdot 5 \cdot x \cdot y$
$\text{LCM} = 2 \cdot 3 \cdot 5 \cdot x \cdot y \cdot z \cdot z = 30xyz^2$

Cumulative Review

47. $2\overline{)19}$ = 9 R 1; 18; 1

$\frac{19}{2} = 9\frac{1}{2}$

48. $4\frac{2}{5} = \frac{(5 \cdot 4) + 2}{5} = \frac{20 + 2}{5} = \frac{22}{5}$

49. $2 + 6(-1) \div 3 = 2 + (-6) \div 3 = 2 + (-2) = 0$

50. $12 - 5 \cdot 2^2 \div 4 = 12 - 5 \cdot 4 \div 4$
$= 12 - 20 \div 4$
$= 12 - 5$
$= 7$

Quick Quiz 5.2

1. $9 = 3 \cdot 3$
$15 = 3 \cdot 5$
$\text{LCM} = 3 \cdot 3 \cdot 5 = 45$

2. $12x = 2 \cdot 2 \cdot 3 \cdot x$
$4x^3 = 2 \cdot 2 \cdot x \cdot x \cdot x$
$9x^2 = 3 \cdot 3 \cdot x \cdot x$
$\text{LCM} = 2 \cdot 2 \cdot 3 \cdot 3 \cdot x \cdot x \cdot x = 36x^3$

3. $3 = 3$
$18 = 2 \cdot 3 \cdot 3$
$30 = 2 \cdot 3 \cdot 5$
$\text{LCM} = 2 \cdot 3 \cdot 3 \cdot 5 = 90$

4. Answers may vary. No, the LCM of $63x^2$ and $75x^3$ has two factors of 3 since $63x^2$ has two factors of 3 and three factors of x since $75x^3$ has three factors of x.

5.3 Exercises

1. When we add two fractions with the same denominator, we add the numerators, and the denominator stays the same.

3. $\frac{4}{5} + \frac{5}{9} \neq \frac{4+5}{5+9} = \frac{9}{14}$
No, we must find a common denominator when we add or subtract fractions with different denominators. Also, we do not add or subtract denominators.

5. $\frac{\boxed{2}}{7} + \frac{3}{7} = \frac{5}{7}$

7. $\frac{\boxed{2}}{4} - \frac{1}{4} = \frac{1}{4}$

9. $\frac{6}{17} + \frac{9}{17} = \frac{6+9}{17} = \frac{15}{17}$

11. $\frac{6}{23} - \frac{5}{23} = \frac{6-5}{23} = \frac{1}{23}$

Copyright © 2012 Pearson Education, Inc.

13. $\frac{-13}{28}+\left(\frac{-11}{28}\right)=\frac{-13+(-11)}{28}$
$=\frac{-24}{28}$
$=\frac{\not{4}(-6)}{\not{4}(7)}$
$=-\frac{6}{7}$

15. $\frac{-31}{51}+\frac{11}{51}=\frac{-31+11}{51}=\frac{-20}{51}=-\frac{20}{51}$

17. $\frac{9}{y}-\frac{8}{y}=\frac{9-8}{y}=\frac{1}{y}$

19. $\frac{31}{a}+\frac{8}{a}=\frac{31+8}{a}=\frac{39}{a}$

21. $\frac{x}{7}-\frac{5}{7}=\frac{x-5}{7}$

23. $\frac{y}{3}+\frac{14}{3}=\frac{y+14}{3}$

25. $5 = 5$
$6 = 2 \cdot 3$
$\text{LCD} = 2 \cdot 3 \cdot 5 = 30$

27. $9 = 3 \cdot 3$
$15 = 3 \cdot 5$
$\text{LCD} = 3 \cdot 3 \cdot 5 = 45$

29. $\frac{1}{5}=\frac{1\cdot 12}{5\cdot 12}=\frac{12}{60}$

31. $\frac{5}{6}=\frac{5\cdot 10}{6\cdot 10}=\frac{50}{60}$

33. LCD = 45
$\frac{11}{15}-\frac{31}{45}=\frac{11\cdot 3}{15\cdot 3}-\frac{31}{45}=\frac{33}{45}-\frac{31}{45}=\frac{2}{45}$

35. LCD = 24
$\frac{17}{24}-\frac{1}{6}=\frac{17}{24}-\frac{1\cdot 4}{6\cdot 4}=\frac{17}{24}-\frac{4}{24}=\frac{13}{24}$

37. LCD = 56
$\frac{3}{8}+\frac{4}{7}=\frac{3\cdot 7}{8\cdot 7}+\frac{4\cdot 8}{7\cdot 8}=\frac{21}{56}+\frac{32}{56}=\frac{53}{56}$

39. LCD = 20
$\frac{-3}{4}+\frac{1}{10}=\frac{-3\cdot 5}{4\cdot 5}+\frac{1\cdot 2}{10\cdot 2}$
$=\frac{-15}{20}+\frac{2}{20}$
$=\frac{-13}{20}$
$=-\frac{13}{20}$

41. LCD = 26
$\frac{-2}{13}+\frac{7}{26}=\frac{-2\cdot 2}{13\cdot 2}+\frac{7}{26}=\frac{-4}{26}+\frac{7}{26}=\frac{3}{26}$

43. LCD = 70
$\frac{-3}{14}+\left(\frac{-1}{10}\right)=\frac{-3\cdot 5}{14\cdot 5}+\left(\frac{-1\cdot 7}{10\cdot 7}\right)$
$=\frac{-15}{70}+\left(\frac{-7}{70}\right)$
$=\frac{-22}{70}$
$=-\frac{\not{2}\cdot 11}{\not{2}\cdot 35}$
$=-\frac{11}{35}$

45. LCD = 70
$\frac{7}{10}-\frac{5}{14}=\frac{7\cdot 7}{10\cdot 7}-\frac{5\cdot 5}{14\cdot 5}$
$=\frac{49}{70}-\frac{25}{70}$
$=\frac{24}{70}$
$=\frac{\not{2}\cdot 12}{\not{2}\cdot 35}$
$=\frac{12}{35}$

47. LCD = 60
$\frac{11}{15}-\frac{5}{12}=\frac{11\cdot 4}{15\cdot 4}-\frac{5\cdot 5}{12\cdot 5}=\frac{44}{60}-\frac{25}{60}=\frac{19}{60}$

49. LCD = $2x$
$\frac{5}{2x}+\frac{8}{x}=\frac{5}{2x}+\frac{8\cdot 2}{x\cdot 2}=\frac{5}{2x}+\frac{16}{2x}=\frac{21}{2x}$

51. LCD = $7x$
$\frac{2}{7x}+\frac{3}{x}=\frac{2}{7x}+\frac{3\cdot 7}{x\cdot 7}=\frac{2}{7x}+\frac{21}{7x}=\frac{23}{7x}$

Copyright © 2012 Pearson Education, Inc.

53. LCD = $6x$

$$\frac{3}{2x}+\frac{5}{6x}=\frac{3\cdot 3}{2x\cdot 3}+\frac{5}{6x}=\frac{9}{6x}+\frac{5}{6x}=\frac{14}{6x}=\frac{\cancel{2}\cdot 7}{\cancel{2}\cdot 3x}=\frac{7}{3x}$$

55. LCD = xy

$$\frac{3}{x}+\frac{4}{y}=\frac{3\cdot y}{x\cdot y}+\frac{4\cdot x}{y\cdot x}=\frac{3y}{xy}+\frac{4x}{xy}=\frac{3y+4x}{xy}$$

57. LCD = ab

$$\frac{4}{a}-\frac{9}{b}=\frac{4\cdot b}{a\cdot b}-\frac{9\cdot a}{b\cdot a}=\frac{4b}{ab}-\frac{9a}{ab}=\frac{4b-9a}{ab}$$

59. LCD = 15

$$\frac{2x}{15}+\frac{3x}{5}=\frac{2x}{15}+\frac{3x\cdot 3}{5\cdot 3}=\frac{2x}{15}+\frac{9x}{15}=\frac{11x}{15}$$

61. LCD = 20

$$\frac{-3x}{10}-\frac{7x}{20}=\frac{-3x\cdot 2}{10\cdot 2}-\frac{7x}{20}=\frac{-6x}{20}-\frac{7x}{20}=\frac{-13x}{20}=-\frac{13x}{20}$$

63. LCD = 12

$$\frac{x}{3}+\left(\frac{-11x}{12}\right)=\frac{x\cdot 4}{3\cdot 4}+\left(\frac{-11x}{12}\right)=\frac{4x}{12}+\left(\frac{-11x}{12}\right)=\frac{-7x}{12}=-\frac{7x}{12}$$

65. LCD = 30

$$\frac{5}{6}+\frac{3}{10}=\frac{5\cdot 5}{6\cdot 5}+\frac{3\cdot 3}{10\cdot 3}=\frac{25}{30}+\frac{9}{30}=\frac{34}{30}=\frac{\cancel{2}\cdot 17}{\cancel{2}\cdot 15}=\frac{17}{15}=1\frac{2}{15}$$

67. LCD = 80

$$\frac{3}{16}+\left(\frac{-9}{20}\right)=\frac{3\cdot 5}{16\cdot 5}+\left(\frac{-9\cdot 4}{20\cdot 4}\right)=\frac{15}{80}+\left(\frac{-36}{80}\right)=\frac{-21}{80}=-\frac{21}{80}$$

69. LCD = xy

$$\frac{9}{y}+\frac{1}{x}=\frac{9\cdot x}{y\cdot x}+\frac{1\cdot y}{x\cdot y}=\frac{9x}{xy}+\frac{y}{xy}=\frac{9x+y}{xy}$$

71. LCD = 60

$$\frac{2x}{15}+\frac{3x}{20}=\frac{2x\cdot 4}{15\cdot 4}+\frac{3x\cdot 3}{20\cdot 3}=\frac{8x}{60}+\frac{9x}{60}=\frac{17x}{60}$$

73. We must add. The LCD is 8.

$$\frac{3}{4}+\frac{7}{8}=\frac{3\cdot 2}{4\cdot 2}+\frac{7}{8}=\frac{6}{8}+\frac{7}{8}=\frac{13}{8}\text{ or }1\frac{5}{8}$$

Pat bought $1\frac{5}{8}$ pounds of fudge.

75. a. We must add. The LCD is 4.

$$\frac{3}{4}+\frac{1}{2}=\frac{3}{4}+\frac{1\cdot 2}{2\cdot 2}=\frac{3}{4}+\frac{2}{4}=\frac{5}{4}\text{ or }1\frac{1}{4}$$

There are $1\frac{1}{4}$ cups of sugar in the recipe.

b. We must subtract. The LCD is 4.

$$\frac{3}{4}-\frac{1}{2}=\frac{3}{4}-\frac{1\cdot 2}{2\cdot 2}=\frac{3}{4}-\frac{2}{4}=\frac{1}{4}$$

There is $\frac{1}{4}$ cup more brown sugar than granulated sugar in the recipe.

Copyright © 2012 Pearson Education, Inc.

77. We must subtract. The LCD is 12.

$$\frac{1}{3}-\frac{1}{4}=\frac{1\cdot 4}{3\cdot 4}-\frac{1\cdot 3}{4\cdot 3}=\frac{4}{12}-\frac{3}{12}=\frac{1}{12}$$

The mother completed $\frac{1}{12}$ more of the job than the son.

79. a. We must add. The LCD is 24.

$$\frac{1}{8}+\frac{1}{12}=\frac{1\cdot 3}{8\cdot 3}+\frac{1\cdot 2}{12\cdot 2}=\frac{3}{24}+\frac{2}{24}=\frac{5}{24}$$

Eric and his cousin inherited $\frac{5}{24}$ of the estate.

b. We must subtract. The LCD is 24.

$$\frac{1}{8}-\frac{1}{12}=\frac{1\cdot 3}{8\cdot 3}-\frac{1\cdot 2}{12\cdot 2}=\frac{3}{24}-\frac{2}{24}=\frac{1}{24}$$

Eric inherited $\frac{1}{24}$ more than his cousin.

81. LCD = 120

$$\begin{aligned}\frac{7}{30}+\frac{3}{40}+\frac{1}{8}&=\frac{7\cdot 4}{30\cdot 4}+\frac{3\cdot 3}{40\cdot 3}+\frac{1\cdot 15}{8\cdot 15}\\&=\frac{28}{120}+\frac{9}{120}+\frac{15}{120}\\&=\frac{52}{120}\\&=\frac{\cancel{4}\cdot 12}{\cancel{4}\cdot 30}\\&=\frac{13}{30}\end{aligned}$$

83. LCD = 12

$$\begin{aligned}\frac{1}{3}+\frac{1}{12}-\frac{1}{6}&=\frac{1\cdot 4}{3\cdot 4}+\frac{1}{12}-\frac{1\cdot 2}{6\cdot 2}\\&=\frac{4}{12}+\frac{1}{12}-\frac{2}{12}\\&=\frac{3}{12}\\&=\frac{\cancel{3}\cdot 1}{\cancel{3}\cdot 4}\\&=\frac{1}{4}\end{aligned}$$

Cumulative Review

85. Replace a with -24 and b with 6.

$$\frac{-a}{b}=\frac{-(a)}{(b)}=\frac{-(-24)}{(6)}=\frac{24}{6}=4$$

86. Replace x with -10 and y with -2.

$$\frac{-x}{-y}=\frac{-(x)}{-(y)}=\frac{-(-10)}{-(-2)}=\frac{10}{2}=5$$

87. Replace a with 3 and b with 1.

$$\frac{(a^2-b)}{-4}=\frac{(3^2-1)}{-4}=\frac{(9-1)}{-4}=\frac{8}{-4}=-2$$

88. Replace x with -2.

$$\begin{aligned}9x-x^2&=9(x)-(x)^2\\&=9(-2)-(-2)^2\\&=9(-2)-4\\&=-18-4\\&=-18+(-4)\\&=-22\end{aligned}$$

89. Total expenses:

$$\begin{array}{r}1295\\469\\387\\+\ 287\\\hline 2438\end{array}$$

Difference:

$$\begin{array}{r}3033\\-\ 2438\\\hline 595\end{array}$$

There is $595 left in the budget.

90. Total expenses:

$$\begin{array}{r}190\\43\\42\\96\\+\ 55\\\hline 426\end{array}$$

Difference:

$$\begin{array}{r}500\\-\ 426\\\hline 74\end{array}$$

Joan has $74 left after expenses.

Quick Quiz 5.3

1. LCD = $9y$

$$\frac{1}{y}-\frac{2}{9y}=\frac{1\cdot 9}{y\cdot 9}-\frac{2}{9y}=\frac{9}{9y}-\frac{2}{9y}=\frac{7}{9y}$$

2. a. LCD = 60

$$\begin{aligned}\frac{5}{12}+\frac{3}{20}&=\frac{5\cdot 5}{12\cdot 5}+\frac{3\cdot 3}{20\cdot 3}\\&=\frac{25}{60}+\frac{9}{60}\\&=\frac{34}{60}\\&=\frac{\cancel{2}\cdot 17}{\cancel{2}\cdot 30}\\&=\frac{17}{30}\end{aligned}$$

Copyright © 2012 Pearson Education, Inc.

b. LCD = 42

$$\frac{3}{14}-\frac{1}{21}=\frac{3\cdot 3}{14\cdot 3}-\frac{1\cdot 2}{21\cdot 2}=\frac{9}{42}-\frac{2}{42}=\frac{7}{42}=\frac{\cancel{7}\cdot 1}{\cancel{7}\cdot 6}=\frac{1}{6}$$

3. LCD = 30

$$\frac{2x}{15}+\frac{5x}{6}=\frac{2x\cdot 2}{15\cdot 2}+\frac{5x\cdot 5}{6\cdot 5}=\frac{4x}{30}+\frac{25x}{30}=\frac{29x}{30}$$

4. a. $20 = 2\cdot 2\cdot 5$

$6 = 2\cdot 3$

$\text{LCD} = 2\cdot 2\cdot 3\cdot 5 = 2^2\cdot 3\cdot 5 = 60$

b. Answers may vary. One possible solution is to multiply each term by a rational number, the numerator and the denominator of which are the quotient of the LCD divided by the denominator of each respective term.

5.4 Understanding the Concept Should We Change to an Improper Fraction?

1. a.

$$25\frac{3\cdot 8}{5\cdot 8}=25\frac{24}{40}$$
$$+32\frac{5\cdot 5}{8\cdot 5}=+32\frac{25}{40}$$
$$57\frac{49}{40}=58\frac{9}{40}$$

b.

$$25\frac{3}{5}=\frac{128}{5}=\frac{1024}{40}$$
$$+32\frac{5}{8}=+\frac{261}{8}=+\frac{1305}{40}$$
$$\frac{2329}{40}\text{ or }58\frac{9}{40}$$

With large numbers, it is easier to keep numbers as mixed numbers even though we must carry.

5.4 Exercises

1. a. Marcy did not change the mixed numbers to improper fractions before she multiplied.

b. $2\frac{2}{3}\cdot 3\frac{4}{5}=\frac{8}{3}\cdot\frac{19}{5}=\frac{152}{15}=10\frac{2}{15}$

3.

$$10\frac{4}{9}$$
$$+11\frac{1}{9}$$
$$21\frac{5}{9}$$

5.

$$5\frac{5}{8}$$
$$+11\frac{1}{8}$$
$$16\frac{6}{8}=16\frac{3}{4}$$

7.

$$5\frac{2\cdot 4}{3\cdot 4}=5\frac{8}{12}$$
$$+8\frac{1\cdot 3}{4\cdot 3}=+8\frac{3}{12}$$
$$13\frac{11}{12}$$

9.

$$14\frac{1\cdot 3}{4\cdot 3}=14\frac{3}{12}$$
$$+6\frac{1\cdot 4}{3\cdot 4}=+6\frac{4}{12}$$
$$20\frac{7}{12}$$

11.

$$7\frac{5\cdot 4}{6\cdot 4}=7\frac{20}{24}$$
$$+4\frac{3\cdot 3}{8\cdot 3}=+4\frac{9}{24}$$
$$11\frac{29}{24}=12\frac{5}{24}$$

13.

$$7\frac{4}{5}$$
$$-2\frac{1}{5}$$
$$5\frac{3}{5}$$

Copyright © 2012 Pearson Education, Inc.

15. $$\begin{array}{rcl} 9\dfrac{2\cdot 2}{3\cdot 2} & = & 9\dfrac{4}{6} \\ -6\dfrac{1}{6} & = & -6\dfrac{1}{6} \\ \hline & & 3\dfrac{3}{6}=3\dfrac{1}{2} \end{array}$$

17. $$\begin{array}{rcl} 11\dfrac{1}{5} & = & 10\dfrac{6}{5} \\ -6\dfrac{3}{5} & = & -6\dfrac{3}{5} \\ \hline & & 4\dfrac{3}{5} \end{array}$$

19. $$\begin{array}{rcccl} 10\dfrac{5\cdot 5}{12\cdot 5} & = & 10\dfrac{25}{60} & = & 9\dfrac{85}{60} \\ -3\dfrac{9\cdot 6}{10\cdot 7} & = & -3\dfrac{54}{60} & = & -3\dfrac{54}{60} \\ \hline & & & & 6\dfrac{31}{60} \end{array}$$

21. $$\begin{array}{rcl} 9 & = & 8\dfrac{4}{4} \\ -2\dfrac{1}{4} & = & -2\dfrac{1}{4} \\ \hline & & 6\dfrac{3}{4} \end{array}$$

23. $$\begin{array}{rcl} 8\dfrac{2\cdot 7}{5\cdot 7} & = & 8\dfrac{14}{35} \\ -6\dfrac{1\cdot 5}{7\cdot 5} & = & -6\dfrac{5}{35} \\ \hline & & 2\dfrac{9}{35} \end{array}$$

25. $$\begin{array}{rcl} 1\dfrac{1\cdot 4}{6\cdot 4} & = & 1\dfrac{4}{24} \\ +\dfrac{3\cdot 3}{8\cdot 3} & = & +\dfrac{9}{24} \\ \hline & & 1\dfrac{13}{24} \end{array}$$

27. $$\begin{array}{rcl} 8\dfrac{1\cdot 3}{4\cdot 3} & = & 8\dfrac{3}{12} \\ +3\dfrac{5\cdot 2}{6\cdot 2} & = & +3\dfrac{10}{12} \\ \hline & & 11\dfrac{13}{12}=12\dfrac{1}{2} \end{array}$$

29. $$\begin{array}{rcl} 32 & = & 31\dfrac{9}{9} \\ -1\dfrac{2}{9} & = & -1\dfrac{2}{9} \\ \hline & & 30\dfrac{7}{9} \end{array}$$

31. $(-2)\cdot 3\dfrac{1}{5}=\dfrac{-2}{1}\cdot\dfrac{16}{5}=-\dfrac{2\cdot 16}{1\cdot 5}=-\dfrac{32}{5}$ or $-6\dfrac{2}{5}$

33. $4\dfrac{1}{3}\cdot 2\dfrac{1}{4}=\dfrac{13}{3}\cdot\dfrac{9}{4}=\dfrac{13\cdot\cancel{3}\cdot 3}{\cancel{3}\cdot 4}=\dfrac{39}{4}$ or $9\dfrac{3}{4}$

35. $-\dfrac{3}{4}\cdot 3\dfrac{5}{7}=-\dfrac{3}{4}\cdot\dfrac{26}{7}=-\dfrac{3\cdot\cancel{2}\cdot 13}{\cancel{2}\cdot 2\cdot 7}=-\dfrac{39}{14}$ or $-2\dfrac{11}{14}$

37. $2\dfrac{1}{4}\div(-4)=\dfrac{9}{4}\div\dfrac{(-4)}{1}=\dfrac{9}{4}\cdot\left(-\dfrac{1}{4}\right)=-\dfrac{9}{16}$

39. $4\dfrac{1}{2}\div 2\dfrac{1}{4}=\dfrac{9}{2}\div\dfrac{9}{4}=\dfrac{9}{2}\cdot\dfrac{4}{9}=\dfrac{\cancel{9}\cdot\cancel{2}\cdot 2}{\cancel{2}\cdot\cancel{9}}=\dfrac{2}{1}=2$

41. $3\dfrac{1}{4}\div\dfrac{3}{8}=\dfrac{13}{4}\div\dfrac{3}{8}=\dfrac{13}{4}\cdot\dfrac{8}{3}=\dfrac{13\cdot\cancel{4}\cdot 2}{\cancel{4}\cdot 3}=\dfrac{26}{3}$ or $8\dfrac{2}{3}$

43. $-6\div\dfrac{1}{4}=\dfrac{(-6)}{1}\cdot\dfrac{4}{1}=\dfrac{-24}{1}=-24$

45. $1\dfrac{1}{4}\cdot 3\dfrac{2}{3}=\dfrac{5}{4}\cdot\dfrac{11}{3}=\dfrac{55}{12}$ or $4\dfrac{7}{12}$

47. $$\begin{aligned} 4\dfrac{1}{2}\div\left(-\dfrac{6}{7}\right)&=\dfrac{9}{2}\cdot\left(-\dfrac{7}{6}\right) \\ &=-\dfrac{\cancel{3}\cdot 3\cdot 7}{2\cdot 2\cdot\cancel{3}} \\ &=-\dfrac{21}{4}\text{ or }-5\dfrac{1}{4} \end{aligned}$$

49. $7\dfrac{1}{2}\div(-8)=\dfrac{15}{2}\div\dfrac{(-8)}{1}=\dfrac{15}{2}\cdot\left(-\dfrac{1}{8}\right)=-\dfrac{15}{16}$

51. We divide.

$26\div 6\dfrac{1}{2}=\dfrac{26}{1}\div\dfrac{13}{2}=\dfrac{26}{1}\cdot\dfrac{2}{13}=\dfrac{\cancel{13}\cdot 2\cdot 2}{1\cdot\cancel{13}}=\dfrac{4}{1}=4$

Andy can cut 4 pieces of rope.

Copyright © 2012 Pearson Education, Inc.

53. We multiply.
$$\frac{1}{3}\cdot 7\frac{1}{5}=\frac{1}{3}\cdot\frac{36}{5}=\frac{1\cdot \cancel{3}\cdot 12}{\cancel{3}\cdot 5}=\frac{12}{5}\text{ or }2\frac{2}{5}$$
He must cut off $2\frac{2}{5}$ feet.

55. We multiply.
$$2\cdot 2\frac{1}{4}=\frac{2}{1}\cdot\frac{9}{4}=\frac{\cancel{2}\cdot 9}{1\cdot\cancel{2}\cdot 2}=\frac{9}{2}\text{ or }4\frac{1}{2}$$
You need $4\frac{1}{2}$ cups of flour.

57. We multiply.
$$4\cdot 2\frac{1}{2}=\frac{4}{1}\cdot\frac{5}{2}=\frac{\cancel{2}\cdot 2\cdot 5}{1\cdot\cancel{2}}=\frac{10}{1}=10$$
You need 10 cups of chocolate chips.

59. We add.
$$\begin{array}{rcl} 9\frac{1\cdot 3}{8\cdot 3} & = & 9\frac{3}{24} \\ 12\frac{1\cdot 8}{3\cdot 8} & = & 12\frac{8}{24} \\ +17\frac{1\cdot 4}{6\cdot 4} & = & +17\frac{4}{24} \\ \hline & & 38\frac{15}{24}=38\frac{5}{8} \end{array}$$
Jeff ran $38\frac{5}{8}$ miles.

61. The value of each number is $\frac{1}{4}$ greater than the preceding number.
The number that follows 1 is $1+\frac{1}{4}=\frac{4}{4}+\frac{1}{4}=\frac{5}{4}$, so $a = 5$. The number that follows $\frac{3}{2}$ is
$\frac{3}{2}+\frac{1}{4}=\frac{3\cdot 2}{2\cdot 2}+\frac{1}{4}=\frac{6}{4}+\frac{1}{4}=\frac{7}{4}$, so $b = 4$.
The number that follows $\frac{11}{4}$ is $\frac{11}{4}+\frac{1}{4}=\frac{12}{4}=3$, so $c = 3$.

Cumulative Review

63. $2 + 9 \cdot 8 = 2 + 72 = 74$

64. $12 - 4 \div 2 = 12 - 2 = 10$

65. $\frac{(5+7)}{(2\cdot 3)}=\frac{12}{6}=2$

66. $\frac{(11-5)}{2}=\frac{6}{2}=3$

67. 19(7) + 28(2) − 2(3) − 1(5) = $178 Jan.
26(7) + 36(2) − 3(3) − 2(5) = $235 Feb.
26(7) + 31(2) − 3(3) − 2(5) = $225 Mar.
178 + 235 + 225 = $638 total

68. The $8 rental fee is a $1 increase over the $7 rental fee, so the total cost will be $1 more for each unit rented.
$638 + $19 + $26 + $26 = $709

Quick Quiz 5.4

1. a.
$$\begin{array}{rcl} 1\frac{7\cdot 4}{9\cdot 4} & = & 1\frac{28}{36} \\ +3\frac{5\cdot 3}{12\cdot 3} & = & +3\frac{15}{36} \\ \hline & & 4\frac{43}{36}=5\frac{7}{36} \end{array}$$

b.
$$\begin{array}{rcccl} 10\frac{1\cdot 4}{3\cdot 4} & = & 10\frac{4}{12} & = & 9\frac{16}{12} \\ -5\frac{5}{12} & = & -5\frac{5}{12} & = & -5\frac{5}{12} \\ \hline & & & & 4\frac{11}{12} \end{array}$$

2. $(-9)\cdot 2\frac{1}{3}=\frac{(-9)}{1}\cdot\frac{7}{3}=-\frac{3\cdot\cancel{3}\cdot 7}{1\cdot\cancel{3}}=-\frac{21}{1}=-21$

3.
$$\begin{aligned} 4\frac{2}{5}\div\left(-1\frac{1}{10}\right)&=\frac{22}{5}\div\left(\frac{-11}{10}\right)\\ &=\frac{22}{5}\cdot\left(-\frac{10}{11}\right)\\ &=-\frac{2\cdot\cancel{11}\cdot 2\cdot\cancel{5}}{\cancel{5}\cdot\cancel{11}}\\ &=-\frac{4}{1}\\ &=-4 \end{aligned}$$

4. Answers may vary. One possible solution is to first express both terms as improper fractions. Multiply the numerators, multiply the denominators, then simplify.

Copyright © 2012 Pearson Education, Inc.

Use Math to Save Money

1. $(4 \times \$8) + \$13 + (5 \times \$10)$
$= \$32 + \$13 + \$50$
$= \$95$
Tracy and Max spend \$95 on lunch every week.

2. $2 \times 5 \times \$3 = 10 \times \$3 = \$30$
The weekly cost of bringing their own lunches is \$30.

3. $\$95 - \$30 = \$65$
They will save \$65 per week by not going out for lunch.

4. $\frac{\$2275}{\$65} = 35$
They will have enough money for the trip after 35 weeks.

How Am I Doing? Sections 5.1–5.4

1. $\frac{16}{9} \cdot \left(\frac{-18}{36}\right) = -\frac{2 \cdot 2 \cdot 2 \cdot \cancel{2} \cdot \cancel{18}}{3 \cdot 3 \cdot \cancel{2} \cdot \cancel{18}} = -\frac{8}{9}$

2. $\frac{-3y^3}{20} \cdot \frac{-12y^2}{21} = \frac{3 \cdot y^3 \cdot \cancel{3} \cdot \cancel{4} \cdot y^2}{\cancel{4} \cdot 5 \cdot \cancel{3} \cdot 7}$
$= \frac{3 \cdot y^{3+2}}{5 \cdot 7}$
$= \frac{3y^5}{35}$

3. $25 \div \frac{5}{7} = \frac{25}{1} \cdot \frac{7}{5} = \frac{5 \cdot \cancel{5} \cdot 7}{1 \cdot \cancel{5}} = \frac{35}{1} = 35$

4. $\frac{3y^4}{20} \div \frac{12y^2}{5} = \frac{3y^4}{20} \cdot \frac{5}{12y^2}$
$= \frac{\cancel{3} \cdot y^4 \cdot \cancel{5}}{2 \cdot 2 \cdot \cancel{5} \cdot 2 \cdot 2 \cdot \cancel{3} \cdot y^2}$
$= \frac{y^{4-2}}{2 \cdot 2 \cdot 2 \cdot 2}$
$= \frac{y^2}{16}$

5. We divide.
$63 \div \frac{3}{4} = \frac{63}{1} \cdot \frac{4}{3} = \frac{\cancel{3} \cdot 3 \cdot 7 \cdot 4}{1 \cdot \cancel{3}} = 84$
Mary Beth will have 84 parcels.

6. $12 = 2 \cdot 2 \cdot 3$
$21 = 3 \cdot 7$
$\text{LCM} = 2 \cdot 2 \cdot 3 \cdot 7 = 84$

7. $7x = 7 \cdot x$
$21 = 3 \cdot 7$
$2x^2 = 2 \cdot x \cdot x$
$\text{LCM} = 2 \cdot 3 \cdot 7 \cdot x \cdot x = 42x^2$

8. $\frac{-3}{7} + \frac{2}{21} = \frac{-3 \cdot 3}{7 \cdot 3} + \frac{2}{21}$
$= \frac{-9+2}{21}$
$= \frac{-7}{21}$
$= -\frac{\cancel{7}}{3 \cdot \cancel{7}}$
$= -\frac{1}{3}$

9. $\frac{11}{16} - \frac{5}{20} = \frac{11 \cdot 5}{16 \cdot 5} - \frac{5 \cdot 4}{20 \cdot 4}$
$= \frac{55}{80} - \frac{20}{80}$
$= \frac{55-20}{80}$
$= \frac{35}{80}$
$= \frac{3 \cdot \cancel{5}}{\cancel{5} \cdot 16}$
$= \frac{7}{16}$

10. $\frac{5x}{8} - \frac{3x}{14} = \frac{5x \cdot 7}{8 \cdot 7} - \frac{3x \cdot 4}{14 \cdot 4} = \frac{35x}{56} - \frac{12x}{56} = \frac{23x}{56}$

11. a. We subtract.
$\frac{1}{4} - \frac{1}{5} = \frac{1 \cdot 5}{4 \cdot 5} - \frac{1 \cdot 4}{5 \cdot 4} = \frac{5}{20} - \frac{4}{20} = \frac{1}{20}$
Abbas completed $\frac{1}{20}$ more of the job the first day.

b. We add.
$\frac{1}{4} + \frac{1}{5} = \frac{1 \cdot 5}{4 \cdot 5} + \frac{1 \cdot 4}{5 \cdot 4} = \frac{5}{20} + \frac{4}{20} = \frac{9}{20}$
Abbas completed $\frac{9}{20}$ of the total job the first two days.

Copyright © 2012 Pearson Education, Inc.

12.
$$\begin{array}{rcr} 2\frac{2\cdot 7}{5\cdot 7} & = & 2\frac{14}{35} \\ +\,4\frac{6\cdot 5}{7\cdot 5} & = & +\,4\frac{30}{35} \\ \hline & & 6\frac{44}{35} = 7\frac{9}{35} \end{array}$$

13.
$$\begin{array}{rcrcr} 6\frac{1\cdot 5}{12\cdot 5} & = & 6\frac{5}{60} & = & 5\frac{65}{60} \\ -\,2\frac{7\cdot 4}{15\cdot 4} & = & -\,2\frac{28}{60} & = & -\,2\frac{23}{60} \\ \hline & & & & 3\frac{37}{60} \end{array}$$

14. $2\frac{2}{3}\cdot 1\frac{5}{16}=\frac{8}{3}\cdot\frac{21}{16}=\frac{\cancel{8}\cdot\cancel{3}\cdot 7}{\cancel{3}\cdot 2\cdot\cancel{8}}=\frac{7}{2}=3\frac{1}{2}$

15.
$$\begin{aligned} 10\frac{2}{9}\div 2\frac{1}{3} &= \frac{92}{9}\div\frac{7}{3} \\ &= \frac{92}{9}\cdot\frac{3}{7} \\ &= \frac{2\cdot 2\cdot 23\cdot\cancel{3}}{\cancel{3}\cdot 3\cdot 7} \\ &= \frac{92}{21} \\ &= 4\frac{8}{21} \end{aligned}$$

5.5 Exercises

1. We multiply before we add.

3.
$$\begin{aligned} \frac{3}{5}-\frac{1}{3}\div\frac{5}{6} &= \frac{3}{5}-\frac{1}{3}\cdot\frac{6}{5} \\ &= \frac{3}{5}-\frac{1\cdot 2\cdot\cancel{3}}{\cancel{3}\cdot 5} \\ &= \frac{3}{5}-\frac{2}{5} \\ &= \frac{1}{5} \end{aligned}$$

5.
$$\begin{aligned} \frac{3}{4}+\frac{1}{4}\cdot\frac{3}{5} &= \frac{3}{4}+\frac{3}{20} \\ &= \frac{3\cdot 5}{4\cdot 5}+\frac{3}{20} \\ &= \frac{15}{20}+\frac{3}{20} \\ &= \frac{18}{20} \\ &= \frac{\cancel{2}\cdot 9}{\cancel{2}\cdot 10} \\ &= \frac{9}{10} \end{aligned}$$

7. $\frac{5}{7}\cdot\frac{1}{3}\div\frac{2}{7}=\frac{5}{21}\div\frac{2}{7}=\frac{5}{21}\cdot\frac{7}{2}=\frac{5\cdot\cancel{7}}{3\cdot\cancel{7}\cdot 2}=\frac{5}{6}$

9.
$$\begin{aligned} \left(\frac{3}{2}\right)^2-\frac{1}{3}+\frac{1}{2} &= \frac{9}{4}-\frac{1}{3}+\frac{1}{2} \\ &= \frac{9\cdot 3}{4\cdot 3}-\frac{1\cdot 4}{3\cdot 4}+\frac{1\cdot 6}{2\cdot 6} \\ &= \frac{27}{12}-\frac{4}{12}+\frac{6}{12} \\ &= \frac{29}{12} \text{ or } 2\frac{5}{12} \end{aligned}$$

11.
$$\begin{aligned} \frac{5}{6}\cdot\frac{1}{2}+\frac{2}{3}\div\frac{4}{3} &= \frac{5}{12}+\frac{2}{3}\div\frac{4}{3} \\ &= \frac{5}{12}+\frac{2}{3}\cdot\frac{3}{4} \\ &= \frac{5}{12}+\frac{1}{2} \\ &= \frac{5}{12}+\frac{1\cdot 6}{2\cdot 6} \\ &= \frac{5}{12}+\frac{6}{12} \\ &= \frac{11}{12} \end{aligned}$$

Copyright © 2012 Pearson Education, Inc.

13. $\frac{2}{9}\cdot\frac{1}{4}+\left(\frac{2}{3}\div\frac{6}{7}\right)=\frac{2}{9}\cdot\frac{1}{4}+\left(\frac{2}{3}\cdot\frac{7}{6}\right)$
$=\frac{2}{9}\cdot\frac{1}{4}+\left(\frac{\cancel{2}\cdot 7}{3\cdot\cancel{2}\cdot 3}\right)$
$=\frac{2}{9}\cdot\frac{1}{4}+\frac{7}{9}$
$=\frac{\cancel{2}\cdot 1}{3\cdot 3\cdot\cancel{2}\cdot 2}+\frac{7}{9}$
$=\frac{1}{18}+\frac{7}{9}$
$=\frac{1}{18}+\frac{7\cdot 2}{9\cdot 2}$
$=\frac{1}{18}+\frac{14}{18}$
$=\frac{15}{18}$
$=\frac{\cancel{3}\cdot 5}{2\cdot\cancel{3}\cdot 3}$
$=\frac{5}{6}$

15. $\frac{3}{4}\cdot\frac{1}{4}+\left(\frac{3}{4}\right)^2=\frac{3}{4}\cdot\frac{1}{4}+\frac{9}{16}$
$=\frac{3}{16}+\frac{9}{16}$
$=\frac{12}{16}$
$=\frac{\cancel{4}\cdot 3}{\cancel{4}\cdot 2\cdot 2}$
$=\frac{3}{4}$

17. $\left(-\frac{2}{5}\right)\cdot\left(\frac{1}{4}\right)^2=\left(-\frac{2}{5}\right)\cdot\frac{1}{16}=-\frac{\cancel{2}\cdot 1}{5\cdot\cancel{2}\cdot 2\cdot 2\cdot 2}=-\frac{1}{40}$

19. $\frac{7+(-3)^2}{\frac{8}{9}}=\frac{[7+(-3)^2]}{\left(\frac{8}{9}\right)}$
$=\frac{(7+9)}{\left(\frac{8}{9}\right)}$
$=\frac{16}{\left(\frac{8}{9}\right)}$
$=16\div\frac{8}{9}$
$=16\cdot\frac{9}{8}$
$=\frac{2\cdot\cancel{8}\cdot 3\cdot 3}{\cancel{8}}$
$=18$

21. $\frac{\frac{4}{7}}{2^3+8}=\frac{\left(\frac{4}{7}\right)}{(2^3+8)}$
$=\frac{\left(\frac{4}{7}\right)}{(8+8)}$
$=\frac{\left(\frac{4}{7}\right)}{16}$
$=\frac{4}{7}\div 16$
$=\frac{4}{7}\cdot\frac{1}{16}$
$=\frac{\cancel{4}\cdot 1}{7\cdot\cancel{4}\cdot 2\cdot 2}$
$=\frac{1}{28}$

23. $\frac{2\cdot 3-1}{\frac{5}{8}}=\frac{(2\cdot 3-1)}{\left(\frac{5}{8}\right)}$
$=\frac{(6-1)}{\left(\frac{5}{8}\right)}$
$=\frac{5}{\left(\frac{5}{8}\right)}$
$=5\div\frac{5}{8}$
$=5\cdot\frac{8}{5}$
$=\frac{\cancel{5}\cdot 2\cdot 2\cdot 2}{1\cdot\cancel{5}}$
$=8$

Copyright © 2012 Pearson Education, Inc.

25. $\dfrac{\frac{6}{7}}{\frac{9}{14}} = \dfrac{6}{7} \div \dfrac{9}{14} = \dfrac{6}{7} \cdot \dfrac{14}{9} = \dfrac{2 \cdot \cancel{3} \cdot 2 \cdot \cancel{7}}{\cancel{7} \cdot \cancel{3} \cdot 3} = \dfrac{4}{3}$

27. $\dfrac{\frac{x^2}{2}}{\frac{x}{4}} = \dfrac{x^2}{2} \div \dfrac{x}{4} = \dfrac{x^2}{2} \cdot \dfrac{4}{x} = \dfrac{x \cdot \cancel{x} \cdot \cancel{2} \cdot 2}{\cancel{2} \cdot \cancel{x}} = 2x$

29. $\dfrac{\frac{x}{3}}{\frac{x^2}{9}} = \dfrac{x}{3} \div \dfrac{x^2}{9} = \dfrac{x}{3} \cdot \dfrac{9}{x^2} = \dfrac{\cancel{x} \cdot \cancel{3} \cdot 3}{\cancel{3} \cdot \cancel{x} \cdot x} = \dfrac{3}{x}$

31.
$$\begin{aligned}
\dfrac{\frac{1}{2}+\frac{3}{4}}{\frac{4}{5}+\frac{1}{10}} &= \dfrac{\left(\frac{1}{2}+\frac{3}{4}\right)}{\left(\frac{4}{5}+\frac{1}{10}\right)} \\
&= \dfrac{\left(\frac{1\cdot 2}{2\cdot 2}+\frac{3}{4}\right)}{\left(\frac{4\cdot 2}{5\cdot 2}+\frac{1}{10}\right)} \\
&= \dfrac{\left(\frac{2}{4}+\frac{3}{4}\right)}{\left(\frac{8}{10}+\frac{1}{10}\right)} \\
&= \dfrac{\frac{5}{4}}{\frac{9}{10}} \\
&= \dfrac{5}{4} \div \dfrac{9}{10} \\
&= \dfrac{5}{4} \cdot \dfrac{10}{9} \\
&= \dfrac{5 \cdot \cancel{2} \cdot 5}{2 \cdot \cancel{2} \cdot 3 \cdot 3} \\
&= \dfrac{25}{18} \text{ or } 1\dfrac{7}{18}
\end{aligned}$$

33.
$$\begin{aligned}
\dfrac{\frac{4}{25}-\frac{3}{50}}{\frac{3}{10}+\frac{5}{20}} &= \dfrac{\left(\frac{4}{25}-\frac{3}{50}\right)}{\left(\frac{3}{10}+\frac{5}{20}\right)} \\
&= \dfrac{\left(\frac{4\cdot 2}{25\cdot 2}-\frac{3}{50}\right)}{\left(\frac{3\cdot 2}{10\cdot 2}+\frac{5}{20}\right)} \\
&= \dfrac{\left(\frac{8}{50}-\frac{3}{50}\right)}{\left(\frac{6}{20}+\frac{5}{20}\right)} \\
&= \dfrac{\frac{5}{50}}{\frac{11}{20}} \\
&= \dfrac{5}{50} \div \dfrac{11}{20} \\
&= \dfrac{5}{50} \cdot \dfrac{20}{11} \\
&= \dfrac{\cancel{5} \cdot 2 \cdot \cancel{10}}{\cancel{5} \cdot \cancel{10} \cdot 11} \\
&= \dfrac{2}{11}
\end{aligned}$$

35. $\dfrac{\frac{x}{10}}{\frac{x^2}{20}} = \dfrac{x}{10} \div \dfrac{x^2}{20} = \dfrac{x}{10} \cdot \dfrac{20}{x^2} = \dfrac{\cancel{x} \cdot \cancel{10} \cdot 2}{\cancel{10} \cdot \cancel{x} \cdot x} = \dfrac{2}{x}$

37.
$$\begin{aligned}
\left(\dfrac{1}{2}\right)^2 + \dfrac{2}{3} \cdot \dfrac{6}{7} &= \dfrac{1}{4} + \dfrac{2}{3} \cdot \dfrac{6}{7} \\
&= \dfrac{1}{4} + \dfrac{2 \cdot 2 \cdot \cancel{3}}{\cancel{3} \cdot 7} \\
&= \dfrac{1}{4} + \dfrac{4}{7} \\
&= \dfrac{1 \cdot 7}{4 \cdot 7} + \dfrac{4 \cdot 4}{7 \cdot 4} \\
&= \dfrac{7}{28} + \dfrac{16}{28} \\
&= \dfrac{23}{28}
\end{aligned}$$

Copyright © 2012 Pearson Education, Inc.

39. $\dfrac{\frac{3}{8}+\frac{1}{4}}{\frac{4}{5}+\frac{3}{4}}=\dfrac{\left(\frac{3}{8}+\frac{1}{4}\right)}{\left(\frac{4}{5}+\frac{3}{4}\right)}$

$=\dfrac{\left(\frac{3}{8}+\frac{1\cdot 2}{4\cdot 2}\right)}{\left(\frac{4\cdot 4}{5\cdot 4}+\frac{3\cdot 5}{4\cdot 5}\right)}$

$=\dfrac{\left(\frac{3}{8}+\frac{2}{8}\right)}{\left(\frac{16}{20}+\frac{15}{20}\right)}$

$=\dfrac{\frac{5}{8}}{\frac{31}{20}}$

$=\dfrac{5}{8}\div\dfrac{31}{20}$

$=\dfrac{5}{8}\cdot\dfrac{20}{31}$

$=\dfrac{5\cdot \not{4}\cdot 5}{\not{4}\cdot 2\cdot 31}$

$=\dfrac{25}{62}$

41. $\dfrac{11+(3)^2}{\frac{2}{3}}=\dfrac{[11+(3)^2]}{\left(\frac{2}{3}\right)}$

$=\dfrac{(11+9)}{\left(\frac{2}{3}\right)}$

$=\dfrac{20}{\left(\frac{2}{3}\right)}$

$=20\div\dfrac{2}{3}$

$=20\cdot\dfrac{3}{2}$

$=\dfrac{\not{2}\cdot 10\cdot 3}{\not{2}}$

$=30$

43. $\dfrac{2\frac{3}{4}\text{ cups of flour}}{20\text{ people}}=\dfrac{x\text{ cups of flour}}{36\text{ people}}$

$36\cdot 2\dfrac{3}{4}=20\cdot x$

$36\cdot\dfrac{11}{4}=20x$

$\dfrac{9\cdot \not{4}\cdot 11}{\not{4}}=20x$

$99=20x$

$\dfrac{99}{20}=\dfrac{20x}{20}$

$4\dfrac{19}{20}=x$

You need $4\dfrac{19}{20}$ cups of sugar to make the recipe for 36 people.

45. $\dfrac{3\frac{3}{4}\text{ bags}}{2\text{ lawns}}=\dfrac{x\text{ bags}}{12\text{ lawns}}$

$12\cdot 3\dfrac{3}{4}=2\cdot x$

$12\cdot\dfrac{15}{4}=2x$

$\dfrac{3\cdot \not{4}\cdot 15}{\not{4}}=2x$

$45=2x$

$\dfrac{45}{2}=\dfrac{2x}{2}$

$22\dfrac{1}{2}=x$

The gardener needs $22\dfrac{1}{2}$ bags for 12 lawns.

47. $\dfrac{2\frac{1}{4}\text{ acres}}{3\text{ houses}}=\dfrac{x\text{ acres}}{28\text{ houses}}$

$28\cdot 2\dfrac{1}{4}=3\cdot x$

$28\cdot\dfrac{9}{4}=3x$

$\dfrac{7\cdot \not{4}\cdot 9}{\not{4}}=3x$

$63=3x$

$\dfrac{63}{3}=\dfrac{3x}{3}$

$21=x$

The developer needs 21 acres to build 28 homes.

Copyright © 2012 Pearson Education, Inc.

49. $\dfrac{\frac{25xy^2}{49}}{\frac{15x^2y}{14}} = \dfrac{25xy^2}{49} \div \dfrac{15x^2y}{14}$

$= \dfrac{25xy^2}{49} \cdot \dfrac{14}{15x^2y}$

$= \dfrac{\cancel{5} \cdot 5 \cdot \cancel{x} \cdot \cancel{y} \cdot y \cdot 2 \cdot \cancel{7}}{\cancel{7} \cdot 7 \cdot 3 \cdot \cancel{5} \cdot \cancel{x} \cdot x \cdot \cancel{y}}$

$= \dfrac{10y}{21x}$

51. $\dfrac{\frac{1}{x} - \frac{1}{y}}{\frac{1}{x} + \frac{1}{y}} = \dfrac{\left(\frac{1}{x} - \frac{1}{y}\right)}{\left(\frac{1}{x} + \frac{1}{y}\right)}$

$= \dfrac{\left(\frac{1 \cdot y}{x \cdot y} - \frac{1 \cdot x}{y \cdot x}\right)}{\left(\frac{1 \cdot y}{x \cdot y} + \frac{1 \cdot x}{y \cdot x}\right)}$

$= \dfrac{\left(\frac{y}{xy} - \frac{x}{xy}\right)}{\left(\frac{y}{xy} + \frac{x}{xy}\right)}$

$= \dfrac{\left(\frac{y-x}{xy}\right)}{\left(\frac{y+x}{xy}\right)}$

$= \dfrac{y-x}{xy} \cdot \dfrac{xy}{y+x}$

$= \dfrac{(y-x) \cdot \cancel{x} \cdot \cancel{y}}{\cancel{x} \cdot \cancel{y} \cdot (y+x)}$

$= \dfrac{y-x}{y+x}$

53. Replace x with $\frac{7}{2}$.

$x - \dfrac{2}{5} \div \dfrac{4}{15} = \dfrac{7}{2} - \dfrac{2}{5} \div \dfrac{4}{15}$

$= \dfrac{7}{2} - \dfrac{2}{5} \cdot \dfrac{15}{4}$

$= \dfrac{7}{2} - \dfrac{\cancel{2} \cdot 3 \cdot \cancel{5}}{\cancel{5} \cdot \cancel{2} \cdot 2}$

$= \dfrac{7}{2} - \dfrac{3}{2}$

$= \dfrac{4}{2}$

$= 2$

55. Replace x with $-\frac{4}{7}$.

$-\dfrac{3}{8} \cdot \dfrac{16}{21} + x = -\dfrac{3}{8} \cdot \dfrac{16}{21} + (x)$

$= -\dfrac{3}{8} \cdot \dfrac{16}{21} + \left(-\dfrac{4}{7}\right)$

$= -\dfrac{\cancel{3} \cdot 2 \cdot \cancel{8}}{\cancel{8} \cdot \cancel{3} \cdot 7} + \left(-\dfrac{4}{7}\right)$

$= -\dfrac{2}{7} + \left(-\dfrac{4}{7}\right)$

$= -\dfrac{6}{7}$

Cumulative Review

57. $(-2)(3)(-1)(-5) = -(2)(3)(1)(5) = -(6)(5) = -30$

58. $(-1)(-3)(-1)(-5) = (1)(3)(1)(5) = (3)(5) = 15$

59. $-50 \div 5 = -10$

60. $-36 \div (-6) = 6$

Quick Quiz 5.5

1. $\left(\dfrac{1}{3}\right)^2 + \dfrac{1}{3} \div \dfrac{5}{6} = \dfrac{1}{9} + \dfrac{1}{3} \div \dfrac{5}{6}$

$= \dfrac{1}{9} + \dfrac{1}{3} \cdot \dfrac{6}{5}$

$= \dfrac{1}{9} + \dfrac{6}{15}$

$= \dfrac{1 \cdot 5}{9 \cdot 5} + \dfrac{6 \cdot 3}{15 \cdot 3}$

$= \dfrac{5}{45} + \dfrac{18}{45}$

$= \dfrac{23}{45}$

2. $\dfrac{\frac{a}{15}}{\frac{a}{10}} = \dfrac{a}{15} \div \dfrac{a}{10} = \dfrac{a}{15} \cdot \dfrac{10}{a} = \dfrac{\cancel{a} \cdot 2 \cdot \cancel{5}}{3 \cdot \cancel{5} \cdot \cancel{a}} = \dfrac{2}{3}$

Copyright © 2012 Pearson Education, Inc.

3. $\frac{\frac{8}{15}-\frac{1}{5}}{\frac{1}{2}+\frac{1}{3}}=\frac{\left(\frac{8}{15}-\frac{1}{5}\right)}{\left(\frac{1}{2}+\frac{1}{3}\right)}$

$=\frac{\left(\frac{8}{15}-\frac{1\cdot 3}{5\cdot 3}\right)}{\left(\frac{1\cdot 3}{2\cdot 3}+\frac{1\cdot 2}{3\cdot 2}\right)}$

$=\frac{\left(\frac{8}{15}-\frac{3}{15}\right)}{\left(\frac{3}{6}+\frac{2}{6}\right)}$

$=\frac{\frac{5}{15}}{\frac{5}{6}}$

$=\frac{5}{15}\div\frac{5}{6}$

$=\frac{5}{15}\cdot\frac{6}{5}$

$=\frac{\cancel{5}\cdot 2\cdot\cancel{3}}{\cancel{3}\cdot 5\cdot\cancel{5}}$

$=\frac{2}{5}$

4. Answers may vary. One possible solution is to first multiply both the numerator and the denominator by 2, eliminating the denominator.

$\frac{1+2\times 3}{\frac{1}{2}}=2(1+2\times 3)$

Next, perform the multiplication operation inside the parentheses, then add that product to 1. Multiply that sum by the 2 outside the parentheses. The expression is completely simplified.

5.6 Exercises

1. At the beginning of week 4, Joan's workout will be the original $\frac{1}{2}$ increased three times by $\frac{1}{4}$ hour.

$\frac{1}{2}+3\cdot\frac{1}{4}=\frac{1}{2}+\frac{3}{4}=\frac{2}{4}+\frac{3}{4}=\frac{5}{4}=1\frac{1}{4}$

Her workout will be $1\frac{1}{4}$ hours.

3. We divide $305\div 4\frac{1}{2}$ to find the unit rate.

$305\div 4\frac{1}{2}=305\div\frac{9}{2}=305\cdot\frac{2}{9}=\frac{610}{9}=67\frac{7}{9}$

Andrew averaged $67\frac{7}{9}$ miles per hour.

5. Water: $6\cdot 1\frac{1}{4}=6\cdot\frac{5}{4}=\frac{\cancel{2}\cdot 3\cdot 5}{\cancel{2}\cdot 2}=\frac{15}{2}=7\frac{1}{2}$ cups

Salt: $6\cdot\frac{1}{8}=\frac{\cancel{2}\cdot 3\cdot 1}{\cancel{2}\cdot 2\cdot 2}=\frac{3}{4}$ tsp

Cereal: $6\cdot\frac{1}{4}=\frac{\cancel{2}\cdot 3\cdot 1}{\cancel{2}\cdot 2}=\frac{3}{2}=1\frac{1}{2}$ cups

7. Each long skirt in a size 8 requires $2\frac{3}{4}$ yards of 45-inch wide fabric.

$5\cdot 2\frac{3}{4}=5\cdot\frac{11}{4}=\frac{55}{4}=13\frac{3}{4}$ yd

You need $13\frac{3}{4}$ yards to make five skirts.

9. a. We subtract.

$$\begin{array}{rcr} 15\frac{2\cdot 2}{3\cdot 2} & = & 15\frac{4}{6} \\ -\,9\frac{1\cdot 3}{2\cdot 3} & = & -\,9\frac{3}{6} \\ \hline & & 6\frac{1}{6}\text{ gal} \end{array}$$

John used $6\frac{1}{6}$ gallons of gasoline.

b. We multiply.

$24\cdot 6\frac{1}{6}=24\cdot\frac{37}{6}=\frac{4\cdot\cancel{6}\cdot 37}{\cancel{6}}=148$ mi

John drove 148 miles.

11. Vacation: $\frac{1}{4}\cdot 2400=\$600$

Pay off bills: $\frac{1}{8}\cdot 2400=\$300$

Savings: $2400 – $600 – $300 = $1500

The Tran family will put $1500 in the savings account.

Copyright © 2012 Pearson Education, Inc.

13. a. The length of the flower bed is the length of the lawn less twice the width of the grass area.

$$\begin{aligned} 40-2\cdot 4\frac{3}{4} &= 40-2\cdot\frac{19}{4} \\ &= 40-\frac{19}{2} \\ &= \frac{80}{2}-\frac{19}{2} \\ &= \frac{61}{2} \\ &= 30\frac{1}{2}\text{ ft} \end{aligned}$$

The width of the flower bed is the width of the lawn less twice the width of the grass area.

$$\begin{aligned} 25-2\cdot 4\frac{3}{4} &= 25-2\cdot\frac{19}{4} \\ &= 25-\frac{19}{2} \\ &= \frac{50}{2}-\frac{19}{2} \\ &= \frac{31}{2} \\ &= 15\frac{1}{2}\text{ ft} \end{aligned}$$

The flower bed is $30\frac{1}{2}$ feet by $15\frac{1}{2}$ feet.

b. Find the perimeter of the flower bed.

$$\begin{aligned} P &= 2L+2W \\ &= 2\left(30\frac{1}{2}\right)+2\left(15\frac{1}{2}\right) \\ &= 2\left(\frac{61}{2}\right)+2\left(\frac{31}{2}\right) \\ &= 61+31 \\ &= 92\text{ ft} \end{aligned}$$

Multiply to find the cost.

$$92\cdot 2\frac{1}{4}=92\cdot\frac{9}{4}=\frac{\cancel{4}\cdot 23\cdot 9}{\cancel{4}}=\$207$$

The fence will cost $207.

15. We multiply the number of shelves per bookcase by the number of bookcases to find the number of shelves.

3 · 4 = 12 shelves

We divide the length of one board by the length of one shelf to determine how many shelves can be cut from one board.

$$8\div 3\frac{3}{4}=8\div\frac{15}{4}=8\cdot\frac{4}{15}=\frac{32}{15}=2\frac{2}{15}\text{ shelves}$$

Thus 2 whole shelves can be cut from each board. Brenda should buy 12 ÷ 2 or 6 boards.

17. a. The length of the pool is the length of the yard less twice the width of the tile area.

$$\begin{aligned} 50-2\cdot 5\frac{1}{2} &= 50-2\cdot\frac{11}{2} \\ &= 50-\frac{22}{2} \\ &= 50-11 \\ &= 39\text{ ft} \end{aligned}$$

The width of the pool is the width of the yard less twice the width of the tile area.

$$\begin{aligned} 30-2\cdot 5\frac{1}{2} &= 30-2\cdot\frac{11}{2} \\ &= 30-\frac{22}{2} \\ &= 30-11 \\ &= 19\text{ ft} \end{aligned}$$

The pool is 39 feet by 19 feet.

b. Find the perimeter of the pool.

$$\begin{aligned} P &= 2L+2W \\ &= 2(39)+2(19) \\ &= 78+38 \\ &= 116\text{ ft} \end{aligned}$$

Multiply to find the cost.

$$116\cdot 2\frac{1}{2}=116\cdot\frac{5}{2}=\frac{58\cdot\cancel{2}\cdot 5}{\cancel{2}}=\$290$$

The tile will cost $290.

19. We divide the length of one piece of wood by the length of one post to determine how many posts can be cut from one piece of wood.

$$10\div 2\frac{3}{4}=10\div\frac{11}{4}=10\cdot\frac{4}{11}=\frac{40}{11}=3\frac{7}{11}\text{ posts}$$

Thus 3 whole posts can be cut from each piece of wood.

We divide the number of posts by 3 to determine the number of pieces of wood.

$$56\div 3=\frac{56}{3}=18\frac{2}{3}$$

Since she must buy whole pieces of wood, Shannon needs 19 pieces of wood.

Copyright © 2012 Pearson Education, Inc.

21. a. We divide the distance by the time.

$$240,000 \div 4\frac{1}{2} = 240,000 \div \frac{9}{2} = 240,000 \cdot \frac{2}{9} = \frac{80,000 \cdot \not{3} \cdot 2}{3 \cdot \not{3}} = \frac{160,000}{3} = 53,333\frac{1}{3} \text{ mi per day}$$

b. We divide the daily rate by 24 to find the hourly rate.

$$53,333\frac{1}{3} \div 24 = \frac{160,000}{3} \div 24 = \frac{160,000}{3} \cdot \frac{1}{24} = \frac{20,000 \cdot \not{8} \cdot 1}{3 \cdot 3 \cdot \not{8}} = \frac{20,000}{9} = 2222\frac{2}{9} \text{ mph}$$

Cumulative Review

23. $3x = 12$

$\frac{3x}{3} = \frac{12}{3}$

$x = 4$

Check: $3x = 12$

$3(4) \stackrel{?}{=} 12$

$12 = 12$ ✓

24. $5x = 45$

$\frac{5x}{5} = \frac{45}{5}$

$x = 9$

Check: $5x = 45$

$5(9) \stackrel{?}{=} 45$

$45 = 45$ ✓

25. $x - 5 = 12$

$x - 5 + 5 = 12 + 5$

$x = 17$

Check: $x - 5 = 12$

$17 - 5 \stackrel{?}{=} 12$

$12 = 12$ ✓

26. $x + 3 = -1$

$x + 3 + (-3) = -1 + (-3)$

$x = -4$

Check: $x + 3 = -1$

$-4 + 3 \stackrel{?}{=} -1$

$-1 = -1$ ✓

Quick Quiz 5.6

1. We multiply the number of shelves per bookcase by the number of bookcases to find the number of shelves.

$3 \cdot 5 = 15$ shelves

We divide the length of one board by the length of one shelf to determine how many shelves can be cut from one board.

$$12 \div 3\frac{1}{3} = 12 \div \frac{10}{3} = 12 \cdot \frac{3}{10} = \frac{6 \cdot \not{2} \cdot 3}{\not{2} \cdot 5} = \frac{18}{5} = 3\frac{3}{5} \text{ shelves}$$

Thus 3 whole shelves can be cut from each board. Alex should purchase 15 ÷ 3 or 5 boards.

2. a. August 1: 134 lb

End of week 1:

$$134 + 2\frac{1}{4} = 136\frac{1}{4} \text{ lb}$$

End of week 2:

$$136\frac{1 \cdot 2}{4 \cdot 2} = 136\frac{2}{8} = 135\frac{10}{8}$$

$$-\ 2\frac{3}{8} = -\ 2\frac{3}{8} = -\ 2\frac{3}{8}$$

$$133\frac{7}{8} \text{ lb}$$

Copyright © 2012 Pearson Education, Inc.

End of week 3:

$$\begin{array}{rcl} 133\frac{7}{8} & = & 133\frac{7}{8} \\ +\ 1\frac{1\cdot 4}{2\cdot 4} & = & +\ 1\frac{4}{8} \\ \hline & & 134\frac{11}{8} = 135\frac{3}{8}\text{ lb} \end{array}$$

Ella weighed $135\frac{3}{8}$ pounds at the end of week 3.

b. We subtract.

$$\begin{array}{r} 135\frac{3}{8} \\ -\ 134 \\ \hline 1\frac{3}{8}\text{ lb} \end{array}$$

Ella weighed $1\frac{3}{8}$ pounds more at the end of week 3 than she weighed on August 1.

3. a. The length of the pavilion is the length of the yard less twice the width of the grass area.

$$\begin{aligned} 32\frac{1}{2}-2\cdot 5\frac{1}{2} &= 32\frac{1}{2}-2\cdot\frac{11}{2} \\ &= 32\frac{1}{2}-11 \\ &= 21\frac{1}{2}\text{ ft} \end{aligned}$$

The width of the pavilion is the width of the yard less twice the width of the grass area.

$$28-2\cdot 5\frac{1}{2}=28-2\cdot\frac{11}{2}=28-11=17\text{ ft}$$

The pavilion is $21\frac{1}{2}$ feet by 17 feet.

b. Find the perimeter of the pavilion.

$$\begin{aligned} P &= 2L+2W \\ &= 2\left(21\frac{1}{2}\right)+2(17) \\ &= 43+34 \\ &= 77\text{ ft} \end{aligned}$$

Multiply to find the cost.

$77\cdot 2=\$154$

The tile will cost \$154.

4. a. Subtract

b. Multiply

c. Divide

5.7 Exercises

1. When you multiply a nonzero fraction by its reciprocal, the product is <u>1</u>.

3. To solve $\frac{x}{6}=2$, we <u>multiply</u> by 6 on both sides of the equation.

5.
$$\begin{aligned} \frac{x}{12} &= -3 \\ \frac{\boxed{12}\cdot x}{12} &= -3\cdot\boxed{12} \\ x &= \boxed{-36} \end{aligned}$$

7.
$$\begin{aligned} \frac{x}{-5} &= 4 \\ \frac{\boxed{-5}\cdot x}{-5} &= 4\cdot\boxed{-5} \\ x &= \boxed{-20} \end{aligned}$$

9.
$$\begin{aligned} \frac{2}{5}x &= -8 \\ \frac{\boxed{5}}{\boxed{2}}\cdot\frac{2}{5}x &= -8\cdot\frac{\boxed{5}}{\boxed{2}} \\ x &= \boxed{-20} \end{aligned}$$

11.
$$\begin{aligned} \frac{-5}{7}x &= 10 \\ \frac{\boxed{7}}{\boxed{-5}}\cdot\frac{(-5)}{7}x &= 10\cdot\frac{\boxed{7}}{\boxed{-5}} \\ x &= \boxed{-14} \end{aligned}$$

13.
$$\begin{aligned} \frac{y}{8} &= 12 \\ \frac{8\cdot y}{8} &= 12\cdot 8 \\ y &= 96 \end{aligned}$$

Check:
$$\begin{aligned} \frac{y}{8} &= 12 \\ \frac{96}{8} &\stackrel{?}{=} 12 \\ 12 &= 12\ \checkmark \end{aligned}$$

15. $\frac{x}{7} = 31$

$\frac{7 \cdot x}{7} = 31 \cdot 7$

$x = 217$

Check: $\frac{x}{7} = 31$

$\frac{217}{7} \stackrel{?}{=} 31$

$31 = 31$ ✓

17. $\frac{m}{13} = -30$

$\frac{13 \cdot m}{13} = -30 \cdot 13$

$m = -390$

Check: $\frac{m}{13} = -30$

$\frac{-390}{13} \stackrel{?}{=} -30$

$-30 = -30$ ✓

19. $\frac{x}{15} = -5$

$\frac{15 \cdot x}{15} = -5 \cdot 15$

$x = -75$

Check: $\frac{x}{15} = -5$

$\frac{-75}{15} \stackrel{?}{=} -5$

$-5 = -5$ ✓

21. $-15 = \frac{a}{4}$

$-15 \cdot 4 = \frac{4 \cdot a}{4}$

$-60 = a$

Check: $-15 = \frac{a}{4}$

$-15 \stackrel{?}{=} \frac{-60}{4}$

$-15 = -15$ ✓

23. $-4 = \frac{a}{-20}$

$-4 \cdot (-20) = \frac{-20 \cdot a}{-20}$

$80 = a$

Check: $-4 = \frac{a}{-20}$

$-4 \stackrel{?}{=} \frac{80}{-20}$

$-4 = -4$ ✓

25. $\frac{x}{3^2} = 2 + 6 \div 3$

$\frac{x}{9} = 2 + 2$

$\frac{x}{9} = 4$

$\frac{9 \cdot x}{9} = 4 \cdot 9$

$x = 36$

Check: $\frac{x}{3^2} = 2 + 6 \div 3$

$\frac{36}{3^2} \stackrel{?}{=} 2 + 6 \div 3$

$\frac{36}{9} \stackrel{?}{=} 2 + 2$

$4 = 4$ ✓

27. $\frac{x}{8} = \frac{1}{4} + \frac{5}{8}$

$\frac{x}{8} = \frac{2}{8} + \frac{5}{8}$

$\frac{x}{8} = \frac{7}{8}$

$\frac{8 \cdot x}{8} = \frac{8 \cdot 7}{8}$

$x = 7$

Check: $\frac{x}{8} = \frac{1}{4} + \frac{5}{8}$

$\frac{7}{8} \stackrel{?}{=} \frac{1}{4} + \frac{5}{8}$

$\frac{7}{8} = \frac{7}{8}$ ✓

29. $\frac{y}{2^3} = \frac{1}{8} + \frac{1}{2}$

$\frac{y}{8} = \frac{1}{8} + \frac{4}{8}$

$\frac{y}{8} = \frac{5}{8}$

$\frac{8 \cdot y}{8} = \frac{8 \cdot 5}{8}$

$y = 5$

Copyright © 2012 Pearson Education, Inc.

Check: $\frac{y}{2^3} = \frac{1}{8} + \frac{1}{2}$

$\frac{5}{2^3} \stackrel{?}{=} \frac{1}{8} + \frac{1}{2}$

$\frac{5}{8} = \frac{5}{8}$ ✓

31. $\frac{3}{5}y = 12$

$\left(\frac{5}{3}\right)\left(\frac{3}{5}\right)y = 12\left(\frac{5}{3}\right)$

$1y = \frac{4 \cdot \cancel{3} \cdot 5}{\cancel{3}}$

$y = 20$

Check: $\frac{3}{5}y = 12$

$\frac{3}{5} \cdot 20 \stackrel{?}{=} 12$

$\frac{3 \cdot 4 \cdot \cancel{5}}{\cancel{5}} \stackrel{?}{=} 12$

$12 = 12$ ✓

33. $\frac{4}{9}x = 12$

$\left(\frac{9}{4}\right)\left(\frac{4}{9}\right)x = 12\left(\frac{9}{4}\right)$

$1x = \frac{3 \cdot \cancel{4} \cdot 9}{\cancel{4}}$

$x = 27$

Check: $\frac{4}{9}x = 12$

$\frac{4}{9} \cdot 27 \stackrel{?}{=} 12$

$\frac{4 \cdot 3 \cdot \cancel{9}}{\cancel{9}} \stackrel{?}{=} 12$

$12 = 12$ ✓

35. $\frac{1}{2}x = -15$

$2\left(\frac{1}{2}\right)x = -15(2)$

$1x = -30$

$x = -30$

Check: $\frac{1}{2}x = -15$

$\frac{1}{2}(-30) \stackrel{?}{=} -15$

$-\frac{\cancel{2} \cdot 15}{\cancel{2}} \stackrel{?}{=} -15$

$-15 = -15$ ✓

37. $\frac{-5}{8}x = 30$

$\left(-\frac{8}{5}\right)\left(-\frac{5}{8}\right)x = 30\left(-\frac{8}{5}\right)$

$1x = -\frac{\cancel{5} \cdot 6 \cdot 8}{\cancel{5}}$

$x = -48$

Check: $\frac{-5}{8}x = 30$

$\frac{-5}{8}(-48) \stackrel{?}{=} 30$

$30 = 30$ ✓

39. $3 = \frac{a}{11}$

$3 \cdot 11 = \frac{11 \cdot a}{11}$

$33 = a$

Check: $3 = \frac{a}{11}$

$3 \stackrel{?}{=} \frac{33}{11}$

$3 = 3$ ✓

41. $\frac{x}{3^2} = 4 + 6 \div 2$

$\frac{x}{9} = 4 + 3$

$\frac{x}{9} = 7$

$\frac{9 \cdot x}{9} = 7 \cdot 9$

$x = 63$

Check: $\frac{x}{3^2} = 4 + 6 \div 2$

$\frac{63}{3^2} \stackrel{?}{=} 4 + 6 \div 2$

$\frac{63}{9} \stackrel{?}{=} 4 + 3$

$7 = 7$ ✓

Copyright © 2012 Pearson Education, Inc.

43.

$$\frac{-3}{4}y = -12$$
$$\left(-\frac{4}{3}\right)\left(-\frac{3}{4}\right)y = -12\left(-\frac{4}{3}\right)$$
$$1y = \frac{4\cdot \cancel{3}\cdot 4}{\cancel{3}}$$
$$y = 16$$

Check:
$$\frac{-3}{4}y = -12$$
$$\frac{-3}{4}\cdot 16 \stackrel{?}{=} -12$$
$$-\frac{3\cdot \cancel{4}\cdot 4}{\cancel{4}} \stackrel{?}{=} -12$$
$$-12 = -12\ ✓$$

45.

$$-1 = \frac{m}{30}$$
$$-1\cdot 30 = \frac{30\cdot m}{30}$$
$$-30 = m$$

Check:
$$-1 = \frac{m}{30}$$
$$-1 \stackrel{?}{=} \frac{-30}{30}$$
$$-1 = -1\ ✓$$

47. a.

$$4x = 52$$
$$\frac{4x}{4} = \frac{52}{4}$$
$$x = 13$$

Check:
$$4x = 52$$
$$4(13) \stackrel{?}{=} 52$$
$$52 = 52\ ✓$$

b.

$$\frac{x}{4} = 52$$
$$\frac{4\cdot x}{4} = 52\cdot 4$$
$$x = 208$$

Check:
$$\frac{x}{4} = 2$$
$$\frac{208}{4} \stackrel{?}{=} 52$$
$$52 = 52\ ✓$$

49. a.

$$4x = -52$$
$$\frac{4x}{4} = \frac{-52}{4}$$
$$x = -13$$

Check:
$$4x = -52$$
$$4(-13) \stackrel{?}{=} -52$$
$$-52 = -52\ ✓$$

b.

$$\frac{x}{4} = -52$$
$$\frac{4\cdot x}{4} = -52\cdot 4$$
$$x = -208$$

Check:
$$\frac{x}{4} = -52$$
$$\frac{-208}{4} \stackrel{?}{=} -52$$
$$-52 = -52\ ✓$$

51. a.

$$x - 7 = 21$$
$$x - 7 + 7 = 21 + 7$$
$$x = 28$$

Check:
$$x - 7 = 21$$
$$28 - 7 \stackrel{?}{=} 21$$
$$21 = 21\ ✓$$

b.

$$x + 12 = 33$$
$$x + 12 + (-12) = 33 + (-12)$$
$$x = 21$$

Check:
$$x + 12 = 33$$
$$21 + 12 \stackrel{?}{=} 33$$
$$33 = 33\ ✓$$

c.

$$5x = 3$$
$$\frac{5x}{5} = \frac{3}{5}$$
$$x = \frac{3}{5}$$

Check:
$$5x = 3$$
$$5\left(\frac{3}{5}\right) \stackrel{?}{=} 3$$
$$3 = 3\ ✓$$

Copyright © 2012 Pearson Education, Inc.

d. $\frac{x}{5} = 11$
$\frac{5 \cdot x}{5} = 11 \cdot 5$
$x = 55$

Check: $\frac{x}{5} = 11$
$\frac{55}{5} \stackrel{?}{=} 11$
$11 = 11$ ✓

53. a. $x - 7 = -12$
$x - 7 + 7 = -12 + 7$
$x = -5$

Check: $x - 7 = -12$
$-5 - 7 \stackrel{?}{=} -12$
$-12 = -12$ ✓

b. $3x = -2$
$\frac{3x}{3} = \frac{-2}{3}$
$x = -\frac{2}{3}$

Check: $3x = -2$
$3\left(-\frac{2}{3}\right) \stackrel{?}{=} -2$
$-2 = -2$ ✓

c. $\frac{x}{6} = -9$
$\frac{6 \cdot x}{6} = -9 \cdot 6$
$x = -54$

Check: $\frac{x}{6} = -9$
$\frac{-54}{6} \stackrel{?}{=} -9$
$-9 = -9$ ✓

d. $x + 11 = -34$
$x + 11 + (-11) = -34 + (-11)$
$x = -45$

Check: $x + 11 = -34$
$-45 + 11 \stackrel{?}{=} -34$
$-34 = -34$ ✓

55. Since $\frac{3}{2}$ is the reciprocal of $\frac{2}{3}$, the fraction $\frac{a}{b}$ is $\frac{2}{3}$.

57. Since $\frac{7}{3}$ is the reciprocal of $\frac{3}{7}$, the fraction $\frac{a}{b}$ is $\frac{3}{7}$.

Cumulative Review

59. $3a + 7b - 9a - 10ab + 2b$
$= (3 - 9)a + (7 + 2)b - 10ab$
$= -6a + 9b - 10ab$

60. $-2x + 3xy - 4y - 6x - 5xy$
$= (-2 - 6)x + (3 - 5)xy - 4y$
$= -8x - 2xy - 4y$

61. $3x^2(x^4 + 8x) = 3x^2(x^4) + 3x^2(8x)$
$= 3x^6 + 24x^3$

62. $-2x^3(x^2 - 6) = -2x^3(x^2) - (-2x^3)(6)$
$= -2x^5 + 12x^3$

63. $A = LW$
$= \left(4\frac{2}{3}\right)\left(3\frac{3}{8}\right)$
$= \left(\frac{14}{3}\right)\left(\frac{27}{8}\right)$
$= \frac{\cancel{2} \cdot 7 \cdot \cancel{3} \cdot 9}{\cancel{3} \cdot \cancel{2} \cdot 4}$
$= \frac{63}{4}$
$= 15\frac{3}{4}\text{ ft}^2$

The area of the stained glass window is $15\frac{3}{4}$ square feet.

64. Multiply the amount of flour in one loaf by 4 to find the amount of flour in 4 loaves.
$4 \cdot 2\frac{1}{2} = 4 \cdot \frac{5}{2} = \frac{2 \cdot \cancel{2} \cdot 5}{\cancel{2}} = 10$ cups

Juan needs 10 cups of flour for 4 loaves of bread.

Copyright © 2012 Pearson Education, Inc.

Quick Quiz 5.7

1. $\frac{x}{5} = 2 - 3^2$
$\frac{x}{5} = 2 - 9$
$\frac{x}{5} = -7$
$\frac{5 \cdot x}{5} = -7 \cdot 5$
$x = -35$

Check: $\frac{x}{5} = 2 - 3^2$
$\frac{-35}{5} \stackrel{?}{=} 2 - 3^2$
$-7 \stackrel{?}{=} 2 - 9$
$-7 = -7$ ✓

2. $\frac{3}{5}a = 18$
$\left(\frac{5}{3}\right)\left(\frac{3}{5}\right)a = 18\left(\frac{5}{3}\right)$
$1a = \frac{\cancel{3} \cdot 6 \cdot 5}{\cancel{3}}$
$a = 30$

Check: $\frac{35}{5}a = 18$
$\frac{3}{5}(30) \stackrel{?}{=} 18$
$\frac{3 \cdot \cancel{5} \cdot 6}{\cancel{5}} \stackrel{?}{=} 18$
$18 = 18$ ✓

3. $\frac{y}{8} = -9$
$\frac{8 \cdot y}{8} = -9 \cdot 8$
$y = -72$

Check: $\frac{y}{8} = -9$
$\frac{-72}{8} \stackrel{?}{=} -9$
$-9 = -9$ ✓

4. Answers may vary. One possible solution is that Amy intended to eliminate the denominator. Amy failed, however, to include the sign of the 5, and this error yielded an answer with the right absolute value, but the wrong sign.

You Try It

1. $\frac{18}{50} \cdot \left(-\frac{20}{21}\right) = -\frac{\cancel{3} \cdot 6 \cdot 2 \cdot \cancel{10}}{5 \cdot \cancel{10} \cdot \cancel{3} \cdot 7} = -\frac{12}{35}$

2. $15x^3 \div \frac{10x^4}{7} = \frac{15x^3}{1} \cdot \frac{7}{10x^4}$
$= \frac{3 \cdot \cancel{5} \cdot \cancel{x} \cdot \cancel{x} \cdot \cancel{x} \cdot 7}{2 \cdot \cancel{5} \cdot \cancel{x} \cdot \cancel{x} \cdot \cancel{x} \cdot x}$
$= \frac{21}{2x}$

3. $20 = 2 \cdot 2 \cdot 5$
$30 = 2 \cdot 3 \cdot 5$
$8 = 2 \cdot 2 \cdot 2$
$\text{LCD} = 2 \cdot 2 \cdot 2 \cdot 3 \cdot 5 = 120$

4. **a.** $\frac{4}{13} - \frac{1}{13} = \frac{4-1}{13} = \frac{3}{13}$

b. $\frac{3}{y} + \frac{6}{y} = \frac{3+6}{y} = \frac{9}{y}$

5. LCD = 90
$\frac{5}{6} + \frac{1}{9} + \frac{7}{15} = \frac{5 \cdot 15}{6 \cdot 15} + \frac{1 \cdot 10}{9 \cdot 10} + \frac{7 \cdot 6}{15 \cdot 6}$
$= \frac{75}{90} + \frac{10}{90} + \frac{42}{90}$
$= \frac{75 + 10 + 42}{90}$
$= \frac{127}{90}$ or $1\frac{37}{90}$

6.
$$\begin{array}{rcr} 8\frac{1 \cdot 3}{3 \cdot 3} & = & 8\frac{3}{9} \\ +\,3\ \frac{8}{9} & = & +\,3\frac{8}{9} \\ \hline & & 11\frac{11}{9} = 12\frac{2}{9} \end{array}$$

7.
$$\begin{array}{rcrcr} 9\frac{2 \cdot 4}{7 \cdot 4} & = & 9\frac{8}{28} & = & 8\frac{36}{28} \\ -\,4\frac{3 \cdot 7}{4 \cdot 7} & = & -\,4\frac{21}{28} & = & -\,4\frac{21}{28} \\ \hline & & & & 4\frac{15}{28} \end{array}$$

Copyright © 2012 Pearson Education, Inc.

8. a. $4\frac{2}{5} \div 2\frac{2}{15} = \frac{22}{5} \div \frac{32}{15}$
$= \frac{22}{5} \cdot \frac{15}{32}$
$= \frac{\cancel{2} \cdot 11 \cdot 3 \cdot \cancel{5}}{\cancel{5} \cdot \cancel{2} \cdot 16}$
$= \frac{33}{16} \text{ or } 2\frac{1}{16}$

b. $5 \cdot 3\frac{1}{3} = \frac{5}{1} \cdot \frac{10}{3} = \frac{5 \cdot 10}{1 \cdot 3} = \frac{50}{3} = 16\frac{2}{3}$

9. $\left(\frac{1}{3} + \frac{2}{7}\right) \cdot \frac{3}{4} - \frac{1}{4} = \left(\frac{7}{21} + \frac{6}{21}\right) \cdot \frac{3}{4} - \frac{1}{4}$
$= \frac{13}{21} \cdot \frac{3}{4} - \frac{1}{4}$
$= \frac{13 \cdot \cancel{3}}{\cancel{3} \cdot 7 \cdot 4} - \frac{1}{4}$
$= \frac{13}{28} - \frac{7}{28}$
$= \frac{6}{28}$
$= \frac{\cancel{2} \cdot 3}{\cancel{2} \cdot 14}$
$= \frac{3}{14}$

10. $\frac{\frac{1}{7} + \frac{9}{14}}{\frac{3}{4} - \frac{1}{2}} = \frac{\frac{2}{14} + \frac{9}{14}}{\frac{3}{4} - \frac{2}{4}}$
$= \frac{\frac{11}{14}}{\frac{1}{4}}$
$= \frac{11}{14} \div \frac{1}{4}$
$= \frac{11}{14} \cdot \frac{4}{1}$
$= \frac{11 \cdot \cancel{2} \cdot 2}{\cancel{2} \cdot 7 \cdot 1}$
$= \frac{22}{7} \text{ or } 3\frac{1}{7}$

11. $\frac{x}{-10} = \frac{1}{3} + \frac{1}{6}$
$\frac{x}{-10} = \frac{2}{6} + \frac{1}{6}$
$\frac{x}{-10} = \frac{3}{6}$
$\frac{-10 \cdot x}{-10} = \frac{-10 \cdot 3}{6}$
$x = -\frac{\cancel{2} \cdot 5 \cdot \cancel{3}}{\cancel{2} \cdot \cancel{3}}$
$x = -\frac{5}{1}$
$x = -5$

Chapter 5 Review Problems

1. LCM: The least common multiple is the smallest common multiple

2. LCD: The least common denominator is the LCM of the denominators

3. Complex Fraction: A fraction that contains at least one fraction in the numerator or denominator

4. **a.** To find the reciprocal, we invert the fraction.
$\frac{4}{9} \to \frac{9}{4}$

b. To find the reciprocal, we write −6 as a fraction and invert the fraction.
$-6 = \frac{-6}{1} \to \frac{1}{-6} = -\frac{1}{6}$

5. **a.** To find the reciprocal, we write 7 as a fraction and then invert the fraction.
$7 = \frac{7}{1} \to \frac{1}{7}$

b. To find the reciprocal, we invert the fraction.
$\frac{1}{11} \to \frac{11}{1} = 11$

6. $\frac{3}{4} \text{ of } \frac{-8}{9} = \frac{3}{4} \cdot \frac{-8}{9} = -\frac{\cancel{3} \cdot 2 \cdot \cancel{4}}{\cancel{4} \cdot \cancel{3} \cdot 3} = -\frac{2}{3}$

7. $\frac{5}{21} \cdot \frac{3}{15} = \frac{\cancel{5} \cdot \cancel{3}}{3 \cdot 7 \cdot \cancel{3} \cdot \cancel{5}} = \frac{1}{21}$

Copyright © 2012 Pearson Education, Inc.

8. $\frac{-4}{35}\cdot\frac{14}{18} = -\frac{2\cdot2\cdot\cancel{2}\cdot\cancel{7}}{5\cdot\cancel{7}\cdot\cancel{2}\cdot3\cdot3} = -\frac{4}{45}$

9. $\frac{9x}{15}\cdot\frac{21}{18x^3} = \frac{\cancel{9}\cdot x\cdot\cancel{3}\cdot7}{\cancel{3}\cdot5\cdot2\cdot\cancel{9}\cdot x^3} = \frac{7}{5\cdot2\cdot x^{3-1}} = \frac{7}{10x^2}$

10.
$$\begin{aligned}\frac{8x^3}{25}\cdot\frac{-45}{18x} &= -\frac{2\cdot2\cdot\cancel{2}\cdot x^3\cdot\cancel{9}\cdot\cancel{5}}{\cancel{5}\cdot5\cdot\cancel{2}\cdot\cancel{9}\cdot x}\\ &= -\frac{2\cdot2\cdot x^{3-1}}{5}\\ &= -\frac{4x^2}{5}\end{aligned}$$

11.
$$\begin{aligned}\frac{-5x^4}{6}\cdot12x^3 &= -\frac{5\cdot x^4\cdot\cancel{6}\cdot2\cdot x^3}{\cancel{6}}\\ &= -5\cdot2\cdot x^{4+3}\\ &= -10x^7\end{aligned}$$

12. $\frac{9}{14}\div\frac{45}{12} = \frac{9}{14}\cdot\frac{12}{45} = \frac{\cancel{9}\cdot\cancel{2}\cdot2\cdot3}{\cancel{2}\cdot7\cdot5\cdot\cancel{9}} = \frac{6}{35}$

13. $\frac{7}{15}\div\frac{-35}{20} = \frac{7}{15}\cdot\frac{20}{-35} = -\frac{\cancel{7}\cdot2\cdot2\cdot\cancel{5}}{3\cdot5\cdot\cancel{5}\cdot\cancel{7}} = -\frac{4}{15}$

14. $\frac{8}{42}\div\frac{-22}{7} = \frac{8}{42}\cdot\frac{7}{-22} = -\frac{\cancel{2}\cdot\cancel{2}\cdot2\cdot\cancel{7}}{\cancel{2}\cdot3\cdot\cancel{7}\cdot\cancel{2}\cdot11} = -\frac{2}{33}$

15.
$$\begin{aligned}\frac{11x^5}{25}\div\frac{3}{5x^2} &= \frac{11x^5}{25}\cdot\frac{5x^2}{3}\\ &= \frac{11\cdot x^5\cdot\cancel{5}\cdot x^2}{5\cdot\cancel{5}\cdot3}\\ &= \frac{11x^7}{15}\end{aligned}$$

16.
$$\begin{aligned}\frac{16x^2}{9}\div\frac{24x^4}{6} &= \frac{16x^2}{9}\cdot\frac{6}{24x^4}\\ &= \frac{\cancel{4}\cdot x\cdot\cancel{x^2}\cdot\cancel{6}}{9\cdot\cancel{4}\cdot\cancel{6}\cdot\cancel{x^2}\cdot x^2}\\ &= \frac{4}{9x^2}\end{aligned}$$

17.
$$\begin{aligned}A &= \frac{1}{2}bh\\ &= \frac{1}{2}\cdot18\text{ m}\cdot7\text{ m}\\ &= \frac{1\cdot18\text{ m}\cdot7\text{ m}}{2\cdot1\cdot1}\\ &= \frac{1\cdot18\cdot7\cdot\text{m}\cdot\text{m}}{2}\\ &= \frac{126\text{ m}^2}{2}\\ &= 63\text{ m}^2\end{aligned}$$

18.
$$\begin{aligned}A &= \frac{1}{2}bh\\ &= \frac{1}{2}\cdot11\text{ in.}\cdot20\text{ in.}\\ &= \frac{1\cdot11\text{ in.}\cdot20\text{ in.}}{2\cdot1\cdot1}\\ &= \frac{1\cdot11\cdot20\cdot\text{in.}\cdot\text{in.}}{2}\\ &= \frac{220\text{ in.}^2}{2}\\ &= 110\text{ in.}^2\end{aligned}$$

19. The word *of* indicates multiplication.
$\frac{2}{7}\cdot\$3500 = \frac{2}{7}\cdot\frac{\$3500}{1} = \$1000$
$1000 is withheld each month.

20. Since we are splitting the flour into 3 equal parts, we divide.
$\frac{1}{2}\div3 = \frac{1}{2}\cdot\frac{1}{3} = \frac{1}{6}$
Les should place $\frac{1}{6}$ pound in each container.

21. $7 = 7$
$21 = 3\cdot7$
$\text{LCM} = 3\cdot7 = 21$

22. $10 = 2\cdot5$
$20 = 2\cdot2\cdot5$
$\text{LCM} = 2\cdot2\cdot5 = 20$

23. $18 = 2\cdot3\cdot3$
$30 = 2\cdot3\cdot5$
$\text{LCM} = 2\cdot3\cdot3\cdot5 = 90$

Copyright © 2012 Pearson Education, Inc.

24. $3 = 3$
$9 = 3 \cdot 3$
$12 = 2 \cdot 2 \cdot 3$
$\text{LCM} = 2 \cdot 2 \cdot 3 \cdot 3 = 36$

25. $4x = 2 \cdot 2 \cdot x$
$8 = 2 \cdot 2 \cdot 2$
$16x = 2 \cdot 2 \cdot 2 \cdot 2 \cdot x$
$\text{LCM} = 2 \cdot 2 \cdot 2 \cdot 2 \cdot x = 16x$

26. $7x^2 = 7 \cdot x \cdot x$
$14x^4 = 2 \cdot 7 \cdot x \cdot x \cdot x \cdot x$
$20x = 2 \cdot 2 \cdot 5 \cdot x$
$\text{LCM} = 2 \cdot 2 \cdot 5 \cdot 7 \cdot x \cdot x \cdot x \cdot x = 140x^4$

27. $18x = 2 \cdot 3 \cdot 3 \cdot x$
$45x^2 = 3 \cdot 3 \cdot 5 \cdot x \cdot x$
$\text{LCM} = 2 \cdot 3 \cdot 3 \cdot 5 \cdot x \cdot x = 90x^2$

28. $20x = 2 \cdot 2 \cdot 5x$
$25x^2 = 5 \cdot 5 \cdot x \cdot x$
$\text{LCM} = 2 \cdot 2 \cdot 5 \cdot 5 \cdot x \cdot x = 100x^2$

29. $\frac{6}{17} - \frac{3}{17} = \frac{6-3}{17} = \frac{3}{17}$

30. $\frac{-23}{27} + \frac{-11}{27} = \frac{-23+(-11)}{27} = \frac{-34}{27} = -\frac{34}{27}$

31. $\frac{7}{x} - \frac{5}{x} = \frac{7-5}{x} = \frac{2}{x}$

32. $\frac{x}{7} - \frac{4}{7} = \frac{x-4}{7}$

33. $\frac{1}{6} + \frac{4}{9} = \frac{1 \cdot 3}{6 \cdot 3} + \frac{4 \cdot 2}{9 \cdot 2} = \frac{3}{18} + \frac{8}{18} = \frac{3+8}{18} = \frac{11}{18}$

34.
$$\begin{aligned}\frac{15}{32} - \frac{7}{28} &= \frac{15 \cdot 7}{32 \cdot 7} - \frac{7 \cdot 8}{28 \cdot 8} \\ &= \frac{105}{224} - \frac{56}{224} \\ &= \frac{49}{224} \\ &= \frac{\cancel{7} \cdot 7}{\cancel{7} \cdot 32} \\ &= \frac{7}{32}\end{aligned}$$

35.
$$\begin{aligned}\frac{-3}{14} + \frac{8}{21} &= \frac{-3 \cdot 3}{14 \cdot 3} + \frac{8 \cdot 2}{21 \cdot 2} \\ &= \frac{-9}{42} + \frac{16}{42} \\ &= \frac{7}{42} \\ &= \frac{1 \cdot \cancel{7}}{6 \cdot \cancel{7}} \\ &= \frac{1}{6}\end{aligned}$$

36. $\frac{5}{2x} + \frac{8}{3x} = \frac{5 \cdot 3}{2x \cdot 3} + \frac{8 \cdot 2}{3x \cdot 2} = \frac{15}{6x} + \frac{16}{6x} = \frac{31}{6x}$

37. $\frac{4x}{15} + \frac{2x}{45} = \frac{4x \cdot 3}{15 \cdot 3} + \frac{2x}{45} = \frac{12x}{45} + \frac{2x}{45} = \frac{14x}{45}$

38.
$$\begin{aligned}\frac{3x}{14} - \frac{5x}{42} &= \frac{3x \cdot 3}{14 \cdot 3} - \frac{5x}{42} \\ &= \frac{9x}{42} - \frac{5x}{42} \\ &= \frac{4x}{42} \\ &= \frac{\cancel{2} \cdot 2x}{\cancel{2} \cdot 21} \\ &= \frac{2x}{21}\end{aligned}$$

39.
$$\begin{array}{rcr} 10\frac{1 \cdot 5}{3 \cdot 5} & = & 10\frac{5}{15} \\ +\ 2\frac{4 \cdot 3}{5 \cdot 3} & = & +\ 2\frac{12}{15} \\ \hline & & 12\frac{17}{15} = 13\frac{2}{15} \end{array}$$

40.
$$\begin{array}{rcr} 12\,\frac{7}{9} & = & 12\frac{7}{9} \\ +\ 6\frac{2 \cdot 3}{3 \cdot 3} & = & +\ 6\frac{6}{9} \\ \hline & & 18\frac{13}{9} = 19\frac{4}{9} \end{array}$$

41.
$$\begin{array}{rcrcr} 11\frac{1 \cdot 5}{5 \cdot 5} & = & 11\frac{5}{25} & = & 10\frac{30}{25} \\ -\ 6\,\frac{11}{25} & = & -6\frac{11}{25} & = & -\ 6\frac{11}{25} \\ \hline & & & & 4\frac{19}{25} \end{array}$$

Copyright © 2012 Pearson Education, Inc.

42.
$$\begin{array}{rcr} 25 & = & 24\frac{14}{14} \\ -16\frac{5}{14} & = & -16\frac{5}{14} \\ \hline & & 8\frac{9}{14} \end{array}$$

43. $-3\cdot\left(4\frac{1}{3}\right)=-\frac{3}{1}\cdot\frac{13}{3}=-\frac{\cancel{3}\cdot 13}{1\cdot\cancel{3}}=-\frac{13}{1}=-13$

44. $2\frac{3}{4}\div 5=\frac{11}{4}\div\frac{5}{1}=\frac{11}{4}\cdot\frac{1}{5}=\frac{11\cdot 1}{4\cdot 5}=\frac{11}{20}$

45. $4\frac{2}{5}\div 8\frac{1}{3}=\frac{22}{5}\div\frac{25}{3}=\frac{22}{5}\cdot\frac{3}{25}=\frac{66}{125}$

46. $-12\div\frac{2}{3}=\frac{-12}{1}\cdot\frac{3}{2}=-\frac{\cancel{2}\cdot 2\cdot 3\cdot 3}{\cancel{2}}=-18$

47. Find the perimeter of the dog run.
$$\begin{aligned} P &= 2L+2W \\ &= 2\left(20\frac{1}{2}\right)+2\left(10\frac{1}{4}\right) \\ &= 2\left(\frac{41}{2}\right)+2\left(\frac{41}{4}\right) \\ &= \frac{82}{2}+\frac{41}{2} \\ &= \frac{123}{2} \text{ or } 61\frac{1}{2} \text{ ft} \end{aligned}$$

We now subtract.
$$\begin{array}{rcr} 75 & = & 74\frac{2}{2} \\ -61\frac{1}{2} & & -61\frac{1}{2} \\ \hline & & 13\frac{1}{2} \text{ ft} \end{array}$$

Shawnee will have $13\frac{1}{2}$ feet of fencing left.

48. We multiply.
$\frac{1}{4}\cdot 7\frac{1}{2}=\frac{1}{4}\cdot\frac{15}{2}=\frac{15}{8}=1\frac{7}{8}$ ft

Mike needs $1\frac{7}{8}$ feet of wood.

49.
$$\begin{aligned} \frac{3}{4}+\frac{1}{2}\cdot\frac{2}{5} &= \frac{3}{4}+\frac{1\cdot\cancel{2}}{\cancel{2}\cdot 5} \\ &= \frac{3}{4}+\frac{1}{5} \\ &= \frac{3\cdot 5}{4\cdot 5}+\frac{1\cdot 4}{5\cdot 4} \\ &= \frac{15}{20}+\frac{4}{20} \\ &= \frac{19}{20} \end{aligned}$$

50.
$$\begin{aligned} \left(\frac{3}{4}\right)^2+\frac{1}{8}\div\frac{1}{2} &= \frac{9}{16}+\frac{1}{8}\div\frac{1}{2} \\ &= \frac{9}{16}+\frac{1}{8}\cdot\frac{2}{1} \\ &= \frac{9}{16}+\frac{1}{4} \\ &= \frac{9}{16}+\frac{1\cdot 4}{4\cdot 4} \\ &= \frac{9}{16}+\frac{4}{16} \\ &= \frac{13}{16} \end{aligned}$$

51.
$$\begin{aligned} \frac{(-2)^2+12}{\frac{4}{7}} &= \frac{4+12}{\frac{4}{7}} \\ &= \frac{16}{\frac{4}{7}} \\ &= 16\div\frac{4}{7} \\ &= \frac{16}{1}\cdot\frac{7}{4} \\ &= 28 \end{aligned}$$

52. $\frac{\frac{4}{3}}{\frac{1}{9}}=\frac{4}{3}\div\frac{1}{9}=\frac{4}{3}\cdot\frac{9}{1}=\frac{2\cdot 2\cdot 3\cdot\cancel{3}}{\cancel{3}\cdot 1}=12$

53. $\frac{\frac{x}{12}}{\frac{x^2}{20}}=\frac{x}{12}\div\frac{x^2}{20}=\frac{\cancel{x}}{\cancelto{3}{12}}\cdot\frac{\cancelto{5}{20}}{x^{\cancel{2}}}=\frac{5}{3x}$

 Copyright © 2012 Pearson Education, Inc.

54. $\dfrac{\frac{1}{2}+\frac{1}{4}}{\frac{2}{3}-\frac{1}{9}}=\dfrac{\left(\frac{1}{2}+\frac{1}{4}\right)}{\left(\frac{2}{3}-\frac{1}{9}\right)}$
$=\dfrac{\left(\frac{1\cdot 2}{2\cdot 2}+\frac{1}{4}\right)}{\left(\frac{2\cdot 3}{3\cdot 3}-\frac{1}{9}\right)}$
$=\dfrac{\left(\frac{2}{4}+\frac{1}{4}\right)}{\left(\frac{6}{9}-\frac{1}{9}\right)}$
$=\dfrac{\frac{3}{4}}{\frac{5}{9}}$
$=\dfrac{3}{4}\div\dfrac{5}{9}$
$=\dfrac{3}{4}\cdot\dfrac{9}{5}$
$=\dfrac{27}{20}$

55. $\dfrac{\frac{1}{3}+\frac{2}{5}}{\frac{1}{5}+\frac{1}{10}}=\dfrac{\left(\frac{1}{3}+\frac{2}{5}\right)}{\left(\frac{1}{5}+\frac{1}{10}\right)}$
$=\dfrac{\left(\frac{1\cdot 5}{3\cdot 5}+\frac{2\cdot 3}{5\cdot 3}\right)}{\left(\frac{1\cdot 2}{5\cdot 2}+\frac{1}{10}\right)}$
$=\dfrac{\left(\frac{5}{15}+\frac{6}{15}\right)}{\left(\frac{2}{10}+\frac{1}{10}\right)}$
$=\dfrac{\frac{11}{15}}{\frac{3}{10}}$
$=\dfrac{11}{15}\div\dfrac{3}{10}$
$=\dfrac{11}{15}\cdot\dfrac{10}{3}$
$=\dfrac{11\cdot 2\cdot \cancel{5}}{3\cdot \cancel{5}\cdot 3}$
$=\dfrac{22}{9}$

56. $\dfrac{3\frac{1}{4}\text{ acres}}{4\text{ houses}}=\dfrac{x\text{ acres}}{32\text{ houses}}$
$32\cdot 3\frac{1}{4}=4\cdot x$
$32\cdot\dfrac{13}{4}=4x$
$\dfrac{8\cdot\cancel{4}\cdot 13}{\cancel{4}}=4x$
$104=4x$
$\dfrac{104}{4}=\dfrac{4x}{4}$
$26=x$
The developer needs 26 acres to build 32 homes.

57. a. The length of the pool is the length of the yard less twice the width of the concrete area.
$38\frac{1}{4}-2\cdot 4\frac{1}{2}=38\frac{1}{4}-2\cdot\frac{9}{2}$
$=38\frac{1}{4}-9$
$=29\frac{1}{4}$ ft
The width of the pool is the width of the yard less twice the width of the concrete area.
$25-2\cdot 4\frac{1}{2}=25-2\cdot\frac{9}{2}=25-9=16$ ft
The pool is $29\frac{1}{4}$ feet by 16 feet.

b. Find the perimeter of the pool.
$P=2L+2W$
$=2\left(29\frac{1}{4}\right)+2(16)$
$=2\left(\frac{117}{4}\right)+32$
$=\frac{117}{2}+32$
$=58\frac{1}{2}+32$
$=90\frac{1}{2}$ ft
Multiply to find the cost.
$90\frac{1}{2}\cdot 3\frac{1}{4}=\frac{181}{2}\cdot\frac{13}{4}=\frac{2353}{8}=\$294\frac{1}{8}$
The tile will cost $\$294\frac{1}{8}$ or \$294.13.

Copyright © 2012 Pearson Education, Inc.

58. We subtract.

$$2\frac{1}{2}-\frac{7}{8}=\frac{5}{2}-\frac{7}{8}$$
$$=\frac{5\cdot 4}{2\cdot 4}-\frac{7}{8}$$
$$=\frac{20}{8}-\frac{7}{8}$$
$$=\frac{13}{8}$$
$$=1\frac{5}{8}\text{ in.}$$

The blade must be lowered $1\frac{5}{8}$ inches.

59.
$$\frac{x}{3}=9$$
$$\frac{3\cdot x}{3}=9\cdot 3$$
$$x=27$$

Check: $\frac{x}{3}=9$
$$\frac{27}{3}\stackrel{?}{=}9$$
$$9=9\checkmark$$

60.
$$\frac{y}{2}=-6$$
$$\frac{2\cdot y}{2}=-6\cdot 2$$
$$y=-12$$

Check: $\frac{y}{2}=-6$
$$\frac{-12}{2}\stackrel{?}{=}-6$$
$$-6=-6\checkmark$$

61.
$$\frac{x}{-6}=-4$$
$$\frac{-6\cdot x}{-6}=-4\cdot(-6)$$
$$x=24$$

Check: $\frac{x}{-6}=-4$
$$\frac{24}{-6}\stackrel{?}{=}-4$$
$$-4=-4\checkmark$$

62.
$$\frac{2}{3}y=22$$
$$\left(\frac{3}{2}\right)\left(\frac{2}{3}\right)y=22\left(\frac{3}{2}\right)$$
$$1y=\frac{\cancel{2}\cdot 11\cdot 3}{\cancel{2}}$$
$$y=33$$

Check: $\frac{2}{3}y=22$
$$\frac{2}{3}\cdot 33\stackrel{?}{=}22$$
$$\frac{2\cdot\cancel{3}\cdot 11}{\cancel{3}}\stackrel{?}{=}22$$
$$22=22\checkmark$$

63.
$$\frac{-3}{4}x=18$$
$$\left(-\frac{4}{3}\right)\left(-\frac{3}{4}\right)x=18\left(-\frac{4}{3}\right)$$
$$1x=-\frac{6\cdot\cancel{3}\cdot 4}{\cancel{3}}$$
$$x=-24$$

Check: $\frac{-3}{4}x=18$
$$\frac{-3}{4}\cdot(-24)\stackrel{?}{=}18$$
$$\frac{3\cdot\cancel{4}\cdot 6}{\cancel{4}}\stackrel{?}{=}18$$
$$18=18\checkmark$$

64.
$$\frac{1}{3}y=4$$
$$3\left(\frac{1}{3}\right)y=4(3)$$
$$1y=12$$
$$y=12$$

Check: $\frac{1}{3}y=4$
$$\frac{1}{3}(12)\stackrel{?}{=}4$$
$$4=4\checkmark$$

Copyright © 2012 Pearson Education, Inc.

65. $\frac{x}{2^3} = 2\cdot 6+1$

$\frac{x}{8} = 12+1$

$\frac{x}{8} = 13$

$\frac{8\cdot x}{8} = 13\cdot 8$

$x = 104$

Check: $\frac{x}{2^3} = 2\cdot 6+1$

$\frac{104}{2^3} \stackrel{?}{=} 2\cdot 6+1$

$\frac{104}{8} \stackrel{?}{=} 12+1$

$13 = 13$ ✓

66. $\frac{y}{3^2} = 1+8\div 2$

$\frac{y}{9} = 1+4$

$\frac{y}{9} = 5$

$\frac{9\cdot y}{9} = 5\cdot 9$

$y = 45$

Check: $\frac{y}{3^2} = 1+8\div 2$

$\frac{45}{3^2} \stackrel{?}{=} 1+8\div 2$

$\frac{45}{9} \stackrel{?}{=} 1+4$

$5 = 5$ ✓

How Am I Doing? Chapter 5 Test

1. $\frac{2}{3}\cdot\frac{9}{10} = \frac{\cancel{2}\cdot\cancel{3}\cdot 3}{\cancel{3}\cdot\cancel{2}\cdot 5} = \frac{3}{5}$

2. $\frac{-1}{4}\cdot\frac{2}{5} = -\frac{1\cdot\cancel{2}}{\cancel{2}\cdot 2\cdot 5} = -\frac{1}{10}$

3. $\frac{7x^2}{8}\cdot\frac{16x}{14} = \frac{\cancel{7}\cdot x^2\cdot\cancel{8}\cdot\cancel{2}\cdot x}{\cancel{8}\cdot\cancel{2}\cdot\cancel{7}} = \frac{x^{2+1}}{1} = x^3$

4. $\frac{-1}{2}\div\frac{1}{4} = \frac{-1}{2}\cdot\frac{4}{1} = -\frac{1\cdot\cancel{2}\cdot 2}{\cancel{2}\cdot 1} = -2$

5. $\frac{2x^8}{5}\div\frac{22x^2}{15} = \frac{2x^8}{5}\cdot\frac{15}{22x^4}$

$= \frac{\cancel{2}\cdot x^8\cdot 3\cdot\cancel{5}}{\cancel{5}\cdot\cancel{2}\cdot 11\cdot x^4}$

$= \frac{3\cdot x^{8-4}}{11}$

$= \frac{3x^4}{11}$

6. $\frac{8x}{15}\cdot\frac{25}{12x^3} = \frac{2\cdot\cancel{4}\cdot x\cdot\cancel{5}\cdot 5}{3\cdot\cancel{5}\cdot 3\cdot\cancel{4}\cdot x^3} = \frac{2\cdot 5}{3\cdot 3\cdot x^{3-1}} = \frac{10}{9x^2}$

7. First six multiples of 6: 6, 12, 18, 24, 30, 36
First six multiples of 9: 9, 18, 27, 36, 45, 54
18 and 36 are common.

8. $14 = 2\cdot 7$
$21 = 3\cdot 7$
$\text{LCM} = 2\cdot 3\cdot 7 = 42$

9. $5a = 5\cdot a$

$10a^4 = 2\cdot 5\cdot a\cdot a\cdot a\cdot a$

$20a^2 = 2\cdot 2\cdot 5\cdot a\cdot a$

$\text{LCM} = 2\cdot 2\cdot 5\cdot a\cdot a\cdot a\cdot a = 20a^4$

10. $5 = 5$
$7 = 7$
$10 = 2\cdot 5$
$\text{LCM} = 2\cdot 5\cdot 7 = 70$

11. $30 = 2\cdot 3\cdot 5$
$4 = 2\cdot 2$
$\text{LCD} = 2\cdot 2\cdot 3\cdot 5 = 60$

12. $\frac{12}{11}+\frac{3}{11} = \frac{12+3}{11} = \frac{15}{11} = 1\frac{4}{11}$

13. $\frac{1}{5a}+\frac{3}{a} = \frac{1}{5a}+\frac{3\cdot 5}{a\cdot 5} = \frac{1}{5a}+\frac{15}{5a} = \frac{16}{5a}$

14. $4\frac{5}{6} = 4\frac{5}{6}$

$+3\frac{1\cdot 2}{3\cdot 2} = +3\frac{2}{6}$

$7\frac{7}{6} = 8\frac{1}{6}$

15. $\frac{2}{21}+\frac{5}{9} = \frac{2\cdot 3}{21\cdot 3}+\frac{5\cdot 7}{9\cdot 7} = \frac{6}{63}+\frac{35}{63} = \frac{41}{63}$

Copyright © 2012 Pearson Education, Inc.

16. $\frac{7}{12}-\frac{2}{15}=\frac{7\cdot5}{12\cdot5}-\frac{2\cdot4}{15\cdot4}$
$=\frac{35}{60}-\frac{8}{60}$
$=\frac{27}{60}$
$=\frac{3\cdot3\cdot\cancel{3}}{2\cdot2\cdot\cancel{3}\cdot5}$
$=\frac{9}{20}$

17.

$$\begin{array}{rcrcr} 10\frac{1\cdot3}{8\cdot3} & = & 10\frac{3}{24} & = & 9\frac{27}{24} \\ -\,2\frac{2\cdot8}{3\cdot8} & = & -\,2\frac{16}{24} & = & -\,2\frac{16}{24} \\ \hline & & & & 7\frac{11}{24} \end{array}$$

18. $\frac{7x}{15}-\frac{x}{20}=\frac{7x\cdot4}{15\cdot4}-\frac{x\cdot3}{20\cdot3}$
$=\frac{20x}{60}-\frac{3x}{60}$
$=\frac{25x}{60}$
$=\frac{5\cdot\cancel{5}\cdot x}{2\cdot2\cdot3\cdot\cancel{5}}$
$=\frac{5x}{12}$

19. $(-4)\cdot5\frac{1}{3}=\frac{-4}{1}\cdot\frac{16}{3}=-\frac{64}{3}=-21\frac{1}{3}$

20. $1\frac{2}{3}\div3=\frac{5}{3}\div3=\frac{5}{3}\cdot\frac{1}{3}=\frac{5}{9}$

21. $2\frac{1}{2}\div\left(-\frac{3}{5}\right)=\frac{5}{2}\div\left(-\frac{3}{5}\right)=-\frac{5}{2}\cdot\frac{5}{3}=-\frac{25}{6}=-4\frac{1}{6}$

22. $\left(\frac{2}{3}\right)^2+\frac{1}{2}\cdot\frac{1}{4}=\frac{4}{9}+\frac{1}{2}\cdot\frac{1}{4}$
$=\frac{4}{9}+\frac{1}{8}$
$=\frac{4\cdot8}{9\cdot8}+\frac{1\cdot9}{8\cdot9}$
$=\frac{32}{72}+\frac{9}{72}$
$=\frac{41}{72}$

23. $\frac{\frac{x}{2}}{\frac{x}{4}}=\frac{x}{2}\div\frac{x}{4}=\frac{x}{2}\cdot\frac{4}{x}=\frac{\cancel{x}\cdot\cancel{2}\cdot2}{\cancel{2}\cdot\cancel{x}}=2$

24. $\frac{\frac{1}{3}+\frac{1}{2}}{\frac{5}{9}-\frac{1}{3}}=\frac{\left(\frac{1}{3}+\frac{1}{2}\right)}{\left(\frac{5}{9}-\frac{1}{3}\right)}$
$=\frac{\left(\frac{1\cdot2}{3\cdot2}+\frac{1\cdot3}{2\cdot3}\right)}{\left(\frac{5}{9}-\frac{1\cdot3}{3\cdot3}\right)}$
$=\frac{\left(\frac{2}{6}+\frac{3}{6}\right)}{\left(\frac{5}{9}-\frac{3}{9}\right)}$
$=\frac{\frac{5}{6}}{\frac{2}{9}}$
$=\frac{5}{6}\div\frac{2}{9}$
$=\frac{5}{6}\cdot\frac{9}{2}$
$=\frac{5\cdot\cancel{3}\cdot3}{2\cdot\cancel{3}\cdot2}$
$=\frac{15}{4}$ or $3\frac{3}{4}$

25. $\frac{x}{-3}=6+2^2$
$\frac{x}{-3}=6+4$
$\frac{x}{-3}=10$
$\frac{-3\cdot x}{-3}=10\cdot(-3)$
$x=-30$

Check: $\frac{x}{-3}=6+2^2$
$\frac{-30}{-3}\stackrel{?}{=}6+2^2$
$10\stackrel{?}{=}6+4$
$10=10$ ✓

26. $\frac{4}{7}y=28$
$\left(\frac{7}{4}\right)\left(\frac{4}{7}\right)y=28\left(\frac{7}{4}\right)$
$1y=\frac{\cancel{4}\cdot7\cdot7}{\cancel{4}}$
$y=49$

Copyright © 2012 Pearson Education, Inc.

Check: $\frac{4}{7}y = 28$

$\frac{4}{7}\cdot 49 \stackrel{?}{=} 28$

$\frac{4\cdot \cancel{7}\cdot 7}{\cancel{7}} \stackrel{?}{=} 28$

$28 = 28$ ✓

27. $\frac{x}{4} = -20$

$\frac{4\cdot x}{4} = -20\cdot 4$

$x = -80$

Check: $\frac{x}{4} = -20$

$\frac{-80}{4} \stackrel{?}{=} -20$

$-20 = -20$ ✓

28. We multiply the number of shelves per bookcase by the number of bookcases to find the number of shelves.

$2 \cdot 5 = 10$ shelves

We divide the length of one board by the length of one shelf to determine how many shelves can be cut from one board.

$10 \div 3\frac{1}{2} = 10 \div \frac{7}{2} = 10\cdot \frac{2}{7} = \frac{20}{7} = 2\frac{6}{7}$ shelves

Thus 2 whole shelves can be cut from each board. Anna needs to buy $10 \div 2$ or 5 boards.

29. a. On Saturday Rudy's temperature was $98\frac{1}{2}° + 3° = 101\frac{1}{2}$°F. On Sunday, his temperature was $101\frac{1}{2}° - 1\frac{1}{2}° = 100$°F.

b. We subtract.

$$\begin{array}{rcr} 100° & = & 99\frac{2}{2}° \\ -\ 98\frac{1}{2}° & = & -\ 98\frac{1}{2}° \\ \hline & & 1\frac{1}{2}°\text{F} \end{array}$$

Rudy's temperature was $1\frac{1}{2}°$ higher than normal.

30. a. The length of the pool is the length of the yard less twice the width of the grass area.

$35\frac{1}{2} - 2\cdot 5 = 35\frac{1}{2} - 10 = 25\frac{1}{2}$ ft

The width of the pool is the width of the yard less twice the width of the grass area.

$22 - 2 \cdot 5 = 22 - 10 = 12$ ft

The pool is $25\frac{1}{2}$ feet by 12 feet.

b. Find the perimeter of the pool.

$P = 2L + 2W$

$= 2\left(25\frac{1}{2}\right) + 2(12)$

$= 2\left(\frac{51}{2}\right) + 24$

$= 51 + 24$

$= 75$ ft

Multiply to find the cost.

$75\cdot 1\frac{1}{2} = 75\cdot \frac{3}{2} = \frac{225}{2} = \$112\frac{1}{2}$

The tile will cost $\$112\frac{1}{2}$.

Copyright © 2012 Pearson Education, Inc.

Chapter 6

6.1 Exercises

1. To subtract two polynomials, we add the opposite of the second polynomial to the first polynomial.

3. The terms of $2z^2+4z-2y^4+3$ are $+2z^2$, $+4z$, $-2y^4$, $+3$.

5. The terms of $6x^6-3x^3-3y-1$ are $+6x^6$, $-3x^3$, $-3y$, -1.

7. $(7y-3)+(-4y+9)=7y-4y-3+9=3y+6$

9. $(2a^2-3a+6)+(4a-2)$
$=2a^2-3a+4a+6-2$
$=2a^2+1a+4$
$=2a^2+a+4$

11. $(5y^2+2y-5)+(-4y^2-8y+2)$
$=5y^2-4y^2+2y-8y-5+2$
$=1y^2-6y-3$
$=y^2-6y-3$

13. $-(5x+2y)=-5x-2y$

15. $-(-8x+4)=8x-4$

17. $-(-3x+6z-5y)=3x-6z+5y$

19. $(10x+7)-(3x+5)=10x+7+(-3x)+(-5)$
$=7x+2$

21. $(7x-3)-(-4x+6)=7x-3+4x+(-6)$
$=11x-9$

23. $(-8a+5)-(4a-3)=-8a+5+(-4a)+3$
$=-12a+8$

25. $(3y^2+4y-5)-(4y^2-6y-8)$
$=3y^2+4y-5+(-4y^2)+6y+8$
$=-1y^2+10y+3$
$=-y^2+10y+3$

27. $(2x^2+6x-5)-(6x^2-4x-8)$
$=2x^2+6x-5+(-6x^2)+4x+8$
$=-4x^2+10x+3$

29. $(-6z^2+9z-1)-(3z^2+8z-7)$
$=-6z^2+9z-1+(-3z^2)+(-8z)+7$
$=-9z^2+z+6$

31. $(2a^2+9a-1)-(-5a^2-8a-4)$
$=2a^2+9a-1+5a^2+8a+4$
$=7a^2+17a+3$

33. $(-7x^2-6x-1)-(2x^2+8x+4)$
$=-7x^2-6x-1+(-2x^2)+(-8x)+(-4)$
$=-9x^2-14x-5$

35. $3x-2(5x^2-6)-(-2x^2+x-1)$
$=3x-10x^2+12-(-2x^2+x-1)$
$=3x-10x^2+12+2x^2-x+1$
$=2x-8x^2+13$
$=-8x^2+2x+13$

37. $6x-(3x^2+8x+2)+2(-x^2-9)$
$=6x-(3x^2+8x+2)-2x^2-18$
$=6x-3x^2-8x-2-2x^2-18$
$=-2x-5x^2-20$
$=-5x^2-2x-20$

39. $(-4x^2+7x+1)-(x^2-5)$
$=-4x^2+7x+1+(-x^2)+5$
$=-5x^2+7x+6$

41. $(4x^2-5x-7)+(3x^2-8x-2)$
$=4x^2+3x^2-5x-8x-7-2$
$=7x^2-13x-9$

43. $(4x^2+6x-2)-(3x^2+7x+1)+(x^2-1)$
$=4x^2+6x-2-3x^2-7x-1+x^2-1$
$=2x^2-x-4$

Copyright © 2012 Pearson Education, Inc.

45. $4(-9x-6)-(3x^2-7x+1)+4x$
$=-36x-24-(3x^2-7x+1)+4x$
$=-36x-24-3x^2+7x-1+4x$
$=-3x^2-25x-25$

47. $(-7x^2+3x-4)-(5x^2+7x-1)+(x^2-8)$
$=-7x^2+3x-4-5x^2-7x+1+x^2-8$
$=-11x^2-4x-11$

49. $5x-2(4x^2+3x-2)-(2x^2+8x-1)$
$=5x-8x^2-6x+4-(2x^2+8x-1)$
$=5x-8x^2-6x+4-2x^2-8x+1$
$=-9x-10x^2+5$
$=-10x^2-9x+5$

51. $7x-2^3(x+3)-(-3)^2(x^2-2x-1)$
$=7x-8(x+3)-9(x^2-2x-1)$
$=7x-8x-24-9x^2+18x+9$
$=17x-9x^2-15$
$=-9x^2+17x-15$

53. $(ax+3)+(2x^2+5x-6)+(8x-2)$
$=2x^2-10x-5$
Simplify the left side of the equation.
$2x^2+(a+5+8)x-5=2x^2-10x-5$
$2x^2+(a+13)x-5=2x^2-10x-5$
Thus, $a+13=-10$
$a+13+(-13)=-10+(-13)$
$a=-23$

55. $(ax^2-bx-7)+(5x^2-2x+3)=9x^2-5x-4$
Simplify the left side of the equation.
$(a+5)x^2-(b+2)x-4=9x^2-5x-4$
Thus $a+5=9$ and
$a+5+(-5)=9+(-5)$
$a=4$
$b+2=5$
$b+2+(-2)=5+(-2)$
$b=3$

Cumulative Review

57. $\dfrac{-6x^8}{2x^2}=-3x^{8-2}=-3x^6$

58. $\dfrac{-8x^6y^2}{2x^2y^7}=-\dfrac{4x^{6-2}}{y^{7-2}}=-\dfrac{4x^4}{y^5}$

59. $(-4x)(2x^2)=(-4)(2)x^{1+2}=-8x^3$

60. $(3y)(-2y)(5y)=(3)(-2)(5)y^{1+1+1}=-30y^3$

61. We subtract.

$$\begin{array}{rcr} 3\frac{1\cdot 7}{2\cdot 7} & = & 3\frac{7}{14} \\ -\,2\frac{2\cdot 2}{7\cdot 2} & = & -\,2\frac{4}{14} \\ \hline & & 1\frac{3}{14}\text{ mi} \end{array}$$

Juan walked $1\frac{3}{14}$ more miles than Maria.

62. We add.
$\frac{3}{4}+\frac{1}{2}=\frac{3}{4}+\frac{1\cdot 2}{2\cdot 2}=\frac{3}{4}+\frac{2}{4}=\frac{5}{4}$ or $1\frac{1}{4}$
There were $\frac{5}{4}$ cups or $1\frac{1}{4}$ cups of sugar in the mixture.

Quick Quiz 6.1

1. a. $(-3x+1)+(-5x-2)=-3x-5x+1-2$
$=-8x-1$

b. $(-6x^2+4x-7)+(8x^2-9x+2)$
$=-6x^2+8x^2+4x-9x-7+2$
$=2x^2-5x-5$

2. a. $(6a-4)-(3a-2)=6a-4+(-3a)+2$
$=3a-2$

b. $(9y^2-3y-8)-(2y^2-4y+8)$
$=9y^2-3y-8+(-2y^2)+4y+(-8)$
$=7y^2+1y-16$
$=7y^2+y-16$

Copyright © 2012 Pearson Education, Inc.

3. a. $-(-2x^2-3x)+(-5x^2+9x+2)-(5x^2-10x+3) = 2x^2+3x-5x^2+9x+2-5x^2+10x-3$
$= -8x^2+22x-1$

b. $9x-(6x^2-2x+3)-3(-4x^2+7x-1) = 9x-(6x^2-2x+3)+12x^2-21x+3$
$= 9x-6x^2+2x-3+12x^2-21x+3$
$= -10x+6x^2+0$
$= 6x^2-10x$

4. Answers may vary. One possible error was not multiplying the negative one coefficient through the entire second expression.

6.2 Exercises

1. The sign of the last term, +4, is wrong. It should be −4.

3. a. A monomial is a polynomial with 1 term.

b. A binomial is a polynomial with 2 terms.

c. A trinomial is a polynomial with 3 terms.

5. First term of the product is: $\boxed{-15y^2}$.
Second term of the product is: $\boxed{+10y}$.
Third term of the product is: $\boxed{-30}$.
Therefore $-5(3y^2-2y+6) = \boxed{-15y^2+10y-30}$.

7. We multiply x times the trinomial: $x\boxed{(x^2+3x+1)} = \boxed{x^3+3x^2+x}$.
Then we multiply −1 times the trinomial: $-1\boxed{(x^2+3x+1)} = \boxed{-x^2-3x-1}$.
Finally we simplify: $(x-1)(x^2+3x+1) = \boxed{x^3+2x^2-2x-1}$.

9. $(x+1)(x-3)$ F = $\boxed{x^2}$
O = $\boxed{-3x}$
I = $\boxed{+1x}$
L = $\boxed{-3}$
$(x+1)(x-3) = \boxed{x^2-2x-3}$

11. $8(2x^2+3x-2) = 8(2x^2)+8(3x)+8(-2)$
$= 16x^2+24x-16$

13. $-4y(2y^2-3y+6) = -4y(2y^2)-4y(-3y)-4y(6)$
$= -8y^3+12y^2-24y$

15. $-3x^2(x-2) = -3x^2(x)-3x^2(-2) = -3x^3+6x^2$

17. $(y^6+9)(-5y^2) = -5y^2(y^6+9) = -5y^8-45y^2$

Copyright © 2012 Pearson Education, Inc.

19. $(x^3-5x-2)(-3x^4) = -3x^4(x^3-5x-2)$
$= -3x^7+15x^5+6x^4$

21. $(x-1)(2x^2-3x-2) = x(2x^2-3x-2)-1(2x^2-3x-2)$
$= x\cdot 2x^2+x(-3x)+x(-2)-1\cdot 2x^2-1(-3x)-1(-2)$
$= 2x^3-3x^2-2x-2x^2+3x+2$
$= 2x^3-5x^2+x+2$

23. $(2y+4)(3y^2+y-5) = 2y(3y^2+y-5)+4(3y^2+y-5)$
$= 2y\cdot 3y^2+2y\cdot y+2y(-5)+4\cdot 3y^2+4\cdot y+4(-5)$
$= 6y^3+2y^2-10y+12y^2+4y-20$
$= 6y^3+14y^2-6y-20$

25. $(3x-1)(x^2+2x+1) = 3x(x^2+2x+1)-1(x^2+2x+1)$
$= 3x\cdot x^2+3x\cdot 2x+3x\cdot 1-1\cdot x^2-1\cdot 2x-1\cdot 1$
$= 3x^3+6x^2+3x-x^2-2x-1$
$= 3x^3+5x^2+x-1$

27. $(x+6)(x+7) = x^2+7x+6x+42$
$= x^2+13x+42$

29. $(x+3)(x+9) = x^2+9x+3x+27$
$= x^2+12x+27$

31. $(a+6)(a+2) = a^2+2a+6a+12 = a^2+8a+12$

33. $(y+4)(y-8) = y^2-8y+4y-32 = y^2-4y-32$

35. $(x+2)(x-4) = x^2-4x+2x-8 = x^2-2x-8$

37. $(x-4)(x+2) = x^2+2x-4x-8 = x^2-2x-8$

39. $(2x+1)(x+2) = 2x^2+4x+x+2 = 2x^2+5x+2$

41. $(3x-3)(x-1) = 3x^2-3x-3x+3 = 3x^2-6x+3$

43. $(2y-1)(y+2) = 2y^2+4y-y-2 = 2y^2+3y-2$

45. $(2y+1)(y-2) = 2y^2-4y+y-2 = 2y^2-3y-2$

47. $-5a(2a-4b-6) = -5a(2a)-5a(-4b)-5a(-6)$
$= -10a^2+20ab+30a$

49. $-7x^3(x-3) = -7x^3(x)-7x^3(-3) = -7x^4+21x^3$

Copyright © 2012 Pearson Education, Inc.

51. $(x-2)(x^2-3x+1) = x(x^2-3x+1)-2(x^2-3x+1)$
$= x\cdot x^2 + x(-3x) + x\cdot 1 - 2\cdot x^2 - 2(-3x) - 2\cdot 1$
$= x^3 - 3x^2 + x - 2x^2 + 6x - 2$
$= x^3 - 5x^2 + 7x - 2$

53. $(z+2)(z-5) = z^2 - 5z + 2z - 10 = z^2 - 3z - 10$

55. $(2x+1)(4x^2+2x-8) = 2x(4x^2+2x-8)+1(4x^2+2x-8)$
$= 2x\cdot 4x^2 + 2x\cdot 2x + 2x(-8) + 1\cdot 4x^2 + 1\cdot 2x + 1(-8)$
$= 8x^3 + 4x^2 - 16x + 4x^2 + 2x - 8$
$= 8x^3 + 8x^2 - 14x - 8$

57. $(y-7)(y+2) = y^2 + 2y - 7y - 14 = y^2 - 5y - 14$

59. **a.** $(z+5)(z+1) = z^2 + z + 5z + 5 = z^2 + 6z + 5$

b. $(z-5)(z-1) = z^2 - z - 5z + 5 = z^2 - 6z + 5$

61. **a.** $(x+3)(x-1) = x^2 - x + 3x - 3$
$= x^2 + 2x - 3$

b. $(x-3)(x+1) = x^2 + x - 3x - 3$
$= x^2 - 2x - 3$

63. $(x+2)(x-1) + 2(3x+3) = x^2 - x + 2x - 2 + 2(3x+3)$
$= x^2 + x - 2 + 6x + 6$
$= x^2 + 7x + 4$

65. $-2x(x^2+3x-1) + (x-2)(x-3) = -2x^2 - 6x^2 + 2x + x^2 - 3x - 2x + 6$
$= -2x^3 - 6x^2 + 2x + x^2 - 5x + 6$
$= -2x^3 - 5x^2 - 3x + 6$

67. $a(2x-3) = -14x + 21$
Simplify on the left side.
$2ax - 3a = -14x + 21$
Thus, $2a = -14$
$\frac{2a}{2} = \frac{-14}{2}$
$a = -7$

Cumulative Review

69. **a.** The number of dimes, D, is three times the number of nickels, N: $D = 3N$.

Copyright © 2012 Pearson Education, Inc.

b. Replace D with 21.

$D = 3N$
$21 = 3N$
$\frac{21}{3} = \frac{3N}{3}$
$7 = N$

There are 7 nickels.

70. a. The length, L, of a piece of wood is double the width: $L = 2W$.

b. Replace L with 40.

$L = 2W$
$40 = 2W$
$\frac{40}{2} = \frac{2W}{2}$
$20 = W$

The width is 20 feet.

71. $\frac{250 \text{ calories}}{2 \text{ cookies}} = \frac{n \text{ calories}}{5 \text{ cookies}}$

$\frac{250}{2} = \frac{n}{5}$
$5 \cdot 250 = 2 \cdot n$
$1250 = 2n$
$\frac{1250}{2} = \frac{2n}{2}$
$625 = n$

There are 625 calories in five Donna Deluxe cookies.

72. $\frac{\$64}{16 \text{ hr}} = \$4$ per hour

She earned \$4 per hour.

Quick Quiz 6.2

1. $(-3z-1)(5z^2) = 5z^2(-3z-1)$
$= 5z^2(-3z) + (5z^2)(-1)$
$= -15z^3 - 5z^2$

2. $(x-1)(4x^2-2x+8)$
$= x(4x^2-2x+8) - 1(4x^2-2x+8)$
$= x \cdot 4x^2 + x(-2x) + x \cdot 8 - 1 \cdot 4x^2 - 1(-2x) - 1 \cdot 8$
$= 4x^3 - 2x^2 + 8x - 4x^2 + 2x - 8$
$= 4x^3 - 6x^2 + 10x - 8$

3. a. $(y-2)(y+1) = y^2 + y - 2y - 2 = y^2 - y - 2$

b. $(2a+3)(a+5) = 2a^2 + 10a + 3a + 15$
$= 2a^2 + 13a + 15$

4. 1. $(x+1)(x+2) = x^2 + 2x + x + 2 = x^2 + 3x + 2$

2. $(x-1)(x-2) = x^2 - 2x - x + 2 = x^2 - 3x + 2$

a. Answers may vary. One possible solution is that in number 1 all products resulting in a first order variable have a positive coefficient, whereas in number 2 the opposite is true.

b. Answers may vary. One possible solution is that all products resulting in a zero order variable are either products of two positive coefficients or two negative coefficients, both resulting in a positive coefficient.

How Am I Doing? Sections 6.1–6.2

1. $(3y^2+5y-2)+(4y-7) = 3y^2+5y+4y-2-7$
$= 3y^2+9y-9$

2. $(-7a+5)-(2a-3) = -7a+5+(-2a)+3$
$= -9a+8$

3. $(-2x^2+4x-7)-(5x^2+3x-4)$
$= -2x^2+4x-7+(-5x^2)+(-3x)+4$
$= -7x^2+1x-3$
$= -7x^2+x-3$

4. $2x-3(5x^2+4)+(-3x^2-x+6)$
$= 2x-15x^2-12+(-3x^2-x+6)$
$= 2x-15x^2-12-3x^2-x+6$
$= 1x-18x^2-6$
$= -18x^2+x-6$

5. $(3x^2-6x-8)-(2x^2+7x+5)+(x^2-7)$
$= 3x^2-6x-8-2x^2-7x-5+x^2-7$
$= 2x^2-13x-20$

6. $-8(2a^2-3a+1) = -8(2a^2)-8(-3a)-8(1)$
$= -16a^2+24a-8$

7. $-2y(-6y+4x-5) = -2y(-6y)-2y(4x)-2y(-5)$
$= 12y^2-8xy+10y$

8. $-4x^2(x^2+6) = -4x^2 \cdot x^2 - 4x^2 \cdot 6$
$= -4x^4 - 24x^2$

Copyright © 2012 Pearson Education, Inc.

9. $(y+2)(4y^2+3y-2)$
$= y(4y^2+3y-2)+2(4y^2+3y-2)$
$= y\cdot 4y^2+y\cdot 3y+y(-2)+2\cdot 4y^2+2\cdot 3y+2(-2)$
$= 4y^3+3y^2-2y+8y^2+6y-4$
$= 4y^3+11y^2+4y-4$

10. $(y+2)(y-4) = y^2-4y+2y-8$
$= y^2-2y-8$

11. $(x-4)(x-1) = x^2-4x-x+4 = x^2-5x+4$

12. $(2y+3)(y+4) = 2y^2+8y+3y+12$
$= 2y^2+11y+12$

6.3 Understanding the Concept Variable Expression or Equation?

1. We can simplify $3x + 1 + 4x$.
2. We can simplify and solve $3x + 1 + 4x = 9$.
3. We can simplify and solve $4 + 2x + 6x = 11$.
4. We can simplify $4 + 2x + 6x$.

6.3 Exercises

1. x = Mark's age and $2x$ = Juan's age.

3. x = Leslie's running record,
$x - 20$ = Alice's running record, and
$x - 50$ = Shannon's running record.

5. Mark drove 500 miles more than Scott. The rest of the answer may vary.

7. Since we are comparing the profit for April to the profit for March, we let the variable represent the profit for March.
Profit for March = x
The profit for April was $8000 less than the profit for March.
Profit for April = $x - 8000$

9. Since we are comparing the profit for the fourth quarter to the profit for the second quarter, we let the variable represent the profit for the second quarter.
Profit for the second quarter = S
The profit for the fourth quarter was $31,100 more than the profit for the second quarter.
Profit for the fourth quarter = $S + 31,100$

11. Since we are comparing the height of the pole to the height of the tree, we let the variable represent the height of the tree.
Height of the tree = t
The height of the pole is one-half the height of the tree.
Height of the pole = $\frac{1}{2}t$

13. Since we are comparing the length of a rectangle to its width, we let the variable represent the width of the rectangle.
Width = W
The length is double the width.
Length = $2W$

15. Since we are comparing the width of a rectangle to its length, we let the variable represent the length of the rectangle.
Length = L
The width is 13 inches shorter than twice the length.
Width = $2L - 13$

17. Since we are comparing the number of DVDs in Carl's collection to the number of DVDs in Toni's collection, we let the variable represent the number of DVDs in Toni's collection.
Number of Toni's DVDs = x
The number of DVDs in Carl's collection is three less than double the number of DVDs in Toni's collection.
Number of Carl's DVDs = $2x - 3$

19. Since we are comparing the length and height of a rectangular box to its width, we let the variable represent the width.
Width = W
The length is double the width.
Length = $2W$
The height is three times the width.
Height = $3W$

21. Since we are comparing the numbers of blue and white cars to the number of red cars, we let the variable represent the number of red cars.
Number of red cars = x
Jim has sixteen more blue cars than red cars.
Number of blue cars = $x + 16$
Jim has seven fewer white cars than red cars.
Number of white cars = $x - 7$

 Copyright © 2012 Pearson Education, Inc.

23. Since we are comparing the lengths of the second and third sides of a triangle to the length of the first side, we let the variable represent the length of the first side.
Length of the first side = F
The second side is 4 inches longer than the first.
Length of the second side = $F + 4$
The third side is 10 inches shorter than two times the first.
Length of the third side = $2F - 10$

25. a. Since we are comparing the height of the building to the height of the tree, we let the variable represent the height of the tree.
Height of the tree = t
The height of the building is four times the height of the tree.
Height of the building = $4t$

b. The difference between the height of the building and the height of the tree is $4t - t$.

c. Combine the like terms.
$4t - t = 3t$

27. a. Since we are comparing the numbers of votes that Jeri and Lee received to the number of votes that Max received, we let the variable represent the number of votes that Max received.
Number of Max's votes = x
Jeri received thirty more votes than Max.
Number of Jeri's votes: $x + 30$
Lee received forty-five fewer votes than Max.
Number of Lee's votes = $x - 45$

b. The number of Max's votes minus the number of Jeri's votes plus the number of Lee's votes is
$x - (x + 30) + (x - 45)$.

c. $$\begin{aligned} x-(x+30)+(x-45) &= x-x-30+x-45 \\ &= x-75 \end{aligned}$$

29. a. Since we are comparing Vu's and Evan's salaries to Sam's salary, we let the variable represent Sam's salary.
Sam's salary = x
Vu's salary is \$125 more than Sam's salary.
Vu's salary = $x + 125$
Evan's salary is \$80 less than Sam's salary.
Evan's salary = $x - 80$

b. Vu's salary plus Sam's salary minus Evan's salary is $(x + 125) + x - (x - 80)$.

c. $$\begin{aligned} (x+125)+x-(x-80) &= x+125+x-x+80 \\ &= x+205 \end{aligned}$$

31. a. Since we are comparing the height and length of a box to its width, we let the variable represent the width.
Width = W
The height is 8 inches longer than the width.
Height = $W + 8$
The length is 2 inches shorter than double the width.

b. The sum of the height, width, and length is $(W + 8) + W + (2W - 2)$.

c. $$\begin{aligned} &(W+8)+W+(2W-2) \\ &= W+8+W+2W-2 \\ &= 4W+6 \end{aligned}$$

33. We can solve $3x + 6$. False; there is no equals sign.

Cumulative Review

35. $$\begin{aligned} 11x &= 44 \\ \frac{11x}{11} &= \frac{44}{11} \\ x &= 4 \end{aligned}$$

36. $$\begin{aligned} y+77 &= -6 \\ y+77+(-77) &= -6+(-77) \\ y &= -83 \end{aligned}$$

37. $$\begin{aligned} \frac{m}{7} &= -5 \\ \frac{7\cdot m}{7} &= -5\cdot 7 \\ m &= -35 \end{aligned}$$

38. $$\begin{aligned} 4x-3x+8 &= 62 \\ x+8 &= 62 \\ + \quad -8 &= -8 \\ \hline x+0 &= 54 \\ x &= 54 \end{aligned}$$

39. $A = LW = (14)(10) = 140$
The area is 140 in.^2.

40. $V = LWH = (6)(4)(5) = 120$
The volume is 120 ft^3.

Copyright © 2012 Pearson Education, Inc.

Quick Quiz 6.3

1. Since we are comparing Tina's salary to Mai's salary, we let the variable represent Mai's salary.
   Mai's salary = x
   Tina's salary is triple Mai's salary.
   Tina's salary = $3x$

2. Since we are comparing Dixie's age and Pumpkin's age to Sugar's age, we let the variable represent Sugar's age.
   Sugar's age = x
   Dixie is 4 years older than Sugar.
   Dixie's age = $x + 4$
   Pumpkin is 3 years younger than twice Sugar's age.
   Pumpkin's age = $2x - 3$

3. **a.** Since we are comparing the costs of the watch and the bracelet to the cost of the ring, we let the variable represent the cost of the ring.
   Cost of the ring = x
   The watch cost seventy-five dollars more than the ring.
   Cost of the watch = $x + 75$
   The bracelet was fifty dollars less than the ring.
   Cost of the bracelet = $x - 50$

   b. The cost of the watch minus the cost of the bracelet plus the cost of the ring is $(x + 75) - (x - 50) + x$.

   c. $$(x+75)-(x-50)+x = x+75-x+50+x = x+125$$

4. **a.** Answers may vary. One possible solution is to let x represent the height because the length and width are represented in terms of height.

   b. x = height
   $3x$ = width
   $2x - 6$ = length

Use Math to Save Money

1. \$440,000 − \$178,000 = \$262,000
   At age 65 Taylor will have \$262,000 more than Doug.

2. \$1,007,000 − \$440,000 = \$567,000
   Ben will have \$567,000 more than Taylor.

3. \$3600 × 10 = \$36,000
   Ben will have contributed \$36,000 more than Taylor.

4. \$567,000 − \$36,000 = \$531,000
   Ben will accumulate \$567,000 more than Taylor; \$531,000 of this amount is due to compound interest.

5. \$280,000 × 3 = \$840,000
   If Sofia saves \$3000 each year, she will have \$840,000 at retirement.

6.4 Exercises

1. The sign in the binomial should be −, not +.

3. $8 = 8 \cdot 1, 8 = 4 \cdot 2$
   $12 = 12 \cdot 1, 12 = 6 \cdot 2, 12 = 4 \cdot 3$

   a. 1, 2, and 4 are the common factors.

   b. The largest common factor is 4, so the GCF of 8 and 12 is 4.

5. $14 = 2 \cdot 2$
   $16 = 2 \cdot 2 \cdot 2 \cdot 2$
   $\text{GCF} = 2 \cdot 2 = 4$

7. $18 = 2 \cdot 3 \cdot 3$
   $27 = 3 \cdot 3 \cdot 3$
   $\text{GCF} = 3 \cdot 3 = 9$

9. $6 = 2 \cdot 3$
   $9 = 3 \cdot 3$
   $15 = 3 \cdot 5$
   $\text{GCF} = 3$

11. $10 = 2 \cdot 5$
    $15 = 3 \cdot 5$
    $20 = 2 \cdot 2 \cdot 5$
    $\text{GCF} = 5$

13. $a^3bc + a^6c$

    a. The variables a and c are common to the terms.

    b. The smaller power on a is 3 and on c is 1.

    c. $\text{GCF} = a^3c$

15. $xy^2 + xy^3$
    Both exponents on x are 1. The smaller exponent on y is 2.
    $\text{GCF} = xy^2$

Copyright © 2012 Pearson Education, Inc.

17. $a^2b^5 + a^3b^4$
The smaller exponent on a is 2, and the smaller exponent on b is 4.
$\text{GCF} = a^2b^4$

19. $a^3bc^2 + ac^3$
The smaller exponent on a is 1, and the smaller exponent on c is 2.
$\text{GCF} = ac^2$

21. $x^3yz^3 + xy^4$
The smaller exponent on x is 1, and the smaller exponent on y is 1.
$\text{GCF} = xy$

23. $9x + 15 = \boxed{3}(3x + 5)$

25. $10xy^2 - 15y = \boxed{5y}(2xy - 3)$

27. $2x + 4 = 2(\boxed{x} + \boxed{2})$

29. $6x^2 - 3x = 3x(\boxed{2x} - \boxed{1})$

31. $18x^3 - 3x^2 - 9x = 3x(\boxed{6x^2} - \boxed{x} - \boxed{3})$

33. a. $2x - 10 = 2(x \boxed{-} 5)$

b. $2x + 10 = 2(x \boxed{+} 5)$

35. $3a - 6$
The GCF is 3.
$3a - 6 = 3 \cdot a - 3 \cdot 2 = 3(a - 2)$
Check: $3(a-2) \stackrel{?}{=} 3a-6$
$3(a-2) = 3 \cdot a - 3 \cdot 2 = 3a - 6$ ✓

37. $5y + 5$
The GCF is 5.
$5y + 5 = 5 \cdot y + 5 \cdot 1 = 5(y + 1)$
Check: $5(y+1) \stackrel{?}{=} 5y+5$
$5(y+1) = 5 \cdot y + 5 \cdot 1 = 5y + 5$ ✓

39. $10a + 4b$
The GCF is 2.
$10a + 4b = 2 \cdot 5a + 2 \cdot 2b = 2(5a + 2b)$
Check: $2(5a+2b) \stackrel{?}{=} 10a+4b$
$2(5a+2b) = 2 \cdot 5a + 2 \cdot 2b = 10a + 4b$ ✓

41. $15m + 3n$
The GCF is 3.
$15m + 3n = 3 \cdot 5m + 3 \cdot n = 3(5m + n)$
Check: $3(5m+n) \stackrel{?}{=} 15m+3n$
$3(5m+n) = 3 \cdot 5m + 3 \cdot n = 15m + 3n$ ✓

43. $7x + 14y + 21$
The GCF is 7.
$7x + 14y + 21 = 7 \cdot x + 7 \cdot 2y + 7 \cdot 3 = 7(x + 2y + 3)$
Check: $7(x+2y+3) \stackrel{?}{=} 7x+14y+21$
$7(x+2y+3) = 7 \cdot x + 7 \cdot 2y + 7 \cdot 3$
$= 7x + 14y + 21$ ✓

45. $8a + 18b - 6$
The GCF is 2.
$8a + 18b - 6 = 2 \cdot 4a + 2 \cdot 9b - 2 \cdot 3$
$= 2(4a + 9b - 3)$
Check: $2(4a+9b-3) \stackrel{?}{=} 8a+18b-6$
$2(4a+9b-3) = 2 \cdot 4a + 2 \cdot 9b - 2 \cdot 3$
$= 8a + 18b - 6$ ✓

47. $2a^2 - 4a$
The GCF is $2a$.
$2a^2 - 4a = 2a \cdot a - 2a \cdot 2 = 2a(a-2)$
Check: $2a(a-2) \stackrel{?}{=} 2a^2 - 4a$
$2a(a-2) = 2a \cdot a - 2a \cdot 2 = 2a^2 - 4a$ ✓

49. $4ab - b^2$
The GCF is b.
$4ab - b^2 = b \cdot 4a - b \cdot b = b(4a - b)$
Check: $b(4a-b) \stackrel{?}{=} 4ab - b^2$
$b(4a-b) = b \cdot 4a - b \cdot b = 4ab - b^2$ ✓

51. $5x + 10xy$
The GCF is $5x$.
$5x + 10xy = 5x \cdot 1 + 5x \cdot 2y = 5x(1 + 2y)$
Check:
$5x(1+2y) \stackrel{?}{=} 5x + 10xy$
$5x(1+2y) = 5x \cdot 1 + 5x \cdot 2y = 5x + 10xy$ ✓

53. $7x^2y - 14xy$
The GCF is $7xy$.
$7x^2y - 14xy = 7xy \cdot x - 7xy \cdot 2 = 7xy(x-2)$
Check:
$7xy(x-2) \stackrel{?}{=} 7x^2y - 14xy$
$7xy(x-2) = 7xy \cdot x - 7xy \cdot 2 = 7x^2y - 14xy$ ✓

Copyright © 2012 Pearson Education, Inc.

55. $12a^2b-6a^2$
The GCF is $6a^2$.
$12a^2b-6a^2=6a^2\cdot 2b-6a^2\cdot 1=6a^2(2b-1)$
Check:
$6a^2(2b-1)\stackrel{?}{=}12a^2b-6a^2$
$6a^2(2b-1)=6a^2\cdot 2b-6a^2\cdot 1=12a^2b-6a^2$ ✓

57. $3x^2-9x+18$
The GCF is 3.
$3x^2-9x+18=3\cdot x^2-3\cdot 3x+3\cdot 6$
$=3(x^2-3x+6)$
Check: $3(x^2-3x+6)\stackrel{?}{=}3x^2-9x+18$
$3(x^2-3x+6)=3\cdot x^2-3\cdot 3x+3\cdot 6$
$=3x^2-9x+18$ ✓

59. $4x^2+8x^3$
The GCF is $4x^2$.
$4x^2+8x^3=4x^2\cdot 1+4x^2\cdot 2x=4x^2(1+2x)$
Check:
$4x^2(1+2x)\stackrel{?}{=}4x^2+8x^3$
$4x^2(1+2x)=4x^2\cdot 1+4x^2\cdot 2x=4x^2+8x^3$ ✓

61. $2x^2y+4xy$
The GCF is $2xy$.
$2x^2y+4xy=2xy\cdot x+2xy\cdot 2=2xy(x+2)$
Check:
$2xy(x+2)\stackrel{?}{=}2x^2y+4xy$
$2xy(x+2)=2xy\cdot x+2xy\cdot 2=2x^2y+4xy$ ✓

63. $4y+2$
The GCF is 2.
$4y+2=2\cdot 2y+2\cdot 1=2(2y+1)$
Check:
$2(2y+1)\stackrel{?}{=}4y+2$
$2(2y+1)=2\cdot 2y+2\cdot 1=4y+2$ ✓

65. $15a-20$
The GCF is 5.
$15a-20=5\cdot 3a-5\cdot 4=5(3a-4)$
Check: $5(3a-4)\stackrel{?}{=}15a-20$
$5(3a-4)=5\cdot 3a-5\cdot 4=15a-20$ ✓

67. $5x-10xy$
The GCF is $5x$.
$5x-10xy=5x\cdot 1-5x\cdot 2y=5x(1-2y)$
Check:
$5x(1-2y)\stackrel{?}{=}5x-10xy$
$5x(1-2y)=5x\cdot 1-5x\cdot 2y=5x-10xy$ ✓

69. $9xy^3-3xy$
The GCF is $3xy$.
$9xy^3-3xy=3xy\cdot 3y^2-3xy\cdot 1=3xy(3y^2-1)$
Check:
$3xy(3y^2-1)\stackrel{?}{=}9xy^3-3xy$
$3xy(3y^2-1)=3xy\cdot 3y^2-3xy\cdot 1=9xy^3-3xy$ ✓

71. $6x-3y+12$
The GCF is 3.
$6x-3y+12=3\cdot 2x-3\cdot y+3\cdot 4=3(2x-y+4)$
Check: $3(2x-y+4)\stackrel{?}{=}6x-3y+12$
$3(2x-y+4)=3\cdot 2x-3\cdot y+3\cdot 4$
$=6x-3y+12$ ✓

73. $4x^2+8x-4$
The GCF is 4.
$4x^2+8x-4=4\cdot x^2+4\cdot 2x-4\cdot 1$
$=4(x^2+2x-1)$
Check: $4(x^2+2x-1)\stackrel{?}{=}4x^2+8x-4$
$4(x^2+2x-1)=4\cdot x^2+4\cdot 2x-4\cdot 1$
$=4x^2+8x-4$ ✓

75. $2x^3y^3-8x^2y^2$
The GCF is $2x^2y^2$.
$2x^3y^3-8x^2y^2=2x^2y^2\cdot xy-2x^2y^2\cdot 4$
$=2x^2y^2(xy-4)$
Check: $2x^2y^2(xy-4)\stackrel{?}{=}2x^3y^3-8x^2y^2$
$2x^2y^2(xy-4)=2x^2y^2\cdot xy-2x^2y^2\cdot 4$
$=2x^3y^3-8x^2y^2$ ✓

77. $6x^2y+2xy+4x$
The GCF is $2x$.
$6x^2y+2xy+4x=2x\cdot 3xy+2x\cdot y+2x\cdot 2$
$=2x(3xy+y+2)$
Check: $2x(3xy+y+2)\stackrel{?}{=}6x^2y+2xy+4x$
$2x(3xy+y+2)=2x\cdot 3xy+2x\cdot y+2x\cdot 2$
$=6x^2y+2xy+4x$ ✓

Copyright © 2012 Pearson Education, Inc.

79. **a.** $-2(x - 5y) = -2 \cdot x - 2(-5y) = -2x + 10y$

b. $2(-x + 5y) = 2(-x) + 2 \cdot 5y = -2x + 10y$

c. The products are the same.

Cumulative Review

81. $3 = 3$
$4 = 2 \cdot 2$
$2 = 2$
$\text{LCD} = 2 \cdot 2 \cdot 3 = 12$

82. $4 = 2 \cdot 2$
$2 = 2$
$5 = 5$
$\text{LCD} = 2 \cdot 2 \cdot 5 = 20$

83. $2x = 2 \cdot x$
$x = x$
$\text{LCD} = 2 \cdot x = 2x$

84. $x = x$
$5x = 5 \cdot x$
$\text{LCD} = 5 \cdot x = 5x$

85. Find the total rainfall for January, February, and March.

$$2\frac{2\cdot 4}{3\cdot 4} = 2\frac{8}{12}$$
$$3\frac{1\cdot 3}{4\cdot 3} = 3\frac{3}{12}$$
$$+\,2\frac{2\cdot 4}{3\cdot 4} = +\,2\frac{8}{12}$$
$$7\frac{19}{12} = 8\frac{7}{12} \text{ in.}$$

Subtract this total from 10 inches.

$$10 = 9\frac{12}{12}$$
$$-\,8\frac{7}{12} = 8\frac{7}{12}$$
$$1\frac{5}{12} = 1\frac{5}{12} \text{ in.}$$

$1\frac{5}{12}$ inches of rain must fall in April.

86. Divide to find the number of servings in 45 pounds.

$45 \div \frac{1}{3} = 45 \times 3 = 135$ servings

Since Louise has 135 servings of potato salad and only 125 guests, she will have enough potato salad.

Quick Quiz 6.4

1. **a.** $12 = 2 \cdot 2 \cdot 3$
$20 = 2 \cdot 2 \cdot 5$
$36 = 2 \cdot 2 \cdot 3 \cdot 3$
$\text{GCF} = 2 \cdot 2 = 4$

b. $x^2yz^2 - x^2y^2$
Both exponents on x are 2. The smaller exponent on y is 1.
$\text{GCF} = x^2y$

2. $4x^2 - 10y + 2$
$\text{GCF} = 2$
$$4x^2 - 10y + 2 = 2 \cdot 2x^2 - 2 \cdot 5y + 2 \cdot 1$$
$$= 2(2x^2 - 5y + 1)$$

3. $5ab^2 - 15ab$
$\text{GCF} = 5ab$
$5ab^2 - 15ab = 5ab \cdot b - 5ab \cdot 3 = 5ab(b - 3)$

4. **a.** Answers may vary. One possible solution is that xy is not part of the GCF because y is not a factor of $16x$.

b. $12xy = 12 \cdot x \cdot y = 3 \cdot 4 \cdot x \cdot y = 2 \cdot 6 \cdot x \cdot y$
$16x = 16 \cdot x = 4 \cdot 4 \cdot x$
$\text{GCF} = 4x$

c. $12xy + 16x = 4x \cdot 3y + 4x \cdot 4 = 4x(3y + 4)$

You Try It

1. $(-5x^2 + 8x - 3) + (3x^2 - 4x - 2)$
$= -5x^2 + 3x^2 + 8x - 4x - 3 - 2$
$= -2x^2 + 4x - 5$

2. $-(6y - 4z + 7) = -6y + 4z - 7$

3. $(7x^2 - 4x - 3) - (4x^2 - 2x - 5)$
$= 7x^2 - 4x - 3 - 4x^2 + 2x + 5$
$= 3x^2 - 2x + 2$

4. $-2a(4a - 5b + 6) = -2a(4a) - 2a(-5b) - 2a(6)$
$= -8a^2 + 10ab - 12a$

Copyright © 2012 Pearson Education, Inc.

5. $(4x+3)(2x^2-x+3)$
$=4x(2x^2-x+3)+3(2x^2-x+3)$
$=4x\cdot 2x^2-4x\cdot x+4x\cdot 3+3\cdot 2x^2-3\cdot x+3\cdot 3$
$=8x^3-4x^2+12x+6x^2-3x+9$
$=8x^3+2x^2+9x+9$

6. $(x+5)(x-2)=x^2-2x+5x-10=x^2+3x-10$

7. a. Since we are comparing the height and length of the box to its width, we let the variable represent the width.
Width = w
The height is triple the width.
Height = $3w$
The length is 3 inches shorter than double the width.
Length = $2w - 3$

b. $3w+w-(2w-3)=3w+w-2w+3=2w+3$

8. $3ab-9a^2b$
The GCF is $3ab$.
$3ab-9a^2b=3ab\cdot 1-3ab\cdot 3a=3ab(1-3a)$

9. $2(-2x^2+2)-(3x^2+5x-6)$
$=-4x^2+4-(3x^2+5x-6)$
$=-4x^2+4-3x^2-5x+6$
$=-7x^2-5x+10$

10. $4x-(x^2+2x)+3(3x^2-6x+4)$
$=4x-x^2-2x+3(3x^2-6x+4)$
$=4x-x^2-2x+9x^2-18x+12$
$=-16x+8x^2+12$
$=8x^2-16x+12$

11. $(-4)(6x^2-8x+5)=-4\cdot 6x^2-4(-8x)-4\cdot 5$
$=-24x^2+32x-20$

12. $(-2y)(y-6)=-2y\cdot y-2y(-6)=-2y^2+12y$

13. $(3x)(9x-3y+2)=3x\cdot 9x+3x(-3y)+3x\cdot 2$
$=27x^2-9xy+6x$

14. $(-5n)(-4n-9m-7)$
$=-5n(-4n)-5n(-9m)-5n(-7)$
$=20n^2+45mn+35n$

15. $(4x^2)(x^4-4)=4x^2\cdot x^4+4x^2(-4)=4x^6-16x^2$

16. $(x^4)(x^5-2x-3)=x^4\cdot x^5+x^4(-2x)+x^4(-3)$
$=x^9-2x^5-3x^4$

17. $(z-4)(5z)=5z(z-4)$
$=5z\cdot z+5z(-4)$
$=5z^2-20z$

18. $(y+10)(-6y)-6y(y+10)=-6y\cdot y-6y\cdot 10$
$=-6y^2-60y$

19. $(x^3-6x)(4x^2)=4x^2(x^3-6x)$
$=4x^2\cdot x^3+4x^2(-6x)$
$=4x^5-24x^3$

20. $(x-2)(2x^2+3x-1)$
$=x(2x^2+3x-1)-2(2x^2+3x-1)$
$=2x^3+3x^2-x-4x^2-6x+2$
$=2x^3-x^2-7x+2$

Chapter 6 Review Problems

1. The terms of $2x^2+5x-3z^3+4$ are $+2x^2$, $+5x$, $-3z^3$, and $+4$.

2. The terms of a^4-2b^2-3b-4 are $+a^4$, $-2b^2$, $-3b$, and -4.

3. $-(2a-3)=-2a+3$

4. $-(-6x+4y-2)=6x-4y+2$

5. $(-3x-9)+(5x-2)=-3x+5x-9-2=2x-11$

6. $(4x+8)-(8x+2)=4x+8+(-8x)+(-2)$
$=-4x+6$

7. $(9a^2-3a+5)-(-4a^2-6a-1)$
$=9a^2-3a+5+4a^2+6a+1$
$=13a^2+3a+6$

8. $(-4x^2-3)-(3x^2+7x+1)+(-x^2-4)$
$=-4x^2-3-3x^2-7x-1-x^2-4$
$=-8x^2-7x-8$

Copyright © 2012 Pearson Education, Inc.

21. $(y+5)(3y^2-2y+3)$
$= y(3y^2-2y+3)+5(3y^2-2y+3)$
$= 3y^3-2y^2+3y+15y^2-10y+15$
$= 3y^3+13y^2-7y+15$

22. $(y-1)(-3y^2+4y+5)$
$= y(-3y^2+4y+5)-1(-3y^2+4y+5)$
$= -3y^3+4y^2+5y+3y^2-4y-5$
$= -3y^3+7y^2+y-5$

23. $(2x+3)(x^2+3x-1)$
$= 2x(x^2+3x-1)+3(x^2+3x-1)$
$= 2x^3+6x^2-2x+3x^2+9x-3$
$= 2x^3+9x^2+7x-3$

24. $(x+2)(x+4) = x^2+4x+2x+8 = x^2+6x+8$

25. $(y+4)(y-7) = y^2-7y+4y-28 = y^2-3y-28$

26. $(x-2)(3x+4) = 3x^2+4x-6x-8 = 3x^2-2x-8$

27. $(x-3)(5x-6) = 5x^2-6x-15x+18$
$= 5x^2-21x+18$

28. Since we are comparing the profit for the third quarter to the profit for the first quarter, we let the variable represent the profit for the first quarter.
Profit for the first quarter = x
The profit for the third quarter is $22,300 more than the profit for the first quarter.
Profit for the third quarter = x + 22,300

29. Since we are comparing the width of a field to its length, we let the variable represent the length of the field.
Length = L
The width is 22 feet shorter than the length.
Width = L – 22

30. Since we are comparing the measures of $\angle a$ and $\angle c$ to the measure of $\angle b$, we let the variable represent the measure of $\angle b$.
Measure of $\angle b = x$
The measure of $\angle a$ is 30° more than the measure of $\angle b$.
Measure of $\angle a = x + 30$
The measure of $\angle c$ is twice the measure of $\angle b$.
Measure of $\angle c = 2x$

31. Since the numbers of carnations and lilies are being compared to the number of roses, we let the variable represent the number of roses.
Number of roses = x
There are three times as many carnations as roses.
Number of carnations = $3x$
There are five more lilies than roses.
Number of lilies = $x + 5$

32. a. Since we are comparing Phoebe's and Kelly's salaries to Erin's salary, we let the variable represent Erin's salary.
Erin's salary = x
Phoebe's salary is $145 more than Erin's salary.
Phoebe's salary = x + 145
Kelly's salary is $60 less than Erin's salary.
Kelly's salary = x – 60

b. Erin's salary plus Phoebe's salary minus Kelly's salary is $x + (x + 145) - (x - 60)$.

c. $x+(x+145)-(x-60) = x+x+145-x+60$
$= x+205$

33. a. Since the numbers of children with brown eyes and with green eyes are being compared to the number of children with blue eyes, we let the variable represent the number of children with blue eyes.
Number of children with blue eyes = x
The number of children with brown eyes is seven more than the number with blue eyes.
Number with brown eyes = $x + 7$
There are nine fewer children with green eyes than with blue eyes.
Number with green eyes = $x - 9$

b. The number of children with blue eyes plus the number of children with brown eyes minus the number of children with green eyes is $x + (x + 7) - (x - 9)$.

c. $x+(x+7)-(x-9) = x+x+7-x+9$
$= x+16$

34. a. Since we are comparing the lengths of the second and third sides of a triangle to the length of the first side, we let the variable represent the length of the first side.
Length of the first side = x
The length of the second side is double the first.
Length of the second side = $2x$
The third side is 10 inches longer than the first.
Length of the third side = $x + 10$

Copyright © 2012 Pearson Education, Inc.

b. The sum of all sides of the triangle is $x + 2x + (x + 10)$.

c. $x+2x+(x+10) = x+2x+x+10 = 4x+10$

35. a. Since we are comparing the length and height of a box to its width, we let the variable represent the width.
Width = W
The length is 7 inches longer than the width.
Length = $W + 7$
The height is 4 inches shorter than three times the width.
Height = $3W - 4$

b. The sum of the height, width, and length is $(3W - 4) + W + (W + 7)$.

c. $(3W-4)+W+(W+7)$
$= 3W-4+W+W+7$
$= 5W+3$

36. $14 = 2 \cdot 7$
$21 = 3 \cdot 7$
GCF = 7

37. $6 = 2 \cdot 3$
$21 = 3 \cdot 7$
GCF = 3

38. $25 = 5 \cdot 5$
$45 = 3 \cdot 3 \cdot 5$
GCF = 5

39. $18 = 2 \cdot 3 \cdot 3$
$36 = 2 \cdot 2 \cdot 3 \cdot 3$
GCF = $2 \cdot 3 \cdot 3 = 18$

40. $8 = 2 \cdot 2 \cdot 2$
$14 = 2 \cdot 7$
$18 = 2 \cdot 3 \cdot 3$
GCF = 2

41. $12 = 2 \cdot 2 \cdot 3$
$16 = 2 \cdot 2 \cdot 2 \cdot 2$
$20 = 2 \cdot 2 \cdot 5$
GCF = $2 \cdot 2 = 4$

42. $a^2bc + ab^3$
The smaller exponent on a is 1, and the smaller exponent on b is 1.
GCF = ab

43. $xy^3z + x^2y^2$
The smaller exponent on x is 1, and the smaller exponent on y is 2.
GCF = xy^2

44. $6x - 14$
GCF = 2
$6x - 14 = 2 \cdot 3x - 2 \cdot 7 = 2(3x - 7)$

45. $5x + 15$
GCF = 5
$5x + 15 = 5 \cdot x + 5 \cdot 3 = 5(x + 3)$

46. $4a + 12b$
GCF = 4
$4a + 12b = 4 \cdot a + 4 \cdot 3b = 4(a + 3b)$

47. $3y - 9z$
GCF = 3
$3y - 9z = 3 \cdot y - 3 \cdot 3z = 3(y - 3z)$

48. $6xy^2 - 12xy$
GCF = $6xy$
$6xy^2 - 12xy = 6xy \cdot y - 6xy \cdot 2 = 6xy(y-2)$

49. $8a^2b - 16ab$
GCF = $8ab$
$8a^2b - 16ab = 8ab \cdot a - 8ab \cdot 2 = 8ab(a-2)$

50. $10x^3y + 5x^2y$
GCF = $5x^2y$
$10x^3y + 5x^2y = 5x^2y \cdot 2x + 5x^2y \cdot 1$
$= 5x^2y(2x+1)$

51. $4y^3 - 6y^2 + 2y$
GCF = $2y$
$4y^3 - 6y^2 + 2y = 2y \cdot 2y^2 - 2y \cdot 3y + 2y \cdot 1$
$= 2y(2y^2 - 3y + 1)$

52. $3a - 6b + 12$
GCF = 3
$3a - 6b + 12 = 3 \cdot a - 3 \cdot 2b + 3 \cdot 4 = 3(a - 2b + 4)$

53. $2x + 4y - 10$
GCF = 2
$2x + 4y - 10 = 2 \cdot x + 2 \cdot 2y - 2 \cdot 5 = 2(x + 2y - 5)$

Copyright © 2012 Pearson Education, Inc.

How Am I Doing? Chapter 6 Test

1. The terms of $x^2y-2x^2+3y-5x$ are $+x^2y$, $-2x^2$, $+3y$, and $-5x$.

2. $-(4x-2y-6)=-4x+2y+6$

3. $(-5x+3)+(-2x+4)=-5x-2x+3+4$
$=-7x+7$

4. $(4y+5)-(2y-3)=4y+5+(-2y)+3=2y+8$

5. $(-7p-2)-(3p+4)=-7p-2-3p+(-4)$
$=-10p-6$

6. $(4x^2+8x-3)+(9x^2-10x+1)$
$=4x^2+9x^2+8x-10x-3+1$
$=13x^2-2x-2$

7. $(-6m^2-3m-8)-(6m^2+3m-4)$
$=-6m^2-3m-8+(-6m^2)+(-3m)+4$
$=-12m^2-6m-4$

8. $(x^2-x+7)+(-2x^2+4x+6)-(x^2+8)$
$=x^2-x+7-2x^2+4x+6-x^2-8$
$=-2x^2+3x+5$

9. $3x-2(7x^2+2x-1)-(3x^2+8x-2)$
$=3x-14x^2-4x+2-(3x^2+8x-2)$
$=3x-14x^2-4x+2-3x^2-8x+2$
$=-9x-17x^2+4$
$=-17x^2-9x+4$

10. $-7a(2a+3b-4)=-7a(2a)-7a(3b)-7a(-4)$
$=-14a^2-21ab+28a$

11. $(-2x^3)(4x^2-3)=-2x^3(4x^2)-2x^3(-3)$
$=-8x^5+6x^3$

12. $(x+5)(x+9)=x^2+9x+5x+45$
$=x^2+14x+45$

13. $(x+3)(x-2)=x^2-2x+3x-6=x^2+x-6$

14. $(2x+1)(x-3)=2x^2-6x+x-3$
$=2x^2-5x-3$

15. $(3x^3-1)(-4x^4)=-4x^4(3x^3-1)$
$=-4x^4(3x^3)-4x^4(-1)$
$=-12x^7+4x^4$

16. $(y-3)(4y^2+2y-6)$
$=y(4y^2+2y-6)-3(4y^2+2y-6)$
$=4y^3+2y^2-6y-12y^2-6y+18$
$=4y^3-10y^2-12y+18$

17. Since we are comparing the width of a piece of wood to its length, we let the variable represent the length of the piece of wood.
Length = L
The width is three inches shorter than the length.
Width = $L-3$

18. Since we are comparing the lengths of the second and third sides of a triangle to the length of the first side, we let the variable represent the length of the first side.
Length of the first side = f
The second side is 6 inches longer than the first.
Length of the second side = $f+6$
The third side is 2 inches shorter than two times the first.
Length of the third side = $2f-2$

19. a. Since we are comparing the numbers of votes that Jason and Nhan received to the number of votes that Lena received, we let the variable represent the number of votes that Lena received.
Number of votes for Lena = x
Jason received 3000 fewer votes than Lena.
Number of votes for Jason: $x-3000$
Nhan received 5100 more votes than Lena.
Number of votes for Nhan = $x+5100$

b. The number of votes received by Nhan plus the number of votes received by Lena minus the number of votes received by Jason is $(x+5100)+x-(x-3000)$.

c. $(x+5100)+x-(x-3000)$
$=x+5100+x-x+3000$
$=x+8100$

20. $9=3\cdot 3$
$21=3\cdot 7$
GCF = 3

Copyright © 2012 Pearson Education, Inc.

21. $8 = 2 \cdot 2 \cdot 2$
$16 = 2 \cdot 2 \cdot 2 \cdot 2$
$20 = 2 \cdot 2 \cdot 5$
$\text{GCF} = 2 \cdot 2 = 4$

22. $x^2yz + x^3z$
The smaller exponent on x is 2, and the smaller exponent on z is 1.
$\text{GCF} = x^2z$

23. $3x + 12$
$\text{GCF} = 3$
$3x + 12 = 3 \cdot x + 3 \cdot 4 = 3(x + 4)$

24. $7x^2 - 14x + 21$
$\text{GCF} = 7$
$$7x^2 - 14x + 21 = 7 \cdot x^2 - 7 \cdot 2x + 7 \cdot 3 = 7(x^2 - 2x + 3)$$

25. $2x^2y - 6xy^2$
$\text{GCF} = 2xy$
$2x^2y - 6xy^2 = 2xy \cdot x - 2xy \cdot 3y = 2xy(x - 3y)$

Cumulative Test for Chapters 1–6

1. $560 + $35 + $410 + $30 + $22 + $120 = $1177
Since $1177 > $1100, she does not have enough.
$1177 − $1100 = $77
She needs $77.

2. $5 \times \$10 + 7 \times \$6 + 4 \times \$7 = \$50 + \$42 + \$28 = \$120$
$\frac{\$120}{2} = \$60$
Assuming Alexandra and Stanley are sharing the cost evenly, Stanley will have to pay $60.

3. $5x - 3x + x + 5 = (5 - 3 + 1)x + 5 = 3x + 5$

4. $-8r + 3 - 5r - 8 = (-8 - 5)r + (3 - 8) = -13r - 5$

5. $8^2 - 10 + 4 = 64 - 10 + 4 = 54 + 4 = 58$

6. $-4 + 2^3 - 9 = -4 + 8 - 9 = 4 - 9 = -5$

7.
$$\begin{aligned} 7 - 24 \div 6(-2)^2 - 3 &= 7 - 24 \div 6(4) - 3 \\ &= 7 - 4(4) - 3 \\ &= 7 - 16 - 3 \\ &= -9 - 3 \\ &= -12 \end{aligned}$$

8.
$$\begin{aligned} 3 - 12 \div (-2) + 4^2 &= 3 - 12 \div (-2) + 16 \\ &= 3 - (-6) + 16 \\ &= 9 + 16 \\ &= 25 \end{aligned}$$

9. $(-10x^2)(5x) = (-10 \cdot 5)x^{2+1} = -50x^3$

10. $(3x^2)(x^3)(x) = 3x^{2+3+1} = 3x^6$

11. $5x(x^2 + 3) = 5x \cdot x^2 + 5x \cdot 3 = 5x^3 + 15x$

12. $\frac{90n^2}{54n} = \frac{\cancel{18} \cdot 5\cancel{n} \cdot n}{\cancel{18} \cdot 3 \cdot \cancel{n}} = \frac{5n}{3}$

13. $\frac{8a^3}{32a^5} = \frac{\cancel{8}a^3}{\cancel{8} \cdot 2 \cdot 2 \cdot a^5} = \frac{1}{4a^{5-3}} = \frac{1}{4a^2}$

14. $\left(\frac{x}{3}\right)^3 = \left(\frac{x^1}{3^1}\right)^3 = \frac{x^{(1)(3)}}{x^{(1)(3)}} = \frac{x^3}{3^3} = \frac{x^3}{27}$

15. a. $A = s^2 = 12^2 = 144$
The area is 144 in.^2.

b. $A = bh = (10)(15) = 150$
The area is 150 ft^2.

16. $V = LWH = (5)(3)(4) = 60$
The volume is 60 yd^3.

17. $4\overline{)37}$ = 9 R 1
```
   9 R 1
4)37
  36
   1
```
$\frac{37}{4} = 9\frac{1}{4}$

18. $3\overline{)40}$ = 13 R 1
```
  13 R 1
3)40
  3
  10
   9
   1
```
$\frac{40}{3} = 13\frac{1}{3}$

19.
```
    550
    / \
  10 · 55
  /\   /\
 2·5  5·11
```
$550 = 2 \cdot 5 \cdot 5 \cdot 11$ or $2 \cdot 5^2 \cdot 11$

Copyright © 2012 Pearson Education, Inc.

20. We divide 310 ÷ 6 to find the unit rate.

$$\begin{array}{r} 51 \text{ R } 4 \\ 6\overline{)310} \\ \underline{30} \\ 10 \\ \underline{6} \\ 4 \end{array} \qquad 51\frac{4}{6} = 51\frac{2}{3}$$

$$\frac{310 \text{ mi}}{6 \text{ hr}} = 51\frac{2}{3} \text{ mph}$$

21. Let x be the amount of carbohydrates in 12 ounces of yogurt. 4 ounces is to 9 grams as 12 ounces is to x grams.

$$\frac{4 \text{ oz}}{9 \text{ g}} = \frac{12 \text{ oz}}{x \text{ g}}$$
$$\frac{4}{9} = \frac{12}{x}$$
$$x \cdot 4 = 9 \cdot 12$$
$$4x = 108$$
$$\frac{4x}{4} = \frac{108}{4}$$
$$x = 27$$

There are 27 grams of carbohydrates in 12 ounces of yogurt.

22. Replace n with −4 and m with 2.

$$\frac{n^2 - 6}{m} = \frac{(-4)^2 - 6}{2} = \frac{16 - 6}{2} = \frac{10}{2} = 5$$

23. **a.**
$$x + 45 = -2$$
$$x + 45 + (-45) = -2 + (-45)$$
$$x = -47$$

b.
$$6x - 5x - 9 = 34$$
$$x - 9 = 34$$
$$x - 9 + 9 = 34 + 9$$
$$x = 43$$

24. **a.**
$$\frac{n}{-8} = 6$$
$$\frac{-8 \cdot n}{-8} = -8 \cdot 6$$
$$n = -48$$

b.
$$-9n = 99$$
$$\frac{-9n}{-9} = \frac{99}{-9}$$
$$n = -11$$

25. $$\frac{2x^4}{5x} \cdot \frac{10x^2}{4} = \frac{\cancel{2} \cdot x^4 \cdot \cancel{2} \cdot \cancel{5} \cdot x^2}{\cancel{5} \cdot x \cdot \cancel{2} \cdot \cancel{2}} = x^{4+2-1} = x^5$$

26. $$-\frac{1}{6} \div \left(-\frac{2}{3}\right) = \frac{1}{6} \cdot \frac{3}{2} = \frac{1 \cdot \cancel{3}}{2 \cdot \cancel{3} \cdot 2} = \frac{1}{4}$$

27. $12 = 2 \cdot 2 \cdot 3$
$28 = 2 \cdot 2 \cdot 7$
$\text{LCD} = 2 \cdot 2 \cdot 3 \cdot 7 = 84$

28. We subtract.

$$\begin{array}{rcr} 12\frac{1 \cdot 3}{2 \cdot 3} & = & 12\frac{3}{6} \\ -\ 9\frac{1 \cdot 2}{3 \cdot 2} & = & -\ 9\frac{2}{6} \\ \hline & & 3\frac{1}{6} \text{ gal} \end{array}$$

Frank used $3\frac{1}{6}$ gallons of gas.

29. We multiply to find the amount Jaci takes to work.

$$\frac{1}{3} \cdot 5\frac{1}{4} = \frac{1}{3} \cdot \frac{21}{4} = \frac{1 \cdot \cancel{3} \cdot 7}{\cancel{3} \cdot 2 \cdot 2} = \frac{7}{4} = 1\frac{3}{4} \text{ lb}$$

We subtract to find how much she has left.

$$\begin{array}{rcr} 5\frac{1}{4} & = & 4\frac{5}{4} \\ -\ 1\frac{3}{4} & = & -\ 1\frac{3}{4} \\ \hline & & 3\frac{2}{4} = 3\frac{1}{2} \text{ lb} \end{array}$$

Jaci has $3\frac{1}{2}$ pounds of candy left.

30. $(-8x^2 + 3x - 6) - (x^2 + 5x)$
$= -8x^2 + 3x - 6 + (-x^2) + (-5x)$
$= -9x^2 - 2x - 6$

31. $(x + 1)(3x^2 - 2x + 6)$
$= x(3x^2 - 2x + 6) + 1(3x^2 - 2x + 6)$
$= 3x^3 - 2x^2 + 6x + 3x^2 - 2x + 6$
$= 3x^3 + x^2 + 4x + 6$

32. $(x + 2)(x + 6) = x^2 + 6x + 2x + 12 = x^2 + 8x + 12$

33. $(2x + 7)(x - 3) = 2x^2 - 6x + 7x - 21$
$= 2x^2 + x - 21$

Copyright © 2012 Pearson Education, Inc.

34. Since we are comparing the height of the building to the height of the tree, we let the variable represent the height of the tree.
Height of the tree = t
The building is 4 feet higher than twice the height of the tree.
Height of the building = $2t + 4$

35. $9a + 18b + 9$
GCF = 9
$9a+18b+9=9\cdot a+9\cdot 2b+9\cdot 1=9(a+2b+1)$

36. $12x^2y+6x^2$
$\text{GCF}=6x^2$
$12x^2y+6x^2=6x^2\cdot 2y+6x^2\cdot 1=6x^2(2y+1)$

Copyright © 2012 Pearson Education, Inc.

Chapter 7

7.1 Exercises

1. To solve the equation $y + 8 = -17$, we add the opposite of +8, which is <u>–8</u>, to both sides of the equation.

3. To solve the equation $x - 15 = 82$, we add <u>15</u> to both sides of the equation.

5. Before we solve the equation $7x + 2 - 6x - 9 = 2 - 8$, we must first <u>simplify</u> each side of the equation.

7. To solve the equation $-9y = -36$, we <u>divide by –9</u> on both sides of the equation.

9. To solve the equation $\frac{y}{-9} = 2$, we <u>multiply by –9</u> on both sides of the equation.

11. We <u>add –3</u> on both sides of the equation when we solve $14 = x + 3$.

13.
$$\begin{aligned} x-7&=-20\\ +\quad +7&=+7\\ \hline x+0&=-13\\ x&=-13 \end{aligned}$$
Check:
$$\begin{aligned} x-7&=-20\\ -13-7&\stackrel{?}{=}-20\\ -20&=-20\ \checkmark \end{aligned}$$

15.
$$\begin{aligned} a+9&=-1\\ +\quad -9&=-9\\ \hline a+0&=-10\\ a&=-10 \end{aligned}$$
Check:
$$\begin{aligned} a+9&=-1\\ -10+9&\stackrel{?}{=}-1\\ -1&=-1\ \checkmark \end{aligned}$$

17.
$$\begin{aligned} -5&=x+5\\ +\quad -5&=\ -5\\ \hline -10&=x+0\\ -10&=x \text{ or } x=-10 \end{aligned}$$
Check:
$$\begin{aligned} -5&=x+5\\ -5&\stackrel{?}{=}-10+5\\ -5&=-5\ \checkmark \end{aligned}$$

19.
$$\begin{aligned} 7-9&=y-4\\ -2&=y-4\\ +\quad +4&=\ +4\\ \hline 2&=y \text{ or } y=2 \end{aligned}$$
Check:
$$\begin{aligned} 7-9&=y-4\\ 7-9&\stackrel{?}{=}2-4\\ -2&=-2\ \checkmark \end{aligned}$$

21.
$$\begin{aligned} 7-12&=x-6+3^2\\ -5&=x-6+3^2\\ -5&=x-6+9\\ -5&=x+3\\ +\quad -3&=\ -3\\ \hline -8&=x \text{ or } x=-8 \end{aligned}$$
Check:
$$\begin{aligned} 7-12&=x-6+3^2\\ 7-12&\stackrel{?}{=}-8-6+3^2\\ -5&\stackrel{?}{=}-8-6+9\\ -5&=-5\ \checkmark \end{aligned}$$

23.
$$\begin{aligned} -8+4^2&=a+6-4\\ -8+16&=a+6-4\\ 8&=a+6-4\\ 8&=a+2\\ +\quad -2&=\ -2\\ \hline 6&=a \text{ or } a=6 \end{aligned}$$
Check:
$$\begin{aligned} -8+4^2&=a+6-4\\ -8+4^2&\stackrel{?}{=}6+6-4\\ -8+16&\stackrel{?}{=}12-4\\ 8&=8\ \checkmark \end{aligned}$$

25.
$$\begin{aligned} 12x-1-11x-1&=-5\\ x-2&=-5\\ +\quad +2&=+2\\ \hline x&=-3 \end{aligned}$$
Check:
$$\begin{aligned} 12x-1-11x-1&=-5\\ 12(-3)-1-11(-3)-1&\stackrel{?}{=}-5\\ -36-1+33-1&\stackrel{?}{=}-5\\ -5&=-5\ \checkmark \end{aligned}$$

27.
$$\begin{aligned} 12-16&=4y+8-3y\\ -4&=4y+8-3y\\ -4&=y+8\\ +\quad -8&=-8\\ \hline -12&=y \text{ or } y=-12 \end{aligned}$$

Copyright © 2012 Pearson Education, Inc.

Check: $12-16=4y+8-3y$
$$12-16 \stackrel{?}{=} 4(-12)+8-3(-12)$$
$$-4 \stackrel{?}{=} -48+8+36$$
$$-4=-4 \checkmark$$

29. $$6x-6-5x+1=-2+4$$
$$x-5=-2+4$$
$$x-5=2$$
$$+5=+5$$
$$x=7$$

Check: $6x-6-5x+1=-2+4$
$$6(7)-6-5(7)+1 \stackrel{?}{=} -2+4$$
$$42-6-35+1 \stackrel{?}{=} 2$$
$$2=2 \checkmark$$

31. $$-16=\frac{x}{3}$$
$$-16\cdot 3=\frac{3\cdot x}{3}$$
$$-48=x \text{ or } x=-48$$

Check: $-16=\frac{x}{3}$
$$-16 \stackrel{?}{=} \frac{-48}{3}$$
$$-16=-16 \checkmark$$

33. $$14=\frac{y}{-2}$$
$$14\cdot(-2)=\frac{-2\cdot y}{-2}$$
$$-28=y \text{ or } y=-28$$

Check: $14=\frac{y}{-2}$
$$14 \stackrel{?}{=} \frac{-28}{-2}$$
$$14=14 \checkmark$$

35. $$\frac{a}{-4}=-6+9$$
$$\frac{a}{-4}=3$$
$$\frac{-4\cdot a}{-4}=3\cdot(-4)$$
$$a=-12$$

Check: $\frac{a}{-4}=-6+9$
$$\frac{-12}{-4} \stackrel{?}{=} -6+9$$
$$3=3 \checkmark$$

37. $$\frac{x}{4}=-8+6$$
$$\frac{x}{4}=-2$$
$$\frac{4\cdot x}{4}=-2\cdot 4$$
$$x=-8$$

Check: $\frac{x}{4}=-8+6$
$$\frac{-8}{4} \stackrel{?}{=} -8+6$$
$$-2=-2 \checkmark$$

39. $$\frac{y}{-2}=4+2^2$$
$$\frac{y}{-2}=4+4$$
$$\frac{y}{-2}=8$$
$$\frac{-2\cdot y}{-2}=8\cdot(-2)$$
$$y=-16$$

Check: $\frac{y}{-2}=4+2^2$
$$\frac{-16}{-2} \stackrel{?}{=} 4+2^2$$
$$8 \stackrel{?}{=} 4+4$$
$$8=8 \checkmark$$

41. $$\frac{3}{5}x=9$$
$$\frac{5}{3}\cdot\frac{3}{5}x=9\cdot\frac{5}{3}$$
$$x=\frac{3\cdot \cancel{3}\cdot 5}{\cancel{3}}$$
$$x=15$$

Check: $\frac{3}{5}x=9$
$$\frac{3}{5}\cdot 15 \stackrel{?}{=} 9$$
$$9=9 \checkmark$$

43. $$\frac{-2}{3}a=18$$
$$\left(-\frac{3}{2}\right)\left(-\frac{2}{3}\right)a=18\left(-\frac{3}{2}\right)$$
$$a=-\frac{9\cdot \cancel{2}\cdot 3}{\cancel{2}}$$
$$a=-27$$

Copyright © 2012 Pearson Education, Inc.

Check: $-\frac{2}{3}a = 18$

$-\frac{2}{3}(-27) \stackrel{?}{=} 18$

$18 = 18$ ✓

45. $\frac{7}{6}y = 4^2 + 5$

$\frac{7}{6}y = 16 + 5$

$\frac{7}{6}y = 21$

$\frac{6}{7} \cdot \frac{7}{6}y = 21 \cdot \frac{6}{7}$

$y = \frac{3 \cdot \cancel{7} \cdot 6}{\cancel{7}}$

$y = 18$

Check: $\frac{7}{6}y = 4^2 + 5$

$\frac{7}{6} \cdot 18 \stackrel{?}{=} 4^2 + 5$

$21 = 21$ ✓

47. $3(2x) = 24$

$6x = 24$

$\frac{6x}{6} = \frac{24}{6}$

$x = 4$

Check: $3(2x) = 24$

$3(2 \cdot 4) \stackrel{?}{=} 24$

$3(8) \stackrel{?}{=} 24$

$24 = 24$ ✓

49. $-6 = 2(-5x)$

$-6 = -10x$

$\frac{-6}{-10} = \frac{-10x}{-10}$

$\frac{\cancel{2} \cdot 3}{\cancel{2} \cdot 5} = x$

$\frac{3}{5} = x$ or $x = \frac{3}{5}$

Check: $-6 = 2(-5x)$

$-6 \stackrel{?}{=} 2\left(-5 \cdot \frac{3}{5}\right)$

$-6 \stackrel{?}{=} 2(-3)$

$-6 = -6$ ✓

51. $8x + 4(4x) = 48$

$8x + 16x = 48$

$24x = 48$

$\frac{24x}{24} = \frac{48}{24}$

$x = 2$

Check: $8x + 4(4x) = 48$

$8 \cdot 2 + 4(4 \cdot 2) \stackrel{?}{=} 48$

$16 + 32 \stackrel{?}{=} 48$

$48 = 48$ ✓

53. $6(-2x) - 3x = -30$

$-12x - 3x = -30$

$-15x = -30$

$\frac{-15x}{-15} = \frac{-30}{-15}$

$x = 2$

Check: $6(-2x) - 3x = -30$

$6(-2 \cdot 2) - 3 \cdot 2 \stackrel{?}{=} -30$

$-24 - 6 \stackrel{?}{=} -30$

$-30 = -30$ ✓

55. $\frac{-10}{2} = 3(4x) + 2x$

$-5 = 3(4x) + 2x$

$-5 = 12x + 2x$

$-5 = 14x$

$\frac{-5}{14} = \frac{14x}{14}$

$-\frac{5}{14} = x$ or $x = -\frac{5}{14}$

Check: $\frac{-10}{2} = 3(4x) + 2x$

$\frac{-10}{2} \stackrel{?}{=} 3\left[4\left(-\frac{5}{14}\right)\right] + 2\left(-\frac{15}{4}\right)$

$-5 \stackrel{?}{=} -\frac{60}{14} - \frac{10}{14}$

$-5 \stackrel{?}{=} -\frac{70}{14}$

$-5 = -5$ ✓

57. $\frac{21}{3} = 5x + 3(-4x)$

$7 = 5x + 3(-4x)$

$7 = 5x - 12x$

$7 = -7x$

$\frac{7}{-7} = \frac{-7x}{-7}$

$-1 = x$ or $x = -1$

Copyright © 2012 Pearson Education, Inc.

Check: $\frac{21}{3}=5x+3(-4x)$
$\frac{21}{3}\stackrel{?}{=}5(-1)+3[-4(-1)]$
$7\stackrel{?}{=}-5+12$
$7=7$ ✓

59. $-x=9$
$-1x=9$
$\frac{-1x}{-1}=\frac{9}{-1}$
$x=-9$
Check: $-x=9$
$-(-9)\stackrel{?}{=}9$
$9=9$ ✓

61. $-x=-4$
$-1x=-4$
$\frac{-1x}{-1}=\frac{-4}{-1}$
$x=4$
Check: $-x=-4$
$-(4)\stackrel{?}{=}-4$
$-4=-4$ ✓

63. a. $2x=-12$
$\frac{2x}{2}=\frac{-12}{2}$
$x=-6$

b. $\frac{x}{2}=-12$
$\frac{2\cdot x}{2}=-12\cdot 2$
$x=-24$

c. $x-2=-12$
$+\ \ +2=+2$
$x=-10$

d. $x+2=-12$
$+\ \ -2=-2$
$x=-14$

65. a. $y-10=9$
$+\ \ +10=+10$
$y=19$

b. $y+10=9$
$+\ \ -10=-10$
$y=-1$

c. $\frac{y}{-10}=9$
$\frac{-10\cdot y}{-10}=9(-10)$
$y=-90$

d. $-10y=9$
$\frac{-10y}{-10}=\frac{9}{-10}$
$y=-\frac{9}{10}$

67. $-15=a+5$
$+\ \ -5=\ -5$
$-20=a$ or $a=-20$

69. $5x+2-4x=9$
$x+2=9$
$+\ \ \ -x=-2$
$x=7$

71. $\frac{x}{7}=-2+3^2$
$\frac{x}{7}=-2+9$
$\frac{x}{7}=7$
$\frac{7\cdot x}{7}=7\cdot 7$
$x=49$

73. $-x=12$
$-1x=12$
$\frac{-1x}{-1}=\frac{12}{-1}$
$x=-12$

75. $\frac{20}{5}=3x+2(-6x)$
$4=3x+2(-6x)$
$4=3x-12x$
$4=-9x$
$\frac{4}{-9}=\frac{-9x}{-9}$
$-\frac{4}{9}=x$ or $x=-\frac{4}{9}$

Copyright © 2012 Pearson Education, Inc.

77.
$$\frac{6}{7}x = 3^2 + 3$$
$$\frac{6}{7}x = 9 + 3$$
$$\frac{6}{7}x = 12$$
$$\frac{7}{6} \cdot \frac{6}{7}x = 12 \cdot \frac{7}{6}$$
$$x = \frac{\cancel{6} \cdot 2 \cdot 7}{\cancel{6}}$$
$$x = 14$$

79. a.
$$-2x = 8$$
$$\frac{-2x}{-2} = \frac{8}{-2}$$
$$x = -4$$

b.
$$-2x = 8$$
$$\left(-\frac{1}{2}\right)(-2)x = 8\left(-\frac{1}{2}\right)$$
$$x = -4$$

c. The answers are the same.

d. Division is defined in terms of multiplication, therefore dividing by −2 is equivalent to multiplying by the reciprocal of −2, which is $-\frac{1}{2}$.

81. a.
$$9 + x + 7 = 18$$
$$x + 16 = 18$$
$$+ \quad -16 = -16$$
$$x = 2$$

b.
$$z + 6 + 8 = 18$$
$$z + 14 = 18$$
$$+ \quad -14 = -14$$
$$z = 4$$

$$y + 10 + 3 = 18$$
$$y + 13 = 18$$
$$+ \quad -13 = -13$$
$$y = 5$$

c.

9	4	5
2	6	10
7	8	3

Sums of diagonals:
9 + 6 + 3 = 18 ✓
7 + 6 + 5 = 18 ✓

83. The sum of the interior angles is 180°.
$$x + 30° + 90° = 180°$$
$$x + 120° = 180°$$
$$+ \quad -120° = -120°$$
$$x = 60°$$
The missing angle is 60°.

85. The sum of the missing angles is 90°.
$$2x + 3x = 90°$$
$$5x = 90°$$
$$\frac{5x}{5} = \frac{90°}{5}$$
$$x = 18°$$
$2x = 2(18°) = 36°$; $3x = 3(18°) = 54°$
The angles measure 36° and 54°.

Cumulative Review

87.
```
     210
     / \
   10 · 21
   /\   /\
  2·5  3·7
```
$210 = 2 \cdot 5 \cdot 3 \cdot 7 = 2 \cdot 3 \cdot 5 \cdot 7$

88.
```
     112
     / \
    4 · 28
   / \  / \
  2·2  4·7
       /\
      2·2
```
$112 = 2 \cdot 2 \cdot 2 \cdot 2 \cdot 7$ or $2^4 \cdot 7$

89. We subtract.
$$\begin{array}{r} 94{,}500{,}000 \\ -\ 91{,}400{,}000 \\ \hline 3{,}100{,}000 \text{ mi} \end{array}$$
The difference is approximately 3,100,000 miles.

Quick Quiz 7.1

1.
$$-3y + 4y + 8 = -9 + 5$$
$$y + 8 = -9 + 5$$
$$y + 8 = -4$$
$$+ \quad -8 = -8$$
$$y = -12$$

Check:
$$-3y + 4y + 8 = -9 + 5$$
$$-3(-12) + 4(-12) + 8 \stackrel{?}{=} -9 + 5$$
$$36 - 48 + 8 \stackrel{?}{=} -4$$
$$-4 = -4 \checkmark$$

Copyright © 2012 Pearson Education, Inc.

2. $\frac{x}{-5} = 3^2 - 7$

$\frac{x}{-5} = 9 - 7$

$\frac{x}{-5} = 2$

$\frac{-5 \cdot x}{-5} = 2(-5)$

$x = -10$

Check: $\frac{x}{-5} = 3^2 - 7$

$\frac{-10}{-5} \stackrel{?}{=} 3^2 - 7$

$2 \stackrel{?}{=} 9 - 7$

$2 = 2$ ✓

3. $2(-7b) + 8b = -24$

$-14b + 8b = -24$

$-6b = -24$

$\frac{-6b}{-6} = \frac{-24}{-6}$

$b = 4$

Check: $2(-7b) + 8b = -24$

$2(-7 \cdot 4) + 8 \cdot 4 \stackrel{?}{=} -24$

$-56 + 32 \stackrel{?}{=} -24$

$-24 = -24$ ✓

4. Answers may vary. Possible solution: Equation (c) may be solved by dividing by 7 on both sides of the equation. This operation may be used to isolate x.

7.2 Exercises

1. If 1 is added to double a number, the result is 5. What is the number? (Answers may vary.)

3. $3x + 6 = 5x + 9$

$+ \boxed{-3x} \quad \boxed{-3x}$

$0 + 6 = 2x + 9$

$+ \boxed{-9} \quad \boxed{-9}$

$-3 = 2x + 0$

$\frac{-3}{\boxed{2}} = \frac{2x}{\boxed{2}}$

$-\frac{3}{2} = x$

5. $6 - 2x = 5 - 9x$

$+ \boxed{+9x} \quad \boxed{+9x}$

$6 + 7x = 5 + 0$

$+ \boxed{-6} \quad \boxed{-6}$

$0 + 7x = -1 + 0$

$\frac{7x}{\boxed{7}} = \frac{-1}{\boxed{7}}$

$x = -\frac{1}{7}$

7. $-4x + 2 + 3x = 13$

$\boxed{-1}x + 2 = 13$

$+ \boxed{-2} \quad \boxed{-2}$

$\frac{-1x}{\boxed{-1}} + 0 = \frac{11}{\boxed{-1}}$

$x = -11$

9. $3x - 9 = 27$

$+ \quad +9 = +9$

$3x = 36$

$\frac{3x}{3} = \frac{36}{3}$

$x = 12$

Check: $3x - 9 = 27$

$3(12) - 9 \stackrel{?}{=} 27$

$36 - 9 \stackrel{?}{=} 27$

$27 = 27$ ✓

11. $5x - 15 = 20$

$+ \quad +15 = +15$

$5x = 35$

$\frac{5x}{5} = \frac{35}{5}$

$x = 7$

Check: $5x - 15 = 20$

$5(7) - 15 \stackrel{?}{=} 20$

$35 - 15 \stackrel{?}{=} 20$

$20 = 20$ ✓

13. $18 = 4x - 10$

$+ +10 = \quad +10$

$28 = 4x$

$\frac{28}{4} = \frac{4x}{4}$

$7 = x$ or $x = 7$

Copyright © 2012 Pearson Education, Inc.

Check:
$$\begin{aligned} 18 &= 4x - 10 \\ 18 &\stackrel{?}{=} 4(7) - 10 \\ 18 &\stackrel{?}{=} 28 - 10 \\ 18 &= 18 \checkmark \end{aligned}$$

15.
$$\begin{aligned} 5x - 1 &= 16 \\ + \quad +1 &= +1 \\ \hline 5x \quad &= 17 \\ \frac{5x}{5} &= \frac{17}{5} \\ x &= \frac{17}{5} \end{aligned}$$

Check:
$$\begin{aligned} 5x - 1 &= 16 \\ 5\left(\frac{17}{5}\right) - 1 &\stackrel{?}{=} 16 \\ 17 - 1 &\stackrel{?}{=} 16 \\ 16 &= 16 \checkmark \end{aligned}$$

17.
$$\begin{aligned} -4y + 9 &= 65 \\ + \quad -9 &= -9 \\ \hline -4y \quad &= 56 \\ \frac{-4y}{-4} &= \frac{56}{-4} \\ y &= -14 \end{aligned}$$

Check:
$$\begin{aligned} -4y + 9 &= 65 \\ -4(-14) + 9 &\stackrel{?}{=} 65 \\ 56 + 9 &\stackrel{?}{=} 65 \\ 65 &= 65 \checkmark \end{aligned}$$

19.
$$\begin{aligned} -6m - 10 &= 88 \\ + \quad +10 &= +10 \\ \hline -6m \quad &= 98 \\ \frac{-6m}{-6} &= \frac{98}{-6} \\ m &= -\frac{49}{3} \end{aligned}$$

Check:
$$\begin{aligned} -6m - 10 &= 88 \\ -6\left(-\frac{49}{3}\right) - 10 &\stackrel{?}{=} 88 \\ 98 - 10 &\stackrel{?}{=} 88 \\ 88 &= 88 \checkmark \end{aligned}$$

21.
$$\begin{aligned} 52 &= -2x - 10 \\ + +10 &= \quad +10 \\ \hline 62 &= -2x \\ \frac{62}{-2} &= \frac{-2x}{-2} \\ -31 &= x \text{ or } x = -31 \end{aligned}$$

Check:
$$\begin{aligned} 52 &= -2x - 10 \\ 52 &\stackrel{?}{=} -2(-31) - 10 \\ 52 &\stackrel{?}{=} 62 - 10 \\ 52 &= 52 \checkmark \end{aligned}$$

23.
$$\begin{aligned} -1 &= -7 - 5y \\ + +7 &= +7 \\ \hline 6 &= \quad -5y \\ \frac{6}{-5} &= \frac{-5y}{-5} \\ -\frac{6}{5} &= y \text{ or } y = -\frac{6}{5} \end{aligned}$$

Check:
$$\begin{aligned} -1 &= -7 - 5y \\ -1 &\stackrel{?}{=} -7 - 5\left(-\frac{6}{5}\right) \\ -1 &\stackrel{?}{=} -7 + 6 \\ -1 &= -1 \checkmark \end{aligned}$$

25.
$$\begin{aligned} 2 &= 4 + 2x \\ + -4 &= -4 \\ \hline -2 &= \quad 2x \\ \frac{-2}{2} &= \frac{2x}{2} \\ -1 &= x \text{ or } x = -1 \end{aligned}$$

Check:
$$\begin{aligned} 2 &= 4 + 2x \\ 2 &\stackrel{?}{=} 4 + 2(-1) \\ 2 &\stackrel{?}{=} 4 - 2 \\ 2 &= 2 \checkmark \end{aligned}$$

27.
$$\begin{aligned} -6 &= 6 - 3y \\ + -6 &= -6 \\ \hline -12 &= \quad -3y \\ \frac{-12}{-3} &= \frac{-3y}{-3} \\ 4 &= y \text{ or } y = 4 \end{aligned}$$

Check:
$$\begin{aligned} -6 &= 6 - 3y \\ -6 &\stackrel{?}{=} 6 - 3(4) \\ -6 &\stackrel{?}{=} 6 - 12 \\ -6 &= -6 \checkmark \end{aligned}$$

29.
$$\begin{aligned} 8y + 6 - 2y &= 18 \\ 6y + 6 &= 18 \\ + \quad -6 &= -6 \\ \hline 6y \quad &= 12 \\ \frac{6y}{6} &= \frac{12}{6} \\ y &= 2 \end{aligned}$$

Copyright © 2012 Pearson Education, Inc.

Check: $8y+6-2y=18$
$8(2)+6-2(2) \stackrel{?}{=} 18$
$16+6-4 \stackrel{?}{=} 18$
$18=18$ ✓

31.
$$\begin{aligned} 9x-2+2x &= 6 \\ 11x-2 &= 6 \\ + \quad +2 &= +2 \\ \hline 11x \quad &= 8 \\ \frac{11x}{11} &= \frac{8}{11} \\ x &= \frac{8}{11} \end{aligned}$$

Check: $9x-2+2x=6$
$9\left(\frac{8}{11}\right)-2+2\left(\frac{8}{11}\right) \stackrel{?}{=} 6$
$\frac{72}{11}-\frac{22}{11}+\frac{16}{11} \stackrel{?}{=} 6$
$\frac{66}{11} \stackrel{?}{=} 6$
$6=6$ ✓

33.
$$\begin{aligned} 7x+8-8x &= 11 \\ -x+8 &= 11 \\ + \quad -8 &= -8 \\ \hline -x \quad &= 3 \\ \frac{-1x}{-1} &= \frac{3}{-1} \\ x &= -3 \end{aligned}$$

Check: $7x+8-8x=11$
$7(-3)+8-8(-3) \stackrel{?}{=} 11$
$-21+8+24 \stackrel{?}{=} 11$
$11=11$ ✓

35.
$$\begin{aligned} 15x &= 9x+48 \\ + \quad -9x &= -9x \\ \hline 6x &= 48 \\ \frac{6x}{6} &= \frac{43}{6} \\ x &= 8 \end{aligned}$$

Check: $15x=9x+48$
$15(8) \stackrel{?}{=} 9(8)+48$
$120 \stackrel{?}{=} 72+48$
$120=120$ ✓

37.
$$\begin{aligned} 4x &= -12x+7 \\ + \quad +12x &= +12x \\ \hline 16x &= 7 \\ \frac{16x}{16} &= \frac{7}{16} \\ x &= \frac{7}{16} \end{aligned}$$

Check: $4x=-12x+7$
$4\left(\frac{7}{16}\right) \stackrel{?}{=} -12\left(\frac{7}{16}\right)+7$
$\frac{7}{4} \stackrel{?}{=} -\frac{21}{4}+\frac{28}{4}$
$\frac{7}{4}=\frac{7}{4}$ ✓

39.
$$\begin{aligned} 8x+2 &= 5x-4 \\ + \quad -5x &= -5x \\ \hline 3x+2 &= -4 \\ + \quad -2 &= -2 \\ \hline 3x \quad &= -6 \\ \frac{3x}{3} &= \frac{-6}{3} \\ x &= -2 \end{aligned}$$

Check: $8x+2=5x-4$
$8(-2)+2 \stackrel{?}{=} 5(-2)-4$
$-16+2 \stackrel{?}{=} -10-4$
$-14=-14$

41.
$$\begin{aligned} 11x+20 &= 12x+2 \\ + \quad -11x &= -11x \\ \hline 20 &= x+2 \\ + \quad -2 &= -2 \\ \hline 18 &= x \text{ or } x=18 \end{aligned}$$

Check: $11x+20=12x+2$
$11(18)+20 \stackrel{?}{=} 12(18)+2$
$198+20 \stackrel{?}{=} 216+2$
$218=218$ ✓

43.
$$\begin{aligned} -2+y+5 &= 3y+9 \\ y+3 &= 3y+9 \\ + \quad -y &= -y \\ \hline 3 &= 2y+9 \\ + \quad -9 &= -9 \\ \hline -6 &= 2y \\ \frac{-6}{2} &= \frac{2y}{2} \\ -3 &= y \text{ or } y=-3 \end{aligned}$$

Copyright © 2012 Pearson Education, Inc.

Check: $-2+y+5=3y+9$
$-2+(-3)+5 \stackrel{?}{=} 3(-3)+9$
$0 \stackrel{?}{=} -9+9$
$0=0$ ✓

45. $13y+9-2y = 6y-8$
$11y+9 = 6y-8$
$+ \quad -6y = -6y$
$5y+9=-8$
$+ \quad -9=-9$
$5y=-17$
$\frac{5y}{5}=\frac{-17}{5}$
$y=-\frac{17}{5}$

Check: $13y+9-2y=6y-8$
$13\left(-\frac{17}{5}\right)+9-2\left(-\frac{17}{5}\right) \stackrel{?}{=} 6\left(\frac{-17}{5}\right)-8$
$-\frac{221}{5}+\frac{45}{5}+\frac{34}{5} \stackrel{?}{=} -\frac{102}{5}-\frac{40}{5}$
$-\frac{142}{5}=-\frac{142}{5}$ ✓

47. $-9-3x+8=-6x+3-3x$
$-3x-1=-6x+3-3x$
$-3x-1=-9x+3$
$+ \quad +9x = +9x$
$6x-1= \quad 3$
$+ \quad +1= \quad +1$
$6x = \quad 4$
$\frac{6x}{6}=\frac{4}{6}$
$x=\frac{2}{3}$

Check: $-9-3x+8=-6x+3-3x$
$-9-3\left(\frac{2}{3}\right)+8 \stackrel{?}{=} -6\left(\frac{2}{3}\right)+3-3\left(\frac{2}{3}\right)$
$-9-2+8 \stackrel{?}{=} -4+3-2$
$-3=-3$ ✓

49. $-2y+6=12$
$+ \quad -6=-6$
$-2y = 6$
$\frac{-2y}{-2}=\frac{6}{-2}$
$y=-3$

Check: $-2y+6=12$
$-3(-3)+6 \stackrel{?}{=} 12$
$6+6 \stackrel{?}{=} 12$
$12=12$ ✓

51. $3x-4=11$
$+ \quad +4=+4$
$3x = 15$
$\frac{3x}{3}=\frac{15}{3}$
$x=5$

Check: $3x-4=11$
$3(5)-4 \stackrel{?}{=} 11$
$15-4 \stackrel{?}{=} 11$
$11=11$ ✓

53. $5x+5-2x=15$
$3x+5=15$
$+ \quad -5=-5$
$3x = 10$
$\frac{3x}{3}=\frac{10}{3}$
$x=\frac{10}{3}$

Check: $5x+5-2x=15$
$5\left(\frac{10}{3}\right)+5-2\left(\frac{10}{3}\right) \stackrel{?}{=} 15$
$\frac{50}{3}+\frac{15}{3}-\frac{20}{3} \stackrel{?}{=} 15$
$\frac{45}{3} \stackrel{?}{=} 15$
$15=15$ ✓

55. $13x = 8x+20$
$+ \quad -8x=-8x$
$5x = \quad 20$
$\frac{5x}{3}=\frac{20}{5}$
$x=4$

Check: $13x=8x+20$
$13(4) \stackrel{?}{=} 8(4)+20$
$52 \stackrel{?}{=} 32+20$
$52=52$ ✓

Copyright © 2012 Pearson Education, Inc.

57.
$$\begin{array}{rl} 4x-24 &= 6x-8 \\ +\ -6x \quad &= -6x \\ \hline -2x-24 &= -8 \\ +\quad +24 &= +24 \\ \hline -2x \quad &= 16 \end{array}$$
$$\frac{-2x}{-2}=\frac{16}{-2}$$
$$x=-8$$

Check:
$$\begin{aligned} 4x-24 &= 6x-8 \\ 4(-8)-24 &\stackrel{?}{=} 6(-8)-8 \\ -32-24 &\stackrel{?}{=} -48-8 \\ -56 &= -56\ \checkmark \end{aligned}$$

59.
$$\begin{array}{rl} 2x-6+2(4x)+13 &= -2-5x+(3^2-3) \\ 2x-6+8x+13 &= -2-5x+(9-3) \\ 10x+7 &= -5x+4 \\ +\quad +5x \quad &= +5x \\ \hline 15x+7 &= 4 \\ +\quad -7 &= -7 \\ \hline 15x \quad &= -3 \end{array}$$
$$\frac{15x}{15}=\frac{-3}{15}$$
$$x=-\frac{1}{5}$$

Check:
$$\begin{aligned} 2x-6+2(4x)+13 &= -2-5x+(3^2-3) \\ 2\left(-\frac{1}{5}\right)-6+2\left[4\left(-\frac{1}{5}\right)\right]+13 &\stackrel{?}{=} -2-5\left(-\frac{1}{5}\right)+(3^2-3) \\ -\frac{2}{5}-\frac{30}{5}-\frac{8}{5}+\frac{65}{5} &\stackrel{?}{=} -2+1+9-3 \\ \frac{25}{5} &\stackrel{?}{=} 5 \\ 5 &= 5\ \checkmark \end{aligned}$$

Cumulative Review

61. $-3(x-4)=-3(x)-3(-4)=-3x+12$

62.
$$\begin{aligned} 2(-3+y) &= 2(-3)+2(y) \\ &= -6+2y \text{ or } 2y-6 \end{aligned}$$

63.
$$\frac{13}{21}=\frac{65}{x}$$
$$x\cdot 13=21\cdot 65$$
$$13x=1365$$
$$\frac{13x}{13}=\frac{1365}{13}$$
$$x=105$$

Copyright © 2012 Pearson Education, Inc.

64.
$$\begin{aligned} \frac{15}{17} &= \frac{x}{85} \\ 85 \cdot 15 &= 17 \cdot x \\ 1275 &= 17x \\ \frac{1275}{17} &= \frac{17x}{17} \\ 75 &= x \text{ or } x = 75 \end{aligned}$$

Quick Quiz 7.2

1.
$$\begin{array}{rrl} & -5a = & 3a+40 \\ + & -3a = & -3a \\ \hline & -8a = & 40 \end{array}$$
$$\begin{aligned} \frac{-8a}{-8} &= \frac{40}{-8} \\ a &= -5 \end{aligned}$$
Check:
$$\begin{aligned} -5a &= 3a+40 \\ -5(-5) &\stackrel{?}{=} 3(-5)+40 \\ 25 &\stackrel{?}{=} -15+40 \\ 25 &= 25 \checkmark \end{aligned}$$

2.
$$\begin{array}{rrl} & 3y+8-6y = & -13 \\ & -3y+8 = & -13 \\ + & -8 = & -8 \\ \hline & -3y = & -21 \end{array}$$
$$\begin{aligned} \frac{-3y}{-3} &= \frac{-21}{-3} \\ y &= 7 \end{aligned}$$
Check:
$$\begin{aligned} 3y+8-6y &= -13 \\ 3(7)+8-6(7) &\stackrel{?}{=} -13 \\ 21+8-42 &= -13 \\ -13 &= -13 \checkmark \end{aligned}$$

3.
$$\begin{array}{rrl} & 4x+9 = & 6x-3 \\ + & -4x = & -4x \\ \hline & 9 = & 2x-3 \\ + & +3 = & +3 \\ \hline & 12 = & 2x \end{array}$$
$$\begin{aligned} \frac{12}{2} &= \frac{2x}{2} \\ 6 &= x \text{ or } x = 6 \end{aligned}$$
Check:
$$\begin{aligned} 4x+9 &= 6x-3 \\ 4(6)+9 &\stackrel{?}{=} 6(6)-3 \\ 24+9 &\stackrel{?}{=} 36-3 \\ 33 &= 33 \checkmark \end{aligned}$$

4. Answers may vary. One possible solution is to first combine the x terms on the left side of the equation. Then subtract 3 from both sides in order to combine the terms without x. Next divide both sides of the equation by the coefficient of x. This last step will isolate the x, and solve the equation.

7.3 Exercises

1. $-5(3x+2)+2x = 16$

$-15x-10+2x=16$	Use distributive property to remove parentheses.
$-13x-10=16$	Combine like terms.
$-13x=26$	Add $+10$ to both sides.
$x=-2$	Divide both sides by -13.

3.
$$\begin{array}{rrl} & -3(2x+1) = & 15 \\ & -6x-3 = & 15 \\ + & +3 & +3 \\ \hline & -6x = & 18 \end{array}$$
$$\begin{aligned} \frac{-6x}{-6} &= \frac{18}{-6} \\ x &= -3 \end{aligned}$$
Check:
$$\begin{aligned} -3(2x+1) &= 15 \\ -3[2(-3)+1] &\stackrel{?}{=} 15 \\ -3[-6+1] &\stackrel{?}{=} 15 \\ -3[-5] &\stackrel{?}{=} 15 \\ 15 &= 15 \checkmark \end{aligned}$$

5.
$$\begin{array}{rrl} & 4(3x-1) = & 12 \\ & 12x-4 = & 12 \\ + & +4 & +4 \\ \hline & 12x = & 16 \end{array}$$
$$\begin{aligned} \frac{12x}{12} &= \frac{16}{12} \\ x &= \frac{4}{3} \end{aligned}$$
Check:
$$\begin{aligned} 4(3x-1) &= 12 \\ 4\left[3\left(\frac{4}{3}\right)-1\right] &\stackrel{?}{=} 12 \\ 4[4-1] &\stackrel{?}{=} 12 \\ 4[3] &\stackrel{?}{=} 12 \\ 12 &= 12 \checkmark \end{aligned}$$

7.
$$\begin{aligned} 36 &= -2(4y+2) \\ 36 &= -8y-4 \\ +{+4} &\quad\ +4 \\ \hline 40 &= -8y \\ \frac{40}{-8} &= \frac{-8y}{-8} \\ -5 &= y \text{ or } y=-5 \end{aligned}$$

Check: $36 = -2(4y+2)$
$36 \stackrel{?}{=} -2[4(-5)+2]$
$36 \stackrel{?}{=} -2[-20+2]$
$36 \stackrel{?}{=} -2[-18]$
$36 = 36$ ✓

9.
$$\begin{aligned} -5(2x+1)+3x &= 37 \\ -10x-5+3x &= 37 \\ -7x-5 &= 37 \\ + \qquad\quad +5 &\quad +5 \\ \hline -7x &= 42 \\ \frac{-7x}{-7} &= \frac{42}{-7} \\ x &= -6 \end{aligned}$$

Check: $-5(2x+1)+3x = 37$
$-5[2(-6)+1]+3(-6) \stackrel{?}{=} 37$
$-5[-12+1]-18 \stackrel{?}{=} 37$
$-5(-11)-18 \stackrel{?}{=} 37$
$55-18 \stackrel{?}{=} 37$
$37 = 37$ ✓

11.
$$\begin{aligned} -2(5y-1)+4y &= -4 \\ -10y+2+4y &= -4 \\ -6y+2 &= -4 \\ + \qquad\quad -2 &\quad -2 \\ \hline -6y &= -6 \\ \frac{-6y}{-6} &= \frac{-6}{-6} \\ y &= 1 \end{aligned}$$

Check: $-2(5y-1)+4y = -4$
$-2[5(1)-1]+4(1) \stackrel{?}{=} -4$
$-2[5-1]+4(1) \stackrel{?}{=} -4$
$-2(4)+4 \stackrel{?}{=} -4$
$-8+4 \stackrel{?}{=} -4$
$-4 = -4$ ✓

13.
$$\begin{aligned} -4(2x+1) &= -5-3x \\ -8x-4 &= -5-3x \\ + \ +3x &\qquad\quad +3x \\ \hline -5x-4 &= -5 \\ + \qquad +4 &\quad +4 \\ \hline -5x \quad &= -1 \\ \frac{-5x}{-5} &= \frac{-1}{-5} \\ x &= \frac{1}{5} \end{aligned}$$

Check: $-4(2x+1) = -5-3x$
$-4\left[2\left(\frac{1}{5}\right)+1\right] \stackrel{?}{=} -5-3\left(\frac{1}{5}\right)$
$-4\left[\frac{2}{5}+\frac{5}{5}\right] \stackrel{?}{=} -\frac{25}{5}-\frac{3}{5}$
$-4\left[\frac{7}{5}\right] \stackrel{?}{=} -\frac{28}{5}$
$-\frac{28}{5} = -\frac{28}{5}$ ✓

15.
$$\begin{aligned} 3(y-4)+6(y+1) &= 57 \\ 3y-12+6y+6 &= 57 \\ 9y-6 &= 57 \\ + \qquad\quad +6 &\quad +6 \\ \hline 9y &= 63 \\ \frac{9y}{9} &= \frac{63}{9} \\ y &= 7 \end{aligned}$$

Check: $3(y-4)+6(y+1) = 57$
$3(7-4)+6(7+1) \stackrel{?}{=} 57$
$3(3)+6(8) \stackrel{?}{=} 57$
$9+48 \stackrel{?}{=} 57$
$57 = 57$ ✓

17.
$$\begin{aligned} 2(x-1)+4(x+2) &= 18 \\ 2x-2+4x+8 &= 18 \\ 6x+6 &= 18 \\ + \qquad\quad -6 &\quad -6 \\ \hline 6x \quad &= 12 \\ \frac{6x}{6} &= \frac{12}{6} \\ x &= 2 \end{aligned}$$

Check: $2(x-1)+4(x+2) = 18$
$2(2-1)+4(2+2) \stackrel{?}{=} 18$
$2(1)+4(4) \stackrel{?}{=} 18$
$2+16 \stackrel{?}{=} 18$
$18 = 18$ ✓

Copyright © 2012 Pearson Education, Inc.

19.
$$\begin{array}{l} 10 = -2(x-2)+6(x-1) \\ 10 = -2x+4+6x-6 \\ 10 = 4x-2 \\ \underline{+\ +2 \qquad +2} \\ 12 = 4x \\ \frac{12}{4} = \frac{4x}{4} \\ 3 = x \text{ or } x = 3 \end{array}$$

Check:
$$\begin{array}{l} 10 = -2(x-2)+6(x-1) \\ 10 \stackrel{?}{=} -2(3-2)+6(3-1) \\ 10 \stackrel{?}{=} -2(1)+6(2) \\ 10 \stackrel{?}{=} -2+12 \\ 10 = 10 \checkmark \end{array}$$

21.
$$\begin{array}{l} 6(x-4) = -9(x+1)+10 \\ 6x-24 = -9x-9+10 \\ 6x-24 = -9x+1 \\ \underline{+\ +9x \qquad +9x} \\ 15x-24 = 1 \\ \underline{+\ \quad +24 \qquad +24} \\ 15x = 25 \\ \frac{15x}{15} = \frac{25}{15} \\ x = \frac{5}{3} \text{ or } 1\frac{2}{3} \end{array}$$

Check:
$$\begin{array}{l} 6(x-4) = -9(x+1)+10 \\ 6\left(\frac{5}{3}-4\right) \stackrel{?}{=} -9\left(\frac{5}{3}+1\right)+10 \\ 6\left(\frac{5}{3}-\frac{12}{3}\right) \stackrel{?}{=} -9\left(\frac{5}{3}+\frac{3}{3}\right)+10 \\ 6\left(-\frac{7}{3}\right) \stackrel{?}{=} -9\left(\frac{8}{3}\right)+10 \\ -14 \stackrel{?}{=} -24+10 \\ -14 = -14 \checkmark \end{array}$$

23.
$$\begin{array}{l} 2(y+3) = -6(y-1)-8 \\ 2y+6 = -6y+6-8 \\ 2y+6 = -6y-2 \\ \underline{+\ +6y \qquad +6y} \\ 8y+6 = -2 \\ \underline{+\ \quad -6 \quad -6} \\ 8y = -8 \\ \frac{8y}{8} = \frac{-8}{8} \\ y = -1 \end{array}$$

Check:
$$\begin{array}{l} 2(y+3) = -6(y-1)-8 \\ 2(-1+3) \stackrel{?}{=} -6(-1-1)-8 \\ 2(2) \stackrel{?}{=} -6(-2)-8 \\ 4 \stackrel{?}{=} 12-8 \\ 4 = 4 \checkmark \end{array}$$

25.
$$\begin{array}{l} (6x^2+4x-1)-(6x^2+9) = 14 \\ 6x^2+4x-1-6x^2-9 = 14 \\ 4x-10 = 14 \\ \underline{+\ \qquad +10 \quad +10} \\ 4x = 24 \\ \frac{4x}{4} = \frac{24}{4} \\ x = 6 \end{array}$$

Check:
$$\begin{array}{l} (6x^2+4x-1)-(6x^2+9) = 14 \\ [6(6)^2+4(6)-1]-[6(6)^2+9] \stackrel{?}{=} 14 \\ [6(36)+4(6)-1]-[6(36)+9] \stackrel{?}{=} 14 \\ [216+24-1]-[216+9] \stackrel{?}{=} 14 \\ 239-225 \stackrel{?}{=} 14 \\ 14 \stackrel{?}{=} 14 \checkmark \end{array}$$

27.
$$\begin{array}{l} (5x^2+x+3)-(5x^2+9) = 4x+1 \\ 5x^2+x+3-5x^2-9 = 4x+1 \\ x-6 = 4x+1 \\ \underline{+\ \qquad -x \quad -x} \\ -6 = 3x+1 \\ \underline{+\ \qquad -1 \quad -1} \\ -7 = 3x \\ \frac{-7}{3} = \frac{3x}{3} \\ -\frac{7}{3} = x \text{ or } x = -\frac{7}{3} \end{array}$$

Check: $(5x^2+x+3)-(5x^2+9)=4x+1$

$$\left[5\left(-\frac{7}{3}\right)^2+\left(-\frac{7}{3}\right)+3\right]-\left[5\left(-\frac{7}{3}\right)^2+9\right]\stackrel{?}{=}4\left(-\frac{7}{3}\right)+1$$

$$\left[5\left(\frac{49}{9}\right)-\frac{7}{3}+3\right]-\left[5\left(\frac{49}{9}\right)+9\right]\stackrel{?}{=}-\frac{28}{3}+1$$

$$\left[\frac{245}{9}-\frac{21}{9}+\frac{27}{9}\right]-\left[\frac{245}{9}+\frac{81}{9}\right]\stackrel{?}{=}-\frac{28}{3}+\frac{3}{3}$$

$$\frac{251}{9}-\frac{326}{9}\stackrel{?}{=}-\frac{25}{3}$$

$$-\frac{75}{9}\stackrel{?}{=}-\frac{25}{9}$$

$$-\frac{25}{3}=-\frac{25}{3}\ \checkmark$$

29. $6(x-1)+2(x+1)=10-5x$

$$6x-6+2x+2=10-5x$$
$$8x-4=10-5x$$
$$+\quad +5x \quad +5x$$
$$13x-4=10$$
$$+\quad +4\quad +4$$
$$13x=14$$
$$\frac{13x}{13}=\frac{14}{13}$$
$$x=\frac{14}{13}$$

Check: $6(x-1)+2(x+1)=10-5x$

$$6\left(\frac{14}{13}-1\right)+2\left(\frac{14}{13}+1\right)\stackrel{?}{=}10-5\left(\frac{14}{13}\right)$$

$$6\left(\frac{14}{13}-\frac{13}{13}\right)+2\left(\frac{14}{13}+\frac{13}{13}\right)\stackrel{?}{=}10-\frac{70}{13}$$

$$6\left(\frac{1}{13}\right)+2\left(\frac{27}{13}\right)\stackrel{?}{=}\frac{130}{13}-\frac{70}{13}$$

$$\frac{6}{13}+\frac{54}{13}\stackrel{?}{=}\frac{60}{13}$$

$$\frac{60}{13}=\frac{60}{13}\ \checkmark$$

31. $4(-6x+2)-6x=68$

$$-24x+8-6x=68$$
$$-30x+8=68$$
$$+\quad -8\quad -8$$
$$-30x=60$$
$$\frac{-30x}{-30}=\frac{60}{-30}$$
$$x=-2$$

Copyright © 2012 Pearson Education, Inc.

Check:
$$\begin{aligned}4(-6x+2)-6x&=68\\4[-6(-2)+2]-6(-2)&\stackrel{?}{=}68\\4[12+2]+12&\stackrel{?}{=}68\\4[14]+12&\stackrel{?}{=}68\\56+12&\stackrel{?}{=}68\\68&=68\ \checkmark\end{aligned}$$

33.
$$\begin{aligned}(4x^2+3x+1)-(4x^2-2)&=5x-2\\4x^2+3x+1-4x^2+2&=5x-2\\3x+3&=5x-2\\+\quad -5x\quad&\ \ -5x\\\hline -2x+3&=-2\\+\quad -3\quad&\ \ -3\\\hline -2x\quad&=-5\\\frac{-2x}{-2}&=\frac{-5}{-2}\\x&=\frac{5}{2}\end{aligned}$$

Check:
$$\begin{aligned}(4x^2+3x+1)-(4x^2-2)&=5x-2\\\left[4\left(\frac{5}{2}\right)^2+3\left(\frac{5}{2}\right)+1\right]-\left[4\left(\frac{5}{2}\right)^2-2\right]&\stackrel{?}{=}5\left(\frac{5}{2}\right)-2\\\left[4\left(\frac{25}{4}\right)+\frac{15}{2}+1\right]-\left[4\left(\frac{25}{4}\right)-2\right]&\stackrel{?}{=}\frac{25}{2}-2\\\left[25+\frac{15}{2}+1\right]-[25-2]&\stackrel{?}{=}\frac{25}{2}-\frac{4}{2}\\\left[\frac{50}{2}+\frac{15}{2}+\frac{3}{2}\right]-23&\stackrel{?}{=}\frac{21}{2}\\\frac{67}{2}-\frac{46}{2}&\stackrel{?}{=}\frac{21}{2}\\\frac{21}{2}&=\frac{21}{2}\ \checkmark\end{aligned}$$

35.
$$\begin{aligned}(2x^2-6x-3)-(x^2+4)&=x^2+6\\2x^2-6x-3-x^2-4&=x^2+6\\x^2-6x-7&=x^2+6\\+\quad +7\quad&\ \ +7\\\hline -6x-7&=\quad 6\\+\quad +7\quad&\ \ +7\\\hline -6x\quad&=\quad 13\\\frac{-6x}{-6}&=\frac{13}{-6}\\x&=-\frac{13}{6}\end{aligned}$$

Copyright © 2012 Pearson Education, Inc.

Check: $(2x^2-6x-3)-(x^2+4)=x^2+6$

$$\left[2\left(-\frac{13}{6}\right)^2-6\left(-\frac{13}{6}\right)-3\right]-\left[\left(-\frac{13}{6}\right)^2+4\right]\stackrel{?}{=}\left(-\frac{13}{6}\right)^2+6$$
$$\left[2\left(\frac{169}{36}\right)+13-3\right]-\left[\frac{169}{36}+4\right]\stackrel{?}{=}\frac{169}{36}+6$$
$$\left[\frac{169}{18}+10\right]-\left[\frac{169}{36}+\frac{144}{36}\right]\stackrel{?}{=}\frac{169}{36}+\frac{216}{36}$$
$$\left[\frac{169}{18}+\frac{180}{18}\right]-\frac{313}{36}\stackrel{?}{=}\frac{385}{36}$$
$$\frac{349}{18}-\frac{313}{36}\stackrel{?}{=}\frac{385}{36}$$
$$\frac{698}{36}-\frac{313}{36}\stackrel{?}{=}\frac{385}{36}$$
$$\frac{385}{36}=\frac{385}{36}\ \checkmark$$

37.
$$\begin{aligned}(x^2+3x-1)-(2x+1)&=x^2-5\\ x^2+3x-1-2x-1&=x^2-5\\ x^2+x-2&=x^2-5\\ +\quad -x^2\quad & \quad -x^2\\ \hline x-2&=-5\\ +\quad +2\quad & \quad +2\\ \hline x&=-3\\ x&=-3\end{aligned}$$

Check: $(x^2+3x-1)-(2x+1)=x^2-5$

$$[(-3)^2+3(-3)-1]-[2(-3)+1]\stackrel{?}{=}(-3)^2-5$$
$$[9-9-1]-[-6+1]\stackrel{?}{=}9-5$$
$$-1-(-5)\stackrel{?}{=}4$$
$$4=4\ \checkmark$$

39.
$$(2x+9)+(3x-2)-5(x+1)=2(x-6)-(x-1)$$
$$[2(19)+9]+[3(19)-2]-5(19+1)\stackrel{?}{=}2(19-6)-(19-1)$$
$$[38+9]+[57-2]-5(20)\stackrel{?}{=}2(13)-18$$
$$47+55-100\stackrel{?}{=}26-18$$
$$2\neq 8$$

Cumulative Review

41. $3=3$
$4=2\cdot 2$
$2=2$
$\text{LCD}=2\cdot 2\cdot 3=12$

42. $4=2\cdot 2$
$2=2$
$5=5$
$\text{LCD}=2\cdot 2\cdot 5=20$

Copyright © 2012 Pearson Education, Inc.

43. $2x = 2 \cdot x$
$x = x$
$\text{LCD} = 2 \cdot x = 2x$

44. $x = x$
$5x = 5 \cdot x$
$\text{LCD} = 5 \cdot x = 5x$

45. The LCD of $\frac{4}{5}$ and $\frac{2}{3}$ is 15.

$\frac{4}{5} = \frac{4 \cdot 3}{5 \cdot 3} = \frac{12}{15};\ \frac{2}{3} = \frac{2 \cdot 5}{3 \cdot 5} = \frac{10}{15}$

Since $\frac{12}{15} > \frac{10}{15}$, we have $\frac{4}{5} > \frac{2}{3}$.

Israel was overcharged.

46. Replace D with $13\frac{1}{4}$.

$$\begin{aligned} P &= 15 + \frac{1}{2}D \\ P &= 15 + \frac{1}{2}\left(13\frac{1}{4}\right) \\ &= 15 + \frac{1}{2}\left(\frac{53}{4}\right) \\ &= 15 + \frac{53}{8} \\ &= \frac{120}{8} + \frac{53}{8} \\ &= \frac{173}{8} \text{ or } 21\frac{5}{8} \end{aligned}$$

The pressure on the diver is $21\frac{5}{8}$ pounds per square inch.

Quick Quiz 7.3

1.
$$\begin{aligned} -4(3x-5) - 2x &= 34 \\ -12x + 20 - 2x &= 34 \\ -14x + 20 &= 34 \\ + \quad -20 &\ \ -20 \\ \hline -14x &= 14 \\ \frac{-14x}{-14} &= \frac{14}{-14} \\ x &= -1 \end{aligned}$$

Check:
$$\begin{aligned} -4(3x-5) - 2x &= 3^4 \\ -4[3(-1)-5] - 2(-1) &\stackrel{?}{=} 34 \\ -4(-3-5) + 2 &\stackrel{?}{=} 34 \\ -4(-8) + 2 &\stackrel{?}{=} 34 \\ 32 + 2 &\stackrel{?}{=} 34 \\ 34 &= 34\ \checkmark \end{aligned}$$

2.
$$\begin{aligned} -8(x+1) &= 4(2x-5) - 20 \\ -8x - 8 &= 8x - 20 - 20 \\ -8x - 8 &= 8x - 40 \\ + \ +8x \quad &\quad +8x \\ \hline -8 &= 16x - 40 \\ + \quad +40 &\quad\quad +40 \\ \hline 32 &= 16x \\ \frac{32}{16} &= \frac{16x}{16} \\ 2 &= x \end{aligned}$$

Check:
$$\begin{aligned} -8(x+1) &= 4(2x-5) - 20 \\ -8(2+1) &\stackrel{?}{=} 4[2(2)-5] - 20 \\ -8(3) &\stackrel{?}{=} 4[4-5] - 20 \\ -24 &\stackrel{?}{=} 4[-1] - 20 \\ -24 &\stackrel{?}{=} -4 - 20 \\ -24 &= -24\ \checkmark \end{aligned}$$

3.
$$\begin{aligned} (4a^2 + 2a - 1) - (4a^2 - 3a + 4) &= 3a + 1 \\ 4a^2 + 2a - 1 - 4a^2 + 3a - 4 &= 3a + 1 \\ 5a - 5 &= 3a + 1 \\ + \quad -3a \quad &\quad -3a \\ \hline 2a - 5 &= \quad 1 \\ + \quad +5 &\quad +5 \\ \hline 2a &= \quad 6 \\ \frac{2a}{2} &= \frac{6}{2} \\ a &= 3 \end{aligned}$$

Check:
$$\begin{aligned} (4a^2 + 2a - 1) - (4a^2 - 3a + 4) &= 3a + 1 \\ [4(3)^2 + 2(3) - 1] - [4(3)^2 - 3(3) + 4] &\stackrel{?}{=} 3(3) + 1 \\ [4(9) + 6 - 1] - [4(9) - 9 + 4] &\stackrel{?}{=} 9 + 1 \\ [36 + 5] - [36 - 5] &\stackrel{?}{=} 10 \\ 41 - 31 &\stackrel{?}{=} 10 \\ 10 &= 10\ \checkmark \end{aligned}$$

4. a. The first step we should perform to solve $2(y - 1) + 2 = -3(y - 2)$ is to simplify the equation using the <u>distributive</u> property.

Copyright © 2012 Pearson Education, Inc.

b. Answers may vary.

$2(y-1)+2=3(y-2)$	Original equation
$2y-2+2=3y-6$	Use the distributive property.
$2y=3y-6$	Combine like terms.
$2y-3y=3y-3y-6$	Subtract $3y$ from both sides to get y in one term only.
$-y=-6$	Simplify.
$y=6$	Divide both sides by -1 to isolate y.

How Am I Doing? Sections 7.1–7.3

1. $-12=a+3$
$+\ -3 \quad -3$
$-15=a$ or $a=-15$

Check: $-12=a+3$
$-12 \stackrel{?}{=} -15+3$
$-12=-12$ ✓

2. $\frac{x}{-2}=4+5^2$
$\frac{x}{-2}=4+25$
$\frac{x}{-2}=29$
$\frac{-2\cdot x}{-2}=29\cdot(-2)$
$x=-58$

Check: $\frac{x}{-2}=4+5^2$
$\frac{-58}{-2} \stackrel{?}{=} 4+5^2$
$29 \stackrel{?}{=} 4+25$
$29=29$ ✓

3. $-x=4$
$-1x=4$
$\frac{-1x}{-1}=\frac{4}{-1}$
$x=-4$

Check: $-x=4$
$-(-4) \stackrel{?}{=} 4$
$4=4$ ✓

4. $\frac{21}{-7}=9x+3(-4x)$
$-3=9x+3(-4x)$
$-3=9x-12x$
$-3=-3x$
$\frac{-3}{-3}=\frac{-3x}{-3}$
$1=x$ or $x=1$

Check: $\frac{21}{-7}=9x+3(-4x)$
$\frac{21}{-7} \stackrel{?}{=} 9(1)+3[-4(1)]$
$-3 \stackrel{?}{=} 9+(-12)$
$-3=-3$ ✓

5. $4x+10=14$
$+\ \ -10\ -10$
$4x=4$
$\frac{4x}{4}=\frac{4}{4}$
$x=1$

Check: $4x+10=14$
$4(1)+10 \stackrel{?}{=} 14$
$4+10 \stackrel{?}{=} 14$
$14=14$ ✓

6. $-5y-9=24$
$+\quad +9\ \ +9$
$-5y=33$
$\frac{-5y}{-5}=\frac{33}{-5}$
$y=-\frac{33}{5}$

Check: $-5y-9=24$
$-5\left(-\frac{33}{5}\right)-9 \stackrel{?}{=} 24$
$33-9 \stackrel{?}{=} 24$
$24=24$ ✓

7. $-3=7-2y$
$+\ -7\ -7$
$-10=\ -2y$
$\frac{-10}{-2}=\frac{-2y}{-2}$
$5=y$ or $y=5$

Check: $-3=7-2y$
$-3 \stackrel{?}{=} 7-2(5)$
$-3 \stackrel{?}{=} 7-10$
$-3=-3$ ✓

Copyright © 2012 Pearson Education, Inc.

8.
$$\begin{aligned} 6x-4&=14x-9\\ +\ -6x\quad &\quad -6x\\ \hline -4&=8x-9\\ +\quad +9\quad &\quad +9\\ \hline 5&=8x\\ \frac{5}{8}&=\frac{8x}{8}\\ \frac{5}{8}&=x \text{ or } x=\frac{5}{8} \end{aligned}$$

Check: $6x-4=14x-9$
$$\begin{aligned} 6\left(\frac{5}{8}\right)-4 &\stackrel{?}{=} 14\left(\frac{5}{8}\right)-9\\ \frac{15}{4}-\frac{16}{4} &\stackrel{?}{=} \frac{35}{4}-\frac{36}{4}\\ -\frac{1}{4}&=-\frac{1}{4}\ \checkmark \end{aligned}$$

9.
$$\begin{aligned} 15y-6+5y&=8y-1\\ 20y-6&=8y-1\\ +\ -8y\quad &\quad -8y\\ \hline 12y-6&=\quad -1\\ +\quad +6\quad &\quad +6\\ \hline 12y\quad &=\quad 5\\ \frac{12y}{12}&=\frac{5}{12}\\ y&=\frac{5}{12} \end{aligned}$$

Check: $15y-6+5y=8y-1$
$$\begin{aligned} 15\left(\frac{5}{12}\right)-6+5\left(\frac{5}{12}\right) &\stackrel{?}{=} 8\left(\frac{5}{12}\right)-1\\ \frac{75}{12}-\frac{72}{12}+\frac{25}{12} &\stackrel{?}{=} \frac{40}{12}-\frac{12}{12}\\ \frac{28}{12}&=\frac{28}{12}\ \checkmark \end{aligned}$$

10.
$$\begin{aligned} 7-8&=2x-5+5x\\ -1&=7x-5\\ +\ +5\quad &\quad +5\\ \hline 4&=7x\\ \frac{4}{7}&=\frac{7x}{7}\\ \frac{4}{7}&=x \text{ or } x=\frac{4}{7} \end{aligned}$$

Check: $7-8=2x-5+5x$
$$\begin{aligned} 7-8 &\stackrel{?}{=} 2\left(\frac{4}{7}\right)-5+5\left(\frac{4}{7}\right)\\ -1 &\stackrel{?}{=} \frac{8}{7}-\frac{35}{7}+\frac{20}{7}\\ -1 &\stackrel{?}{=} -\frac{7}{7}\\ -1&=-1\ \checkmark \end{aligned}$$

11.
$$\begin{aligned} -3(2x+1)+4x&=11\\ -6x-3+4x&=11\\ -2x-3&=11\\ +\quad +3\quad &\ +3\\ \hline -2x&=14\\ \frac{-2x}{-2}&=\frac{14}{-2}\\ x&=-7 \end{aligned}$$

Check: $-3(2x+1)+4x=11$
$$\begin{aligned} -3[2(-7)+1]+4(-7) &\stackrel{?}{=} 11\\ -3[-14+1]+(-28) &\stackrel{?}{=} 11\\ -3(-13)-28 &\stackrel{?}{=} 11\\ 39-28 &\stackrel{?}{=} 11\\ 11&=11\ \checkmark \end{aligned}$$

12.
$$\begin{aligned} 2(x+1)&=-3(x+2)+10\\ 2x+2&=-3x-6+10\\ 2x+2&=-3x+4\\ +\ +3x\quad &\quad +3x\\ \hline 5x+2&=\quad 4\\ +\quad -2\quad &\quad -2\\ \hline 5x\quad &=\quad 2\\ \frac{5x}{5}&=\frac{2}{5}\\ x&=\frac{2}{5} \end{aligned}$$

Check: $2(x+1)=-3(x+2)+10$
$$\begin{aligned} 2\left(\frac{2}{5}+1\right) &\stackrel{?}{=} -3\left(\frac{2}{5}+2\right)+10\\ 2\left(\frac{2}{5}+\frac{5}{5}\right) &\stackrel{?}{=} -3\left(\frac{2}{5}+\frac{10}{5}\right)+10\\ 2\left(\frac{7}{5}\right) &\stackrel{?}{=} -3\left(\frac{12}{5}\right)+10\\ \frac{14}{5} &\stackrel{?}{=} -\frac{36}{5}+\frac{50}{5}\\ \frac{14}{5}&=\frac{14}{5}\ \checkmark \end{aligned}$$

Copyright © 2012 Pearson Education, Inc.

13. $(3y^2+8y-1)-(3y^2+2)=5$

$$\begin{aligned} 3y^2+8y-1-3y^2-2&=5 \\ 8y-3&=5 \\ +\quad\quad +3&\;+3 \\ \hline 8y&=8 \\ \frac{8y}{8}&=\frac{8}{8} \\ y&=1 \end{aligned}$$

Check:

$$\begin{aligned} (3y^2+8y-1)-(3y^2+2)&=5 \\ [3(1)^2+8(1)-1]-[3(1)^2+2]&\stackrel{?}{=}5 \\ [3+8-1]-[3+2]&\stackrel{?}{=}5 \\ 10-5&\stackrel{?}{=}5 \\ 5&=5\ \checkmark \end{aligned}$$

7.4 Exercises

1. To solve $\frac{x}{3}+\frac{x}{5}=10$, we multiply each term by <u>15</u>, so that we clear the fractions.

3.

$$\begin{aligned} \frac{x}{4}+\frac{x}{3}&=7 \\ \boxed{12}\cdot\frac{x}{4}+\boxed{12}\cdot\frac{x}{3}&=\boxed{12}\cdot 7 \\ \boxed{3}x+\boxed{4}x&=84 \\ \boxed{7}x&=84 \\ x&=\boxed{12} \end{aligned}$$

5. $\frac{x}{6}+\frac{x}{2}=8$

LCD = 12

$$\begin{aligned} 12\left(\frac{x}{6}\right)+12\left(\frac{x}{2}\right)&=12(8) \\ 2x+6x&=96 \\ 8x&=96 \\ \frac{8x}{8}&=\frac{96}{8} \\ x&=12 \end{aligned}$$

Check:

$$\begin{aligned} \frac{x}{6}+\frac{x}{2}&=8 \\ \frac{12}{6}+\frac{12}{2}&\stackrel{?}{=}8 \\ 2+6&=8\ \checkmark \end{aligned}$$

7. $\frac{x}{6}+\frac{x}{4}=5$

LCD = 12

$$\begin{aligned} 12\left(\frac{x}{6}\right)+12\left(\frac{x}{4}\right)&=12(5) \\ 2x+3x&=60 \\ 5x&=60 \\ \frac{5x}{5}&=\frac{60}{5} \\ x&=12 \end{aligned}$$

Check:

$$\begin{aligned} \frac{x}{6}+\frac{x}{4}&=5 \\ \frac{12}{6}+\frac{12}{4}&\stackrel{?}{=}5 \\ 2+3&=5\ \checkmark \end{aligned}$$

9. $\frac{x}{2}-\frac{x}{5}=6$

LCD = 10

$$\begin{aligned} 10\left(\frac{x}{2}\right)-10\left(\frac{x}{5}\right)&=10(6) \\ 5x-2x&=60 \\ 3x&=60 \\ \frac{3x}{3}&=\frac{60}{3} \\ x&=20 \end{aligned}$$

Check:

$$\begin{aligned} \frac{x}{2}-\frac{x}{5}&=6 \\ \frac{20}{2}-\frac{20}{5}&\stackrel{?}{=}6 \\ 10-4&=6\ \checkmark \end{aligned}$$

11. $3x+\frac{2}{3}=\frac{9}{2}$

LCD = 6

$$\begin{aligned} 6(3x)+6\left(\frac{2}{3}\right)&=6\left(\frac{9}{2}\right) \\ 18x+4&=27 \\ +\quad\quad -4&\;-4 \\ \hline 18x\quad\;&=23 \\ \frac{18x}{18}&=\frac{23}{18} \\ x&=\frac{23}{18} \end{aligned}$$

Copyright © 2012 Pearson Education, Inc.

Check:
$$3x+\frac{2}{3}=\frac{9}{2}$$
$$3\left(\frac{23}{18}\right)+\frac{2}{3}\stackrel{?}{=}\frac{9}{2}$$
$$\frac{23}{6}+\frac{4}{6}\stackrel{?}{=}\frac{9}{2}$$
$$\frac{27}{6}\stackrel{?}{=}\frac{9}{2}$$
$$\frac{9}{2}=\frac{9}{2}\ \checkmark$$

13. $5x+\frac{1}{8}=\frac{3}{4}$

LCD = 8
$$8(5x)+8\left(\frac{1}{8}\right)=8\left(\frac{3}{4}\right)$$
$$40x+1=6$$
$$+\quad -1\ -1$$
$$40x\quad =5$$
$$\frac{40x}{40}=\frac{5}{40}$$
$$x=\frac{1}{8}$$

Check:
$$5x+\frac{1}{8}=\frac{3}{4}$$
$$5\left(\frac{1}{8}\right)+\frac{1}{8}\stackrel{?}{=}\frac{3}{4}$$
$$\frac{5}{8}+\frac{1}{8}\stackrel{?}{=}\frac{3}{4}$$
$$\frac{6}{8}\stackrel{?}{=}\frac{3}{4}$$
$$\frac{3}{4}=\frac{3}{4}\ \checkmark$$

15. $5x-\frac{1}{2}=\frac{1}{8}$

LCD = 8
$$8(5x)-8\left(\frac{1}{2}\right)=8\left(\frac{1}{8}\right)$$
$$40x-4=1$$
$$+\quad +4\ +4$$
$$40x\quad =5$$
$$\frac{40x}{40}=\frac{5}{40}$$
$$x=\frac{1}{8}$$

Check:
$$5x-\frac{1}{2}=\frac{1}{8}$$
$$5\left(\frac{1}{8}\right)-\frac{1}{2}\stackrel{?}{=}\frac{1}{8}$$
$$\frac{5}{8}-\frac{4}{8}\stackrel{?}{=}\frac{1}{8}$$
$$\frac{1}{8}=\frac{1}{8}\ \checkmark$$

17. $-2x+\frac{1}{2}=\frac{3}{7}$

LCD = 14
$$14(-2x)+14\left(\frac{1}{2}\right)=14\left(\frac{3}{7}\right)$$
$$-28x+7=\ 6$$
$$+\quad -7\ -7$$
$$-28x\quad =-1$$
$$\frac{-28x}{-28}=\frac{-1}{-28}$$
$$x=\frac{1}{28}$$

Check:
$$-2x+\frac{1}{2}=\frac{3}{7}$$
$$-2\left(\frac{1}{28}\right)+\frac{1}{2}\stackrel{?}{=}\frac{3}{7}$$
$$-\frac{1}{14}+\frac{7}{14}\stackrel{?}{=}\frac{3}{7}$$
$$\frac{6}{14}\stackrel{?}{=}\frac{3}{7}$$
$$\frac{3}{7}=\frac{3}{7}\ \checkmark$$

19. $-4x+\frac{2}{3}=\frac{1}{6}$

LCD = 6
$$6(-4x)+6\left(\frac{2}{3}\right)=6\left(\frac{1}{6}\right)$$
$$-24x+4=\ 1$$
$$+\quad -4\ -4$$
$$-24\quad x=-3$$
$$\frac{-24x}{-24}=\frac{-3}{-29}$$
$$x=\frac{1}{8}$$

Copyright © 2012 Pearson Education, Inc.

Check: $-4x+\frac{2}{3}=\frac{1}{6}$

$-4\left(\frac{1}{8}\right)+\frac{2}{3}\stackrel{?}{=}\frac{1}{6}$

$-\frac{1}{2}+\frac{2}{3}\stackrel{?}{=}\frac{1}{6}$

$-\frac{3}{6}+\frac{4}{6}\stackrel{?}{=}\frac{1}{6}$

$\frac{1}{6}=\frac{1}{6}$ ✓

21. $\frac{x}{3}+x=8$

LCD = 3

$3\left(\frac{x}{3}\right)+3(x)=3(8)$

$x+3x=24$

$4x=24$

$\frac{4x}{4}=\frac{24}{4}$

$x=6$

Check: $\frac{x}{3}+6=8$

$\frac{6}{3}+6\stackrel{?}{=}8$

$2+6=8$ ✓

23. $\frac{x}{3}+x=6$

LCD = 3

$3\left(\frac{x}{3}\right)+3(x)=3(6)$

$x+3x=18$

$4x=18$

$\frac{4x}{4}=\frac{18}{4}$

$x=\frac{9}{2}$

Check: $\frac{x}{3}+x=6$

$\frac{\frac{9}{2}}{3}+\frac{9}{2}\stackrel{?}{=}6$

$\frac{9}{2}\cdot\frac{1}{3}+\frac{9}{2}\stackrel{?}{=}6$

$\frac{3}{2}+\frac{9}{2}\stackrel{?}{=}6$

$\frac{12}{2}\stackrel{?}{=}6$

$6=6$ ✓

25. $\frac{x}{2}+x=6$

LCD = 2

$2\left(\frac{x}{2}\right)+2(x)=2(6)$

$x+2x=12$

$3x=12$

$\frac{3x}{3}=\frac{12}{3}$

$x=4$

Check: $\frac{x}{2}+x=6$

$\frac{4}{2}+4\stackrel{?}{=}6$

$2+4=6$ ✓

27. $\frac{x}{4}-2x=3$

LCD = 4

$4\left(\frac{x}{4}\right)-4(2x)=4(3)$

$x-8x=12$

$-7x=12$

$\frac{-7x}{-7}=\frac{12}{-7}$

$x=-\frac{12}{7}$

Check: $\frac{x}{4}-2x=3$

$\frac{\left(-\frac{12}{7}\right)}{4}-2\left(-\frac{12}{7}\right)\stackrel{?}{=}3$

$-\frac{12}{7}\cdot\frac{1}{4}+\frac{24}{7}\stackrel{?}{=}3$

$-\frac{3}{7}+\frac{24}{7}\stackrel{?}{=}3$

$\frac{21}{7}\stackrel{?}{=}3$

$3=3$ ✓

Copyright © 2012 Pearson Education, Inc.

29. $2x+\frac{1}{3}=\frac{1}{6}$

LCD = 6

$$6(2x)+6\left(\frac{1}{3}\right)=6\left(\frac{1}{6}\right)$$
$$12x+2=1$$
$$+\quad -2 \quad -2$$
$$12x=-1$$
$$\frac{12x}{12}=\frac{-1}{12}$$
$$x=-\frac{1}{12}$$

Check: $2x+\frac{1}{3}=\frac{1}{6}$
$$2\left(-\frac{1}{12}\right)+\frac{1}{3}\stackrel{?}{=}\frac{1}{6}$$
$$-\frac{1}{6}+\frac{2}{6}\stackrel{?}{=}\frac{1}{6}$$
$$\frac{1}{6}=\frac{1}{6}\ \checkmark$$

31. $\frac{x}{3}+x=8$

LCD = 3

$$3\left(\frac{x}{3}\right)+3(x)=3(8)$$
$$x+3x=24$$
$$4x=24$$
$$\frac{4x}{4}=\frac{24}{4}$$
$$x=6$$

Check: $\frac{x}{3}+x=8$
$$\frac{6}{3}+6\stackrel{?}{=}8$$
$$2+6=8\ \checkmark$$

33. $\frac{x}{2}+\frac{x}{5}=7$

LCD = 10

$$10\left(\frac{x}{2}\right)+10\left(\frac{x}{5}\right)=10(7)$$
$$5x+2x=70$$
$$7x=70$$
$$\frac{7x}{7}=\frac{70}{7}$$
$$x=10$$

Check: $\frac{x}{2}+\frac{x}{5}=7$
$$\frac{10}{2}+\frac{10}{5}\stackrel{?}{=}7$$
$$5+2=7\ \checkmark$$

35. $\frac{5}{2}+\frac{x}{3}=\frac{1}{6}$

LCD = 6

$$6\left(\frac{5}{2}\right)+6\left(\frac{x}{3}\right)=6\left(\frac{1}{6}\right)$$
$$15+2x=1$$
$$+\quad -15 \quad -15$$
$$2x=-14$$
$$\frac{2x}{2}=\frac{-14}{2}$$
$$x=-7$$

Check: $\frac{5}{2}+\frac{x}{3}=\frac{1}{6}$
$$\frac{5}{2}+\frac{-7}{3}\stackrel{?}{=}\frac{1}{6}$$
$$\frac{15}{6}-\frac{14}{6}\stackrel{?}{=}\frac{1}{6}$$
$$\frac{1}{6}=\frac{1}{6}\ \checkmark$$

37. $\frac{3}{2}+\frac{x}{10}=\frac{1}{5}$

LCD = 10

$$10\left(\frac{3}{2}\right)+10\left(\frac{x}{10}\right)=10\left(\frac{1}{5}\right)$$
$$15+x=2$$
$$+\quad -15 \quad -15$$
$$x=-13$$

Check: $\frac{3}{2}+\frac{x}{10}=\frac{1}{5}$
$$\frac{3}{2}+\frac{-13}{10}\stackrel{?}{=}\frac{1}{5}$$
$$\frac{15}{10}-\frac{13}{10}\stackrel{?}{=}\frac{1}{5}$$
$$\frac{2}{10}\stackrel{?}{=}\frac{1}{5}$$
$$\frac{1}{5}=\frac{1}{5}\ \checkmark$$

Copyright © 2012 Pearson Education, Inc.

39. $\frac{x}{2}-\frac{2}{6}=-\frac{5}{6}$

LCD = 6

$$6\left(\frac{x}{2}\right)-6\left(\frac{2}{6}\right)=6\left(-\frac{5}{6}\right)$$
$$3x-2=-5$$
$$\underline{+\quad +2\quad +2}$$
$$3x=-3$$
$$\frac{3x}{3}=\frac{-3}{3}$$
$$x=-1$$

Check: $\frac{x}{2}-\frac{2}{6}=-\frac{5}{6}$
$$\frac{-1}{2}-\frac{2}{6}\stackrel{?}{=}-\frac{5}{6}$$
$$-\frac{3}{6}-\frac{2}{6}\stackrel{?}{=}-\frac{5}{6}$$
$$-\frac{5}{6}=-\frac{5}{6}\ \checkmark$$

41. $x+\frac{2}{3}+2=\frac{3}{2}+\frac{1}{4}$

LCD = 12

$$12(x)+12\left(\frac{2}{3}\right)+12(2)=12\left(\frac{3}{2}\right)+12\left(\frac{1}{4}\right)$$
$$12x+8+24=18+3$$
$$12x+32=21$$
$$\underline{+\quad -32\quad -32}$$
$$12x=-11$$
$$\frac{12x}{12}=\frac{-11}{12}$$
$$x=-\frac{11}{12}$$

43. $4+\frac{6}{x}+\frac{2}{5}=\frac{3}{2x}$

LCD = $10x$

$$10x(4)+10x\left(\frac{6}{x}\right)+10x\left(\frac{2}{5}\right)=10x\left(\frac{3}{2x}\right)$$
$$40x+60+4x=15$$
$$44x+60=15$$
$$\underline{+\quad -60\quad -60}$$
$$44x=-45$$
$$\frac{44x}{44}=\frac{-45}{44}$$
$$x=-\frac{45}{44}$$

45. $\frac{1}{3}\left(\frac{x}{2}+3\right)+\frac{1}{4}=\frac{5}{6}$
$$\frac{x}{6}+1+\frac{1}{4}=\frac{5}{6}$$

LCD = 12

$$12\left(\frac{x}{6}\right)+12(1)+12\left(\frac{1}{4}\right)=12\left(\frac{5}{6}\right)$$
$$2x+12+3=10$$
$$2x+15=10$$
$$\underline{+\quad -15\ -15}$$
$$2x=-5$$
$$\frac{2x}{2}=\frac{-5}{2}$$
$$x=-\frac{5}{2}$$

Cumulative Review

47. Six more than twice a number: $6 + 2x$ or $2x + 6$

48. Twelve less than some number: $x - 12$

49. The sum of 4 and x: $4 + x$

50. The sum of 5 and y is multiplied by 2: $2(5 + y)$

51. $\frac{1}{3}+\frac{1}{4}=\frac{1\cdot 4}{3\cdot 4}+\frac{1\cdot 3}{4\cdot 3}=\frac{4}{12}+\frac{3}{12}=\frac{7}{12}$

The first and second partners own $\frac{7}{12}$ of the restaurant.

$$1-\frac{7}{12}=\frac{12}{12}-\frac{7}{12}=\frac{5}{12}$$

The third partner owns $\frac{5}{12}$ of the restaurant.

52.
$$4\frac{1}{4}\ =\ 4\frac{1}{4}$$
$$\underline{+\ 3\frac{1\cdot 2}{2\cdot 2}}\ =\ \underline{+\ 3\frac{2}{4}}$$
$$7\frac{3}{4}\text{ lb}$$

$$11\ =\ 10\frac{4}{4}$$
$$\underline{-\ 7\frac{3}{4}}\ =\ \underline{-\ 7\frac{3}{4}}$$
$$3\frac{1}{4}\text{ lb}$$

The boxer needs to lose $3\frac{1}{4}$ pounds the third week.

Copyright © 2012 Pearson Education, Inc.

Quick Quiz 7.4

1. $\frac{x}{3}+\frac{x}{2}=20$

LCD = 6

$$6\left(\frac{x}{3}\right)+6\left(\frac{x}{2}\right)=6(20)$$
$$2x+3x=120$$
$$5x=120$$
$$\frac{5x}{5}=\frac{120}{5}$$
$$x=24$$

Check: $\frac{x}{3}+\frac{x}{2}=20$
$$\frac{24}{3}+\frac{24}{2}\stackrel{?}{=}20$$
$$8+12=20\ \checkmark$$

2. $3a+\frac{2}{3}=\frac{4}{5}$

LCD = 15

$$15(3a)+15\left(\frac{2}{3}\right)=15\left(\frac{4}{5}\right)$$
$$45a+10=12$$
$$+\quad -10\ -10$$
$$45a\quad =2$$
$$\frac{45a}{45}=\frac{2}{45}$$
$$a=\frac{2}{45}$$

Check: $3a+\frac{2}{3}=\frac{4}{5}$
$$3\left(\frac{2}{45}\right)+\frac{2}{3}\stackrel{?}{=}\frac{4}{5}$$
$$\frac{2}{15}+\frac{10}{15}\stackrel{?}{=}\frac{4}{5}$$
$$\frac{12}{15}\stackrel{?}{=}\frac{4}{5}$$
$$\frac{4}{5}=\frac{4}{5}\ \checkmark$$

3. $\frac{y}{4}+y=25$

LCD = 4

$$4\left(\frac{y}{4}\right)+4(y)=4(25)$$
$$y+4y=100$$
$$5y=100$$
$$\frac{5y}{5}=\frac{100}{5}$$
$$y=20$$

Check: $\frac{y}{4}+y=25$
$$\frac{20}{4}+20\stackrel{?}{=}25$$
$$5+20=25\ \checkmark$$

4. Answers may vary. One possible solution to solving the equation follows:
Identify the LCD as 4, and multiply all terms.
$$-2x+\frac{3}{4}=\frac{1}{2}\to 4(-2x)+4\left(\frac{3}{4}\right)=4\left(\frac{1}{2}\right)$$
Multiply through to eliminate fractions.
$$4(-2x)+4\left(\frac{3}{4}\right)=4\left(\frac{1}{2}\right)\to -8x+3=2$$
Subtract 3 from both sides of the equation.
$$-8x+3-3=2-3\to -8x=-1$$
Divide both sides by −8 to isolate x.
$$\frac{-8x}{-8}=\frac{-1}{-8}\to x=\frac{1}{8}$$

7.5 Understanding the Concept Forming Equations

1. The next two numbers in the sequence x are 7 and 8.
x is: 1, 2, 3, 4, 5, 6, 7, 8, ...
We evaluate $7^2=49$ and $8^2=64$ to find the next two values of y.
y is: 1, 4, 9, 16, 25, 36, 49, 64, ...

7.5 Exercises

1. a. If three times a number is increased by nine, the result is fifteen: $3n + 9 = 15$.

b.
$$3n+9=15$$
$$+\ -9\ -9$$
$$3n\quad =6$$
$$\frac{3n}{3}=\frac{6}{3}$$
$$n=2$$

3. a. If triple a number is decreased by four, the result is five: $3n - 4 = 5$.

b.
$$\begin{aligned} 3n - 4 &= 5 \\ + \ +4 &\ +4 \\ \hline 3n &= 9 \\ \frac{3n}{3} &= \frac{9}{3} \\ n &= 3 \end{aligned}$$

5. a. If the sum of 4 and a number is multiplied by 2, the result is 12: $2(4 + n) = 12$.

b.
$$\begin{aligned} 2(4+n) &= 12 \\ 8+2n &= 12 \\ + \ -8 \quad &\ -8 \\ \hline 2n &= 4 \\ \frac{2n}{2} &= \frac{4}{2} \\ n &= 2 \end{aligned}$$

7.
$$\begin{aligned} P &= 2L+2W \\ 60 &= 2(x+15)+2(12) \\ 60 &= 2x+30+24 \\ 60 &= 2x+54 \\ + -54 = &\quad -54 \\ \hline 6 &= 2x \\ \frac{6}{2} &= \frac{2x}{2} \\ 3 &= x \end{aligned}$$
The value of x is 3 meters.

9. The perimeter is the sum of the lengths of the sides.
$$\begin{aligned} x+(x+1)+(x+2) &= 12 \\ 3x+3 &= 12 \\ + \qquad -3 &= -3 \\ \hline 3x &= 9 \\ \frac{3x}{3} &= \frac{9}{3} \\ x &= 3 \end{aligned}$$
First side: $x = 3$ cm
Second side: $x + 1 = 3 + 1 = 4$ cm
Third side: $x + 2 = 3 + 2 = 5$ cm

11.
$$\begin{aligned} A &= LW \\ 90 &= 15(2x) \\ 90 &= 30x \\ \frac{90}{30} &= \frac{30x}{30} \\ 3 &= x \end{aligned}$$
The value of x is 3 inches.

13. a. $L = 15$, $W = x + 12$, $P = 70$
$$\begin{aligned} P &= 2L+2W \\ 70 &= 2(15)+2(x+12) \\ 70 &= 30+2x+24 \\ 70 &= 2x+54 \\ + -54 = &\quad -54 \\ \hline 16 &= 2x \\ \frac{16}{2} &= \frac{2x}{2} \\ 8 &= x \end{aligned}$$
The width should be enlarged 8 feet.

b. $W = x + 12 = 8 + 12 = 20$
The width will be 20 feet.

15. a. $L = 9 + x$, $W = 7$, $A = 105$
$$\begin{aligned} A &= LW \\ 105 &= (9+x)(7) \\ 105 &= 63+7x \\ + \ -63 &\ -63 \\ \hline 42 &= 7x \\ \frac{42}{7} &= \frac{7x}{7} \\ 6 &= x \end{aligned}$$
The length should be enlarged 6 feet.

b. $L = 9 + x = 9 + 6 = 15$
The length will be 15 feet.

17. a. We are comparing the tutor's salary to Mark's salary, so we let the variable M represent Mark's salary.
Mark's salary = M
The tutor earns $11,400 less than Mark.
Tutor's salary = $M - 11,400$

b. The sum of Mark's salary and the tutor's salary is $28,890.
$M + (M - 11,400) = 28,890$

c.
$$\begin{aligned} M+(M-11,400) &= 28,890 \\ 2M-11,400 &= 28,890 \\ + \qquad 11,400 &\quad 11,400 \\ \hline 2M &= 40,290 \\ \frac{2M}{2} &= \frac{40,290}{2} \\ M &= 20,145 \end{aligned}$$
$M - 11,400 = 20,145 - 11,400 = 8745$
Mark earns $20,145, and the tutor earns $8745.

d.
$$\begin{aligned} \$20,145+\$8745 &\stackrel{?}{=} \$28,890 \\ \$28,890 &= \$28,890 \ \checkmark \end{aligned}$$

Copyright © 2012 Pearson Education, Inc.

19. a. We are comparing the distance Cal drove the second day to the distance he drove the first day, so we let the variable x represent the distance Cal drove the first day.
Miles Cal drove the first day = x
He drove 165 more miles the second day than the first day.
Miles he drove the second day = $x + 165$

b. In two days Cal drove 825 miles.
$x + (x + 165) = 825$

c.
$$\begin{aligned} x+(x+165) &= 825 \\ 2x+165 &= 825 \\ + \quad -165 &\ \ -165 \\ \hline 2x &= 660 \\ \frac{2x}{2} &= \frac{660}{2} \\ x &= 330 \end{aligned}$$
$x + 165 = 330 + 165 = 495$
Cal drove 330 miles the first day and 495 miles the second day.

d. $330 \text{ mi} + 495 \text{ mi} \stackrel{?}{=} 825 \text{ mi}$
$825 \text{ mi} = 825 \text{ mi}$ ✓

21. a. We are comparing the time for the first flight to the time for the second flight, so we let the variable s represent the time for the second flight.
Flight time for the second flight = s
The time for the first flight is half of the second.
Flight time for the first flight $= \frac{1}{2}s$

b. The total flying time for the two flights is 15 hr.
$s + \left(\frac{1}{2}s\right) = 15$ or $s + \frac{s}{2} = 15$

c. $s + \frac{s}{2} = 15$
LCD = 2
$$\begin{aligned} 2(s) + 2\left(\frac{s}{2}\right) &= 2(15) \\ 2s + s &= 30 \\ 3s &= 30 \\ \frac{3s}{3} &= \frac{30}{3} \\ s &= 10 \end{aligned}$$
$\frac{1}{2}s = \frac{1}{2}(10) = 5$
The second flight is 10 hours, and the first flight is 5 hours.

d. $5 \text{ hr} + 10 \text{ hr} \stackrel{?}{=} 15 \text{ hr}$
$15 \text{ hr} = 15 \text{ hr}$ ✓

23. a. We are comparing the lengths of the second and third sides to the length of the first side, so we let the variable L be the length of the first side.
Length of first side = L
The length of the second side is double the first.
Length of second side = $2L$
The third side is 12 meters longer than the first side.
Length of third side = $L + 12$

b. The perimeter is 120 meters.
$L + 2L + (L + 12) = 120$

c.
$$\begin{aligned} L+2L+(L+12) &= 120 \\ 4L+12 &= 120 \\ + \quad -12 &\ \ -12 \\ \hline 4L &= 108 \\ \frac{4L}{4} &= \frac{108}{4} \\ L &= 27 \end{aligned}$$
$2L = 2(27) = 54$
$L + 12 = 27 + 12 = 39$
The first side is 27 meters, the second side is 54 meters, and the third side is 39 meters.

d. $27 \text{ m} + 54 \text{ m} + 39 \text{ m} \stackrel{?}{=} 120 \text{ m}$
$120 \text{ m} = 120 \text{ m}$ ✓

25. a. We are comparing the length to the width, so we let the variable W represent the width.
Width = W
The length is 2 meters less than triple the width.
Length = $3W - 2$

b. The perimeter is 68 meters.
$2(3W - 2) + 2W = 68$

c.
$$\begin{aligned} 2(3W-2)+2W &= 68 \\ 6W-4+2W &= 68 \\ 8W-4 &= 68 \\ +\quad 4\ & \quad 4 \\ \hline 8W &= 72 \\ \frac{8W}{8} &= \frac{72}{8} \\ W &= 9 \end{aligned}$$

$3W - 2 = 3(9) - 2 = 27 - 2 = 25$
The rectangle is 25 meters long by 9 meters wide.

d.
$$\begin{aligned} 2(25\text{ m})+2(9\text{ m}) &\stackrel{?}{=} 68\text{ m} \\ 50\text{ m}+18\text{ m} &\stackrel{?}{=} 68\text{ m} \\ 68\text{ m} &= 68\text{ m}\ \checkmark \end{aligned}$$

27. a. We are comparing the number of students in the spring and summer to the number of students in the fall, so we let the variable x represent the number of students in the fall.
Number in fall = x
There were 95 more students in the spring than in the fall.
Number in spring = $x + 95$
There were 75 fewer students in the summer than in the fall.
Number in summer = $x - 75$

b. A total of 395 students took English.
$x + (x + 95) + (x - 75) = 395$

c.
$$\begin{aligned} x+(x+95)+(x-75) &= 395 \\ 3x+20 &= 395 \\ +\quad -20\ & \quad -20 \\ \hline 3x &= 375 \\ \frac{3x}{3} &= \frac{375}{3} \\ x &= 125 \end{aligned}$$

$x + 95 = 125 + 95 = 220$
$x - 75 = 125 - 75 = 50$
125 students took English in the fall, 220 in the spring and 50 in the summer.

d.
$$\begin{aligned} 125+220+50 &\stackrel{?}{=} 395 \\ 395 &= 395\ \checkmark \end{aligned}$$

29. The sum of the angle measures is 180°.
$$\begin{aligned} (3x-1)+(5x-3) &= 180 \\ 8x-4 &= 180 \\ +\quad 4\ & \quad 4 \\ \hline 8x &= 184 \\ \frac{8x}{8} &= \frac{184}{8} \\ x &= 23 \end{aligned}$$
The value of x is 23.

31. $P = 4s,\ A = s^2$
If the perimeter and area are equal, then $4s = s^2$, or $4 \cdot s = s \cdot s$. If $s = 4$, the statement is true since $4 \cdot 4 = 4 \cdot 4$. The length of a side is 4 units.

33. a. The values of x are consecutive integers. The next two values of x are 7 and 8.

b. $3(1) = 3,\ 3(2) = 6,\ 3(3) = 9, \ldots$
We multiply each x by 3 to get y.

c. $3x = y$

d. $3(6) = 18,\ 3(7) = 21,\ 3(8) = 24$, and $3(30) = 90$
The missing values of y are 18, 21, 24, and 90.

35. a. The values of x are consecutive integers. the missing numbers are 4 and 6.

b. $3(0) + 1 = 1,\ 3(1) + 1 = 4,\ 3(2) + 1 = 7,\ 3(3) + 1 = 10, \ldots$
We multiply each x by 3 and then add 1 to obtain y.

c. $3x + 1 = y$

d. $3(4) + 1 = 13,\ 3(6) + 1 = 19,\ 3(45) + 1 = 136$
The missing values of y are 13, 19, and 136.

Cumulative Review

37. The length will not change: $13\frac{1}{2}$ in.
The width will be one-half of the current width.
$\frac{1}{2} \cdot 13\frac{1}{2} = \frac{1}{2} \cdot \frac{27}{2} = \frac{27}{4} = 6\frac{3}{4}$ in.
The height will be double the current height.

Copyright © 2012 Pearson Education, Inc.

$2 \cdot \frac{5}{8} = \frac{5}{4} = 1\frac{1}{4}$ in.

The dimensions are $13\frac{1}{2}$ in. by $6\frac{3}{4}$ in. by $1\frac{1}{4}$ in.

38. a. The length is the length of 4 bricks and 3 grout seams.

$$4\left(8\frac{1}{4}\right)+3\left(\frac{3}{4}\right)=4\left(\frac{33}{4}\right)+\frac{9}{4}$$
$$=33+2\frac{1}{4}$$
$$=35\frac{1}{4}\text{ in.}$$

The width is the width of 3 bricks and 3 grout seams.

$$3\left(3\frac{3}{4}\right)+3\left(\frac{3}{4}\right)=3\left(\frac{15}{4}\right)+\frac{9}{4}$$
$$=\frac{45}{4}+\frac{9}{4}$$
$$=\frac{54}{4}$$
$$=\frac{27}{2}$$
$$=13\frac{1}{2}\text{ in.}$$

The length is $35\frac{1}{4}$ inches, and the width is $13\frac{1}{2}$ inches.

b. $P = 2L + 2W$

$$=2\left(35\frac{1}{4}\right)+2\left(13\frac{1}{2}\right)$$
$$=2\left(\frac{141}{4}\right)+2\left(\frac{27}{2}\right)$$
$$=\frac{141}{2}+27$$
$$=70\frac{1}{2}+27$$
$$=97\frac{1}{2}\text{ in.}$$

Quick Quiz 7.5

1. We are comparing the first and second sides to the third side, so we let the variable x represent the length of the third side.
Length of the third side = x
The first side is 5 meters longer than the third side.
Length of first side = $x + 5$
The second side is twice the length of the third side.
Length of second side = $2x$

2. The perimeter is 29 meters.
$x + (x + 5) + 2x = 29$

3.
$$x+(x+5)+2x=29$$
$$4x+5=29$$
$$+\quad -5 \quad -5$$
$$4x = 24$$
$$\frac{4x}{4}=\frac{24}{4}$$
$$x=6$$

$x + 5 = 6 + 5 = 11$
$2x = 2(6) = 12$
The third side is 6 meters, the first side is 11 meters, and the second side is 12 meters.

4. a. Answers may vary. One possible solution is that Eduardo mistakenly left out Sam's contribution to the total laps.

b. Let S = laps completed by Sam, then A = laps completed by Alicia, and M = laps completed by Miguel.
Given: $S + A + M = 29$
$A = S - 4$
$M = S + 2$
Combining equations:
$$S+(S-4)+(S+2)=29$$
$$3S-2=29$$
$$S=\frac{31}{3}=10\frac{1}{3}$$
$$M=S+2=10\frac{1}{3}+2=12\frac{1}{3}$$

Miguel completed $12\frac{1}{3}$ laps.

Use Math to Save Money

1. \$315 + \$22 + \$15 = \$352
It will cost \$352 to buy this suit.

Copyright © 2012 Pearson Education, Inc.

2. \$335 + \$13 = \$348
It will cost \$348 to buy the second suit.

3. Yes, Arnold can buy the second suit since its total cost is less than \$350.

4. \$315 + \$22 = \$337
With free tailoring, the first suit would cost \$337.

5. If tailoring were free, Arnold should buy the first suit.

You Try It

1.
$$\begin{aligned} -1-6 &= 9x+5-8x-2 \\ -7 &= x+3 \\ +\quad -3 &\quad -3 \\ \hline -10 &= x \end{aligned}$$

Check:
$$\begin{aligned} -1-6 &= 9x+5-8x-2 \\ -1-6 &\stackrel{?}{=} 9(-10)+5-8(-10)-2 \\ -7 &\stackrel{?}{=} -90+5+80-2 \\ -7 &= -7 \checkmark \end{aligned}$$

2.
$$\begin{aligned} \frac{y}{-3} &= 2^2-8 \\ \frac{y}{-3} &= 4-8 \\ \frac{y}{-3} &= -4 \\ \frac{-3y}{-3} &= -4(-3) \\ y &= 12 \end{aligned}$$

3.
$$\begin{aligned} -\frac{15}{5} &= -2(4x)+10x \\ -3 &= -8x+10x \\ -3 &= 2x \\ \frac{-3}{2} &= \frac{2x}{2} \\ -\frac{3}{2} &= x \end{aligned}$$

4.
$$\begin{aligned} 6x-4 &= 3+8x+7 \\ 6x-4 &= 8x+10 \\ +\ -6x &\quad -6x \\ \hline -4 &= 2x+10 \\ +\ -10 &\quad -10 \\ \hline -14 &= 2x \\ \frac{-14}{2} &= \frac{2x}{2} \\ -7 &= x \end{aligned}$$

Check:
$$\begin{aligned} 6x-4 &= 3+8x+7 \\ 6(-7)-4 &\stackrel{?}{=} 3+8(-7)+7 \\ -42-4 &\stackrel{?}{=} 3+(-56)+7 \\ -46 &= -46 \checkmark \end{aligned}$$

5.
$$\begin{aligned} -3(8x-2) &= 2(7x+1)+23 \\ -24x+6 &= 14x+2+23 \\ -24x+6 &= 14x+25 \\ +\quad +24x &\quad +24x \\ \hline 6 &= 38x+25 \\ +\quad -25 &\quad -25 \\ \hline -19 &= 38x \\ \frac{-19}{38} &= \frac{38x}{38} \\ -\frac{1}{2} &= x \end{aligned}$$

6. $2x+\frac{1}{3}=\frac{2}{7}$
The LCD is 21.
$$\begin{aligned} (21)2x+(21)\cdot\frac{1}{3} &= (21)\cdot\frac{2}{7} \\ 42x+7 &= 6 \\ +\quad -7 &\quad -7 \\ \hline 42x &= -1 \\ \frac{42x}{42} &= \frac{-1}{42} \\ x &= -\frac{1}{42} \end{aligned}$$

Chapter 7 Review Problems

1.
$$\begin{aligned} x-5 &= 37 \\ +\quad 5 &\quad 5 \\ \hline x &= 42 \end{aligned}$$

Check:
$$\begin{aligned} x-5 &= 37 \\ 42-5 &\stackrel{?}{=} 37 \\ 37 &= 37 \checkmark \end{aligned}$$

2.
$$\begin{aligned} 2(3x) &= -36 \\ 6x &= -36 \\ \frac{6x}{6} &= \frac{-36}{6} \\ x &= -6 \end{aligned}$$

Check:
$$\begin{aligned} 2(3x) &= -36 \\ 2[3(-6)] &\stackrel{?}{=} -36 \\ 2[-18] &\stackrel{?}{=} -36 \\ -36 &= -36 \checkmark \end{aligned}$$

Copyright © 2012 Pearson Education, Inc.

3. $-15-20=3x+13-2x$

$-35=3x+13-2x$

$-35=x+13$

$+\quad -13 \qquad -13$

$-48=x$ or $x=-48$

Check: $-15-20=3x+13-2x$

$-15-20 \stackrel{?}{=} 3(-48)+13-2(-48)$

$-35 \stackrel{?}{=} -144+13+96$

$-35=-35$ ✓

4. $\frac{y}{-2}=-8+3^2$

$\frac{y}{-2}=-8+9$

$\frac{y}{-2}=1$

$\frac{-2\cdot y}{-2}=1\cdot(-2)$

$y=-2$

Check: $\frac{y}{-2}=-8+3^2$

$\frac{-2}{-2} \stackrel{?}{=} -8+3^2$

$1 \stackrel{?}{=} -8+9$

$1=1$ ✓

5. $-5+2^2=\frac{x}{-5}$

$-5+4=\frac{x}{-5}$

$-1=\frac{x}{-5}$

$-1(-5)=\frac{-5\cdot x}{-5}$

$5=x$ or $x=5$

Check: $-5+2^2=\frac{x}{-5}$

$-5+2^2 \stackrel{?}{=} \frac{5}{-5}$

$-5+4 \stackrel{?}{=} -1$

$-1=-1$ ✓

6. $\frac{6}{7}a=2^2-1$

$\frac{6}{7}a=4-1$

$\frac{6}{7}a=3$

$\frac{7}{6}\cdot\left(\frac{6}{7}a\right)=\frac{7}{6}\cdot 3$

$a=\frac{7}{2}$

Check: $\frac{6}{7}a=2^2-1$

$\frac{6}{7}\left(\frac{7}{2}\right) \stackrel{?}{=} 2^2-1$

$\frac{6}{2} \stackrel{?}{=} 4-1$

$3=3$ ✓

7. $-3x+2(4x)=\frac{18}{-2}$

$-3x+8x=\frac{18}{-2}$

$5x=\frac{18}{-2}$

$5x=-9$

$\frac{5x}{5}=\frac{-9}{5}$

$x=-\frac{9}{5}$

Check: $-3x+2(4x)=\frac{18}{-2}$

$-3\left(-\frac{9}{5}\right)+2\left[4\left(-\frac{9}{5}\right)\right] \stackrel{?}{=} \frac{18}{-2}$

$\frac{27}{5}-\frac{72}{5} \stackrel{?}{=} -9$

$-\frac{45}{5} \stackrel{?}{=} -9$

$-9=-9$ ✓

8. $-x=-18$

$-1x=-18$

$\frac{-1x}{-1}=\frac{-18}{-1}$

$x=18$

Check: $-x=-18$

$-(18) \stackrel{?}{=} -18$

$-18=-18$ ✓

Copyright © 2012 Pearson Education, Inc.

9.
$$-y = \frac{3}{4}$$
$$-1y = \frac{3}{4}$$
$$\frac{-1y}{-1} = \frac{\frac{3}{4}}{-1}$$
$$y = -\frac{3}{4}$$

Check:
$$-y = \frac{3}{4}$$
$$-\left(-\frac{3}{4}\right) \stackrel{?}{=} \frac{3}{4}$$
$$\frac{3}{4} = \frac{3}{4} \checkmark$$

10.
$$\begin{array}{r} 6x - 8 = 34 \\ + \quad 8 \quad 8 \\ \hline 6x \quad = 42 \end{array}$$
$$\frac{6x}{6} = \frac{42}{6}$$
$$x = 7$$

Check:
$$6x - 8 = 34$$
$$6(7) - 8 \stackrel{?}{=} 34$$
$$42 - 8 \stackrel{?}{=} 34$$
$$34 = 34 \checkmark$$

11.
$$\begin{array}{r} -2y + 16 = 58 \\ + \quad -16 \ -16 \\ \hline -2y \quad = 42 \end{array}$$
$$\frac{-2y}{-2} = \frac{42}{-2}$$
$$y = -21$$

Check:
$$-2y + 16 = 58$$
$$-2(-21) + 16 \stackrel{?}{=} 58$$
$$42 + 16 \stackrel{?}{=} 58$$
$$58 = 58 \checkmark$$

12.
$$\begin{array}{r} -20 = 8 - 7y \\ + -8 \ -8 \quad \quad \\ \hline -28 = \ -7y \end{array}$$
$$\frac{-28}{-7} = \frac{-7y}{-7}$$
$$4 = y \text{ or } y = 4$$

Check:
$$-20 = 8 - 7y$$
$$-20 \stackrel{?}{=} 8 - 7(4)$$
$$-20 \stackrel{?}{=} 8 - 28$$
$$-20 = -20 \checkmark$$

13.
$$\begin{array}{r} 14 = 10 - 4x \\ + -10 \ -10 \quad \quad \\ \hline 4 = \ -4x \end{array}$$
$$\frac{4}{-4} = \frac{-4x}{-4}$$
$$-1 = x \text{ or } x = -1$$

Check:
$$14 = 10 - 4x$$
$$14 \stackrel{?}{=} 10 - 4(-1)$$
$$14 \stackrel{?}{=} 10 + 4$$
$$14 = 14 \checkmark$$

14.
$$\begin{array}{r} 1 - 4y = 19 \\ + -1 \quad \quad -1 \\ \hline -4y = 18 \end{array}$$
$$\frac{-4y}{-4} = \frac{18}{-4}$$
$$y = -\frac{9}{2}$$

Check:
$$1 - 4y = 19$$
$$1 - 4\left(-\frac{9}{2}\right) \stackrel{?}{=} 19$$
$$1 + 18 = 19$$
$$19 = 19 \checkmark$$

15.
$$8x - 7 - 5x = 15$$
$$\begin{array}{r} 3x - 7 = 15 \\ + \quad \quad 7 \quad 7 \\ \hline 3x = 22 \end{array}$$
$$\frac{3x}{3} = \frac{22}{3}$$
$$x = \frac{22}{3}$$

Check:
$$8x - 7 - 5x = 15$$
$$8\left(\frac{22}{3}\right) - 7 - 5\left(\frac{22}{3}\right) \stackrel{?}{=} 15$$
$$\frac{176}{3} - \frac{21}{3} - \frac{110}{3} \stackrel{?}{=} 15$$
$$\frac{45}{3} \stackrel{?}{=} 15$$
$$15 = 15 \checkmark$$

16.
$$\begin{array}{r} -6x = 9x + 36 \\ + -9x \ -9x \quad \quad \\ \hline -15x = \quad \quad 36 \end{array}$$
$$\frac{-15x}{-15} = \frac{36}{-15}$$
$$x = -\frac{12}{5}$$

Copyright © 2012 Pearson Education, Inc.

Check: $-6x = 9x + 36$

$$-6\left(-\frac{12}{5}\right) \stackrel{?}{=} 9\left(-\frac{12}{5}\right) + 36$$

$$\frac{72}{5} \stackrel{?}{=} -\frac{108}{5} + \frac{180}{5}$$

$$\frac{72}{5} = \frac{72}{5} \checkmark$$

17.

$$\begin{array}{rcl} 5x &=& 2x + 30 \\ + \; -2x & & -2x \\ \hline 3x &=& 30 \end{array}$$

$$\frac{3x}{3} = \frac{30}{3}$$

$$x = 10$$

Check: $5x = 2x + 30$

$$5(10) \stackrel{?}{=} 2(10) + 30$$

$$50 \stackrel{?}{=} 20 + 30$$

$$50 = 50 \checkmark$$

18.

$$3x - 8 - 5x = 6$$

$$\begin{array}{rcl} -2x - 8 &=& 6 \\ + \quad\; 8 & & 8 \\ \hline -2x &=& 14 \end{array}$$

$$\frac{-2x}{-2} = \frac{14}{-2}$$

$$x = -7$$

Check: $3x - 8 - 5x = 6$

$$3(-7) - 8 - 5(-7) \stackrel{?}{=} 6$$

$$-21 - 8 + 35 \stackrel{?}{=} 6$$

$$6 = 6 \checkmark$$

19.

$$-3 + 2y + 6 = -4y + 12$$

$$\begin{array}{rcl} 2y + 3 &=& -4y + 12 \\ + \quad 4y & & 4y \\ \hline 6y + 3 &=& 12 \\ + \quad -3 & & -3 \\ \hline 6y &=& 9 \end{array}$$

$$\frac{6y}{6} = \frac{9}{6}$$

$$y = \frac{3}{2}$$

Check: $-3 + 2y + 6 = -4y + 12$

$$-3 + 2\left(\frac{3}{2}\right) + 6 \stackrel{?}{=} -4\left(\frac{3}{2}\right) + 12$$

$$-3 + 3 + 6 \stackrel{?}{=} -6 + 12$$

$$6 = 6 \checkmark$$

Copyright © 2012 Pearson Education, Inc.

20.
$$\begin{array}{rl} 6y-8+2y &= -6+9y \\ 8y-8 &= -6+9y \\ +\ -9y \qquad & \qquad -9y \\ \hline -y-8 &= -6 \\ +\qquad 8 & \quad 8 \\ \hline -y \qquad &= 2 \\ \frac{-y}{-1} &= \frac{2}{-1} \\ y &= -2 \end{array}$$

Check:
$$\begin{aligned} 6y-8+2y &= -6+9y \\ 6(-2)-8+2(-2) &\stackrel{?}{=} -6+9(-2) \\ -12-8-4 &\stackrel{?}{=} -6-18 \\ -24 &= -24\ \checkmark \end{aligned}$$

21.
$$\begin{array}{rl} 8x-9-5x+18 &= -3x-2+2x \\ 3x+9 &= -x-2 \\ +\qquad x \qquad & \quad x \\ \hline 4x+9 &= \quad -2 \\ +\qquad -9 & \quad -9 \\ \hline 4x \qquad &= -11 \\ \frac{4x}{4} &= \frac{-11}{4} \\ x &= -\frac{11}{4} \end{array}$$

Check:
$$\begin{aligned} 8x-9-5x+18 &= -3x-2+2x \\ 8\left(-\frac{11}{4}\right)-9-5\left(-\frac{11}{4}\right)+18 &\stackrel{?}{=} -3\left(-\frac{11}{4}\right)-2+2\left(-\frac{11}{4}\right) \\ -\frac{88}{4}-\frac{36}{4}+\frac{55}{4}+\frac{72}{4} &\stackrel{?}{=} \frac{33}{4}-\frac{8}{4}-\frac{22}{4} \\ \frac{3}{4} &= \frac{3}{4}\ \checkmark \end{aligned}$$

22.
$$\begin{array}{rl} -3x-2-9x+5 &= 2x+8-7x \\ -12x+3 &= -5x+8 \\ +\qquad 5x \qquad & \quad 5x \\ \hline -7x+3 &= \quad 8 \\ +\qquad -3 & \quad -3 \\ \hline -7x \qquad &= 5 \\ \frac{-7x}{-7} &= \frac{5}{-7} \\ x &= -\frac{5}{7} \end{array}$$

Check:
$$\begin{aligned} -3x-2-9x+5 &= 2x+8-7x \\ -3\left(-\frac{5}{7}\right)-2-9\left(-\frac{5}{7}\right)+5 &\stackrel{?}{=} 2\left(-\frac{5}{7}\right)+8-7\left(-\frac{5}{7}\right) \\ \frac{15}{7}-\frac{14}{7}+\frac{45}{7}+\frac{35}{7} &\stackrel{?}{=} -\frac{10}{7}+\frac{56}{7}+\frac{35}{7} \\ \frac{81}{7} &= \frac{81}{7}\ \checkmark \end{aligned}$$

Copyright © 2012 Pearson Education, Inc.

23.
$$\begin{aligned} -3(x+5) &= 21 \\ -3x-15 &= 21 \\ +\quad 15 &\quad 15 \\ \hline -3x &= 36 \\ \frac{-3x}{-3} &= \frac{36}{-3} \\ x &= -12 \end{aligned}$$

Check:
$$\begin{aligned} -3(x+5) &= 21 \\ -3(-12+5) &\stackrel{?}{=} 21 \\ -3(-7) &\stackrel{?}{=} 21 \\ 21 &= 21 \checkmark \end{aligned}$$

24.
$$\begin{aligned} 4(2x+9)+5x &= -3 \\ 8x+36+5x &= -3 \\ 13x+36 &= -3 \\ +\quad -36 &\quad -36 \\ \hline 13x &= -39 \\ \frac{13x}{13} &= \frac{-39}{13} \\ x &= -3 \end{aligned}$$

Check:
$$\begin{aligned} 4(2x+9)+5x &= -3 \\ 4[2(-3)+9]+5(-3) &\stackrel{?}{=} -3 \\ 4(-6+9)-15 &\stackrel{?}{=} -3 \\ 4(3)-15 &\stackrel{?}{=} -3 \\ 12-15 &\stackrel{?}{=} -3 \\ -3 &= -3 \checkmark \end{aligned}$$

25.
$$\begin{aligned} 7(x+1)+3(x+1) &= -10 \\ 7x+7+3x+3 &= -10 \\ 10x+10 &= -10 \\ +\quad -10 &\quad -10 \\ \hline 10x &= -20 \\ \frac{10x}{10} &= \frac{-20}{10} \\ x &= -2 \end{aligned}$$

Check:
$$\begin{aligned} 7(x+1)+3(x+1) &= -10 \\ 7(-2+1)+3(-2+1) &\stackrel{?}{=} -10 \\ 7(-1)+3(-1) &\stackrel{?}{=} -10 \\ -7-3 &\stackrel{?}{=} -10 \\ -10 &= -10 \checkmark \end{aligned}$$

Copyright © 2012 Pearson Education, Inc.

26.

$$\begin{aligned}
-3(x-7) &= 5(x+6)-10 \\
-3x+21 &= 5x+30-10 \\
-3x+21 &= 5x+20 \\
+\ -5x \quad & \quad -5x \\
\hline
-8x+21 &= \quad 20 \\
+\quad -21 & \quad\ \ 21 \\
\hline
-8x \quad &= \quad -1 \\
\frac{-8x}{-8} &= \frac{-1}{-8} \\
x &= \frac{1}{8}
\end{aligned}$$

Check:

$$\begin{aligned}
-3(x-7) &= 5(x+6)-10 \\
-3\left(\frac{1}{8}-7\right) &\stackrel{?}{=} 5\left(\frac{1}{8}+6\right)-10 \\
-3\left(\frac{1}{8}-\frac{56}{8}\right) &\stackrel{?}{=} 5\left(\frac{1}{8}+\frac{48}{8}\right)-\frac{80}{8} \\
-3\left(-\frac{55}{8}\right) &\stackrel{?}{=} 5\left(\frac{49}{8}\right)-\frac{80}{8} \\
\frac{165}{8} &\stackrel{?}{=} \frac{245}{8}-\frac{80}{8} \\
\frac{165}{8} &= \frac{165}{8} \checkmark
\end{aligned}$$

27.

$$\begin{aligned}
(9y^2+8y-2)-(9y^2+5y-1) &= 14 \\
9y^2+8y-2-9y^2-5y+1 &= 14 \\
3y-1 &= 14 \\
+ \qquad\qquad\qquad\qquad 1 & \quad\ \ 1 \\
\hline
3y \quad &= 15 \\
\frac{3y}{3} &= \frac{15}{3} \\
y &= 5
\end{aligned}$$

Check:

$$\begin{aligned}
(9y^2+8y-2)-(9y^2+5y-1) &= 14 \\
(9\cdot 5^2+8\cdot 5-2)-(9\cdot 5^2+5\cdot 5-1) &\stackrel{?}{=} 14 \\
(9\cdot 25+40-2)-(9\cdot 25+25-1) &\stackrel{?}{=} 14 \\
(225+40-2)-(225+25-1) &\stackrel{?}{=} 14 \\
263-249 &\stackrel{?}{=} 14 \\
14 &= 14 \checkmark
\end{aligned}$$

Copyright © 2012 Pearson Education, Inc.

28. $(5x^2 + x - 2) - (5x^2 - 5) = 6x + 9$

$$\begin{aligned}
5x^2 + x - 2 - 5x^2 + 5 &= 6x + 9 \\
x + 3 &= 6x + 9 \\
+ \quad -6x \quad &\quad -6x \\
-5x + 3 &= 9 \\
+ \quad -3 \quad &\quad -3 \\
-5x &= 6 \\
\frac{-5x}{-5} &= \frac{6}{-5} \\
x &= -\frac{6}{5}
\end{aligned}$$

Check: $(5x^2 + x - 2) - (5x^2 - 5) = 6x + 9$

$$\left[5\left(-\frac{6}{5}\right)^2 + \left(-\frac{6}{5}\right) - 2\right] - \left[5\left(-\frac{6}{5}\right)^2 - 5\right] \stackrel{?}{=} 6\left(-\frac{6}{5}\right) + 9$$

$$\left[5\left(\frac{36}{25}\right) - \frac{6}{5} - 2\right] - \left[5\left(\frac{36}{25}\right) - 5\right] \stackrel{?}{=} -\frac{36}{5} + \frac{45}{5}$$

$$\left(\frac{36}{5} - \frac{6}{5} - \frac{10}{5}\right) - \left(\frac{36}{5} - \frac{25}{5}\right) \stackrel{?}{=} \frac{9}{5}$$

$$\frac{20}{5} - \frac{11}{5} \stackrel{?}{=} \frac{9}{5}$$

$$\frac{9}{5} = \frac{9}{5} \checkmark$$

29. $\frac{x}{3} + \frac{x}{4} = 7$

LCD = 12

$$\begin{aligned}
12\left(\frac{x}{3}\right) + 12\left(\frac{x}{4}\right) &= 12(7) \\
4x + 3x &= 84 \\
7x &= 84 \\
\frac{7x}{7} &= \frac{84}{7} \\
x &= 12
\end{aligned}$$

Check: $\frac{x}{3} + \frac{x}{4} = 7$

$$\frac{12}{3} + \frac{12}{4} \stackrel{?}{=} 7$$

$$4 + 3 \stackrel{?}{=} 7$$

$$7 = 7 \checkmark$$

Copyright © 2012 Pearson Education, Inc.

30. $\frac{x}{5}-\frac{x}{2}=6$

LCD = 10

$$10\left(\frac{x}{5}\right)-10\left(\frac{x}{2}\right)=10(6)$$
$$2x-5x=60$$
$$-3x=60$$
$$\frac{-3x}{-3}=\frac{60}{-3}$$
$$x=-20$$

Check: $\frac{x}{5}-\frac{x}{2}=6$
$$\frac{-20}{5}-\frac{-20}{2}\stackrel{?}{=}6$$
$$-4+10\stackrel{?}{=}6$$
$$6=6\ \checkmark$$

31. $y+\frac{y}{5}=-12$

LCD = 5

$$5(y)+5\left(\frac{y}{5}\right)=5(-12)$$
$$5y+y=-60$$
$$6y=-60$$
$$\frac{6y}{6}=\frac{-60}{6}$$
$$y=-10$$

Check: $y+\frac{y}{5}=-12$
$$-10+\frac{-10}{5}\stackrel{?}{=}-12$$
$$-10-2\stackrel{?}{=}-12$$
$$-12=-12\ \checkmark$$

32. $2y-\frac{y}{4}=7$

LCD = 4

$$4(2y)-4\left(\frac{y}{4}\right)=4(7)$$
$$8y-y=28$$
$$7y=28$$
$$\frac{7y}{7}=\frac{28}{7}$$
$$y=4$$

Check: $2y-\frac{y}{4}=7$
$$2(4)-\frac{4}{4}\stackrel{?}{=}7$$
$$8-1\stackrel{?}{=}7$$
$$7=7\ \checkmark$$

33. $2x-\frac{3}{4}=\frac{1}{3}$

LCD = 12

$$12(2x)-12\left(\frac{3}{4}\right)=12\left(\frac{1}{3}\right)$$
$$\begin{array}{rr} & 24x-9=\ 4 \\ + & 9\quad 9 \\ \hline & 24x\quad =13 \end{array}$$
$$\frac{24x}{24}=\frac{13}{24}$$
$$x=\frac{13}{24}$$

Check: $2x-\frac{3}{4}=\frac{1}{3}$
$$2\left(\frac{13}{24}\right)-\frac{3}{4}\stackrel{?}{=}\frac{1}{3}$$
$$\frac{13}{12}-\frac{9}{12}\stackrel{?}{=}\frac{1}{3}$$
$$\frac{4}{12}\stackrel{?}{=}\frac{1}{3}$$
$$\frac{1}{3}=\frac{1}{3}\ \checkmark$$

34. $2x-\frac{3}{4}=\frac{1}{2}$

LCD = 4

$$4(2x)-4\left(\frac{3}{4}\right)=4\left(\frac{1}{2}\right)$$
$$\begin{array}{rr} & 8x-3=2 \\ + & 3\quad 3 \\ \hline & 8x\quad =5 \end{array}$$
$$\frac{8x}{8}=\frac{5}{8}$$
$$x=\frac{5}{8}$$

Copyright © 2012 Pearson Education, Inc.

Check: $2x - \frac{3}{4} = \frac{1}{2}$

$$2\left(\frac{5}{8}\right) - \frac{3}{4} \stackrel{?}{=} \frac{1}{2}$$
$$\frac{5}{4} - \frac{3}{4} \stackrel{?}{=} \frac{1}{2}$$
$$\frac{2}{4} \stackrel{?}{=} \frac{1}{2}$$
$$\frac{1}{2} = \frac{1}{2} \checkmark$$

35. $-\frac{1}{3} = 3y + \frac{1}{2}$

LCD = 6

$$6\left(-\frac{1}{3}\right) = 6(3y) + 6\left(\frac{1}{2}\right)$$
$$-2 = 18y + 3$$
$$+ \quad -3 \qquad -3$$
$$-5 = 18y$$
$$\frac{-5}{18} = \frac{18y}{18}$$
$$-\frac{5}{18} = y \text{ or } y = -\frac{5}{18}$$

Check: $-\frac{1}{3} = 3y + \frac{1}{2}$

$$-\frac{1}{3} \stackrel{?}{=} 3\left(-\frac{5}{18}\right) + \frac{1}{2}$$
$$-\frac{1}{3} \stackrel{?}{=} -\frac{5}{6} + \frac{3}{6}$$
$$-\frac{1}{3} \stackrel{?}{=} -\frac{2}{6}$$
$$-\frac{1}{3} = -\frac{1}{3} \checkmark$$

36. a. Four times a number decreased by nine is fifteen: $4n - 9 = 15$.

b.
$$4n - 9 = 15$$
$$+ \quad 9 \quad 9$$
$$4n = 24$$
$$\frac{4n}{4} = \frac{24}{4}$$
$$n = 6$$

The number is 6.

37. a. $P = 2L + 2W$

$54 = 2(16) + 2(x + 3)$

b.
$$54 = 2(16) + 2(x + 3)$$
$$54 = 32 + 2x + 6$$
$$54 = 38 + 2x$$
$$+ -38 \quad -38$$
$$16 = 2x$$
$$\frac{16}{2} = \frac{2x}{2}$$
$$8 = x$$

$W = x + 3 = 8 + 3 = 11$ ft

The width is 11 feet.

38. a.
$$P = 2L + 2W$$
$$56 = 2(15 + x) + 2(8)$$
$$56 = 30 + 2x + 16$$
$$56 = 46 + 2x$$
$$+ -46 \quad -46$$
$$10 = 2x$$
$$\frac{10}{2} = \frac{2x}{2}$$
$$5 = x$$

The garden should be lengthened 5 feet.

b. $L = 15 + x = 15 + 5 = 20$ ft

The length will be 20 feet.

39. a. We are comparing Sara's age to Tara's age, so we let the variable T represent Tara's age.

Tara's age = T

Sara is seven years younger than Tara.

Sara's age = $T - 7$

b. The sum of their ages is twenty-nine.

$T + (T - 7) = 29$

c.
$$T + (T - 7) = 29$$
$$2T - 7 = 29$$
$$+ \qquad 7 \quad 7$$
$$2T = 36$$
$$\frac{2T}{2} = \frac{36}{2}$$
$$T = 18$$

$T - 7 = 18 - 7 = 11$

Tara is 18 years old, and Sara is 11 years old.

d.
$$18 + 11 \stackrel{?}{=} 29$$
$$29 = 29 \checkmark$$

Copyright © 2012 Pearson Education, Inc.

40. a. We are comparing the enrollments in the spring and summer to the enrollment in the fall, so we let the variable L represent the fall enrollment.
Number of fall students = L
85 more students were enrolled in the spring than in the fall.
Number of spring students = $L + 85$
65 fewer students were enrolled in the summer than in the fall.
Number of summer students = $L - 65$

b. A total of 491 students took Spanish.
$L + (L + 85) + (L - 65) = 491$

c.
$$\begin{aligned} L+(L+85)+(L-65) &= 491 \\ 3L+20 &= 491 \\ + \quad -20 &\ \ -20 \\ \hline 36 &= 471 \\ \frac{3L}{3} &= \frac{471}{3} \\ L &= 157 \end{aligned}$$

$L + 85 = 157 + 85 = 242$
$L - 65 = 157 - 65 = 92$
157 students took Spanish in the fall, 242 in the spring, and 92 in the summer.

d.
$$\begin{aligned} 157+242+92 &\stackrel{?}{=} 491 \\ 491 &= 491 \checkmark \end{aligned}$$

How Am I Doing? Chapter 7 Test

1.
$$\begin{aligned} 5(2x) &= -30 \\ 10x &= -30 \\ \frac{10x}{10} &= \frac{-30}{10} \\ x &= -3 \end{aligned}$$

2.
$$\begin{aligned} 7x+3-6x &= -4 \\ x+3 &= -4 \\ + \quad -3 &\ \ -3 \\ \hline x &= -7 \end{aligned}$$

3.
$$\begin{aligned} \frac{2}{3}a &= 3^2+1 \\ \frac{2}{3}a &= 9+1 \\ \frac{2}{3}a &= 10 \\ \frac{3}{2}\cdot\left(\frac{2}{3}a\right) &= \frac{3}{2}\cdot 10 \\ a &= 15 \end{aligned}$$

4.
$$\begin{aligned} -x &= 12 \\ -1x &= 12 \\ -1\cdot(-1x) &= -1\cdot 12 \\ x &= -12 \end{aligned}$$

5.
$$\begin{aligned} 5x-11 &= -1 \\ + \quad 11 &\ \ \ 11 \\ \hline 5x &= 10 \\ \frac{5x}{5} &= \frac{10}{5} \\ x &= 2 \end{aligned}$$

6.
$$\begin{aligned} \frac{x}{5} &= -2+4^2 \\ \frac{x}{5} &= -2+16 \\ \frac{x}{5} &= 14 \\ \frac{5\cdot x}{5} &= 14\cdot 5 \\ x &= 70 \end{aligned}$$

7.
$$\begin{aligned} 4(-2y)+5y &= 12 \\ -8y+5y &= 12 \\ -3y &= 12 \\ \frac{-3y}{-3} &= \frac{12}{-3} \\ y &= -4 \end{aligned}$$

8.
$$\begin{aligned} 12y+6-11y-1 &= -8+10 \\ y+5 &= -8+10 \\ y+5 &= 2 \\ + \quad -5 &\ \ -5 \\ \hline y &= -3 \end{aligned}$$

9.
$$\begin{aligned} -3 &= 7-4y \\ + -7 &\ \ -7 \\ \hline -10 &= -4y \\ \frac{-10}{-4} &= \frac{-4y}{-4} \\ \frac{5}{2} &= y \text{ or } y = \frac{5}{2} \end{aligned}$$

Copyright © 2012 Pearson Education, Inc.

10.
$$\begin{aligned} 6x+4-9x &= 15 \\ -3x+4 &= 15 \\ + \quad -4 & \quad -4 \\ \hline -3x &= 11 \\ \frac{-3x}{-3} &= \frac{11}{-3} \\ x &= -\frac{11}{3} \end{aligned}$$

11.
$$\begin{aligned} 14x &= -2x+16 \\ +2x & \quad 2x \\ \hline 16x &= 16 \\ \frac{16x}{16} &= \frac{16}{16} \\ x &= 1 \end{aligned}$$

12.
$$\begin{aligned} 2x+4(3x-6) &= 4-(6x+2) \\ 2x+12x-24 &= 4-6x-2 \\ 14x-24 &= 2-6x \\ + \quad 6x & \quad\quad 6x \\ \hline 20x-24 &= 2 \\ + \quad 24 & \quad 24 \\ \hline 20x &= 26 \\ \frac{20x}{20} &= \frac{26}{20} \\ x &= \frac{13}{10} \end{aligned}$$

13.
$$\begin{aligned} 3(2x+6)+3x &= -27 \\ 6x+18+3x &= -27 \\ 9x+18 &= -27 \\ + \quad -18 & \quad -18 \\ \hline 9x &= -45 \\ \frac{9x}{9} &= \frac{-45}{9} \\ x &= -5 \end{aligned}$$

14.
$$\begin{aligned} 4(x-1) &= -6(x+2)+48 \\ 4x-4 &= -6x-12+48 \\ 4x-4 &= -6x+36 \\ +6x & \quad 6x \\ \hline 10x-4 &= 36 \\ + \quad 4 & \quad 4 \\ \hline 10x &= 40 \\ \frac{10x}{10} &= \frac{40}{10} \\ x &= 4 \end{aligned}$$

15.
$$\begin{aligned} (3x^2-2x+1)+(-3x^2-10) &= 5x+5 \\ 3x^2-2x+1-3x^2-10 &= 5x+5 \\ -2x-9 &= 5x+5 \\ + \quad -5x & \quad -5x \\ \hline -7x-9 &= 5 \\ + \quad 9 & \quad 9 \\ \hline -7x &= 14 \\ \frac{-7x}{-7} &= \frac{14}{-7} \\ x &= -2 \end{aligned}$$

16. $\frac{x}{5}+\frac{x}{2}=7$

LCD = 10

$$\begin{aligned} 10\left(\frac{x}{5}\right)+10\left(\frac{x}{2}\right) &= 10(7) \\ 2x+5x &= 70 \\ 7x &= 70 \\ \frac{7x}{7} &= \frac{70}{7} \\ x &= 10 \end{aligned}$$

17. $2x+\frac{1}{3}=\frac{1}{2}$

LCD = 6

$$\begin{aligned} 6(2x)+6\left(\frac{1}{3}\right) &= 6\left(\frac{1}{2}\right) \\ 12x+2 &= 3 \\ + \quad -2 & \quad -2 \\ \hline 12x &= 1 \\ \frac{12x}{12} &= \frac{1}{12} \\ x &= \frac{1}{12} \end{aligned}$$

18. $\frac{x}{6}+x=14$

LCD = 6

$$\begin{aligned} 6\left(\frac{x}{6}\right)+6(x) &= 6(14) \\ x+6x &= 84 \\ 7x &= 84 \\ \frac{7x}{7} &= \frac{84}{7} \\ x &= 12 \end{aligned}$$

Copyright © 2012 Pearson Education, Inc.

19. The first and third sides are being compared to the second side, so we let the variable s represent the length of the second side.
Length of second side = s
The first side is 6 feet longer than the second.
Length of first side = $s + 6$
The length of the third side is triple the second.
Length of third side = $3s$

20. The perimeter is 26 feet.
$s + (s + 6) + 3s = 26$

21.
$$\begin{aligned} s+(s+6)+3s &= 26 \\ 5s+6 &= 26 \\ +\quad -6 &\quad -6 \\ \hline 5s &= 20 \\ \frac{5s}{5} &= \frac{20}{5} \\ s &= 4 \end{aligned}$$
$s + 6 = 4 + 6 = 10$; $3s = 3(4) = 12$
The first side is 10 feet, the second side is 4 feet, and the third side is 12 feet.

22. We are comparing the sales clerk's salary to Anna's salary, so we let the variable A represent Anna's salary.
Anna's annual salary = A
The sales clerk earns $4000 per year less than Anna.
Sales clerk's annual salary = $A - 4000$

23. The sum of their salaries is $61,200.
$A + (A - 4000) = 61{,}200$

24.
$$\begin{aligned} A+(A-4000) &= 61{,}200 \\ 2A-4000 &= 61{,}200 \\ +\quad 4000 &\quad 4{,}000 \\ \hline 2A &= 65{,}200 \\ \frac{2A}{2} &= \frac{65{,}200}{2} \\ A &= 32{,}600 \end{aligned}$$
$A - 4000 = 32{,}600 - 4000 = 28{,}600$
Anna earns $32,600 per year, and the sales clerk earns $28,600 per year.

 Copyright © 2012 Pearson Education, Inc.

Chapter 8

8.1 Exercises

1. When we change 9 to 10, we write 0 and carry 1 to the 2, the next place value to the left. Thus the 2 changes to 3.

3. 5.32
The last digit is in the hundredths place.
Five and thirty-two hundredths.

5. 0.428
The last digit is in the thousandths place.
Four hundred twenty-eight thousandths

7. Three hundred twenty-four thousandths: 0.324

9. Fifteen and three hundred forty-six ten thousandths: 15.0346

11. Twenty-five and 54/100

13. One hundred forty-three and 56/100

15. There is one place after the decimal so there is one zero in the denominator.

$$0.7 = \frac{7}{10}$$

17. There are two places after the decimal so there are two zeros in the denominator.

$$4.17 = 4\frac{17}{100}$$

19. There are three places after the decimal so there are three zeros in the denominator.

$$32.081 = 32\frac{81}{1000}$$

21. There are four places after the decimal so there are four zeros in the denominator.

$$0.5731 = \frac{5731}{10,000}$$

23. There is one zero in the denominator so we move the decimal in the numerator one place to the left.

$$6\frac{7}{10} = 6.7$$

25. There are three zeros in the denominator so we move the decimal in the numerator three places to the left. We need to insert a zero so that we can move the decimal three places.

$$12\frac{37}{1000} = 12.037$$

27. There are two zeros in the denominator so we move the decimal in the numerator two places to the left. We need to insert a zero so that we can move the decimal two places.

$$\frac{1}{100} = 0.01$$

29. There are three zeros in the denominator so we move the decimal in the numerator three places to the left. We need to insert two zeros so that we can move the decimal three places.

$$\frac{1}{1000} = 0.001$$

31. 0.426 ? 0.429
The thousandths digits differ. Since $6 < 9$,
$0.426 < 0.429$.

33. 0.09 ? 0.11
The tenths digits differ. Since $0 < 1$, $0.09 < 0.11$.

35. 0.36 ? 0.366
0.360 ? 0.366
The thousandths digits differ. Since $0 < 6$,
$0.360 < 0.366$.

37. 0.7431 ? 0.743
0.7431 ? 0.7430
The ten thousandths digits differ. Since $1 > 0$,
$0.7431 > 0.7430$.

39. 0.3 ? 0.05
The tenths digits differ. Since $3 > 0$, $0.3 > 0.05$.

41. 0.502 ? 0.52
The hundredths digits differ. Since $0 < 2$
$0.502 < 0.52$.

43. 523.7235
The round-off place digit is in the hundredths place: $523.7\underline{2}35$. The digit to the right is less than 5. Do not change the round-off place digit. Drop all digits to the right.
523.72

Copyright © 2012 Pearson Education, Inc.

45. 43.961
The round-off place digit is in the hundredths place: 43.9$\underline{6}$1. The digit to the right is less than 5. Do not change the round-off place digit. Drop all digits to the right.
43.96

47. 9.0546
The round-off place digit is in the tenths place: 9.$\underline{0}$546. The digit to the right is greater than or equal to 5. Increase the round-off place digit by 1. Drop all digits to the right.
9.1

49. 462.931
The round-off place digit is in the tenths place: 462.$\underline{9}$31. The digit to the right is less than 5. Do not change the round-off place digit. Drop all digits to the right.
462.9

51. 312.95144
The round-off place digit is in the thousandths place: 312.95$\underline{1}$44. The digit to the right is less than 5. Do not change the round-off place digit. Drop all digits to the right.
312.951

53. 1286.3496
The round-off place digit is in the thousandths place: 1286.34$\underline{9}$6. The digit to the right is greater than or equal to 5. We increase 9 to 10 by changing 9 to 0 followed by changing 4 to 5. We must include the zero on the right because we were asked to round to the nearest thousandth.
1286.350

55. 0.063148
The round-off place digit is in the ten thousandths place: 0.063$\underline{1}$48. The digit to the right is less than 5. Do not change the round-off place digit. Drop all digits to the right.
0.0631

57. 0.047362
The round-off place digit is in the ten thousandths place: 0.047$\underline{3}$62. The digit to the right is greater than or equal to 5. Increase the round-off place digit by 1. Drop all digits to the right.
0.0474

59. 42.5 million miles
The round-off place digit is in the ones place: 4$\underline{2}$.5. The digit to the right is greater than or equal to 5. Increase the round-off place digit by 1.
43 million miles.

34.6 million miles
The round-off place digit is in the ones place: 3$\underline{4}$.6. The digit to the right is greater than or equal to 5. Increase the round-off place digit by 1.
35 million miles

61. New York steak: $15.25
The round-off place digit is in the ones place: 1$\underline{5}$.25. The digit to the right is less than 5. Do not change the round-off place digit.
$15

63.

Stuffed pork chops:	$13.75 → $14
Lobster:	$18.25 → $18
Prime rib:	$15.75 → $16
Add:	$48

65. 0.0069, 0.73, $\frac{7}{10}$, 0.007, 0.071

$\frac{7}{10} = 0.7 = 0.70$: Since 0.73 and 0.70 differ in the hundredths place and $3 > 0$, $0.73 > \frac{7}{10}$. Since 0.70 and 0.071 differ in the tenths place and $7 > 0$, $\frac{7}{10} > 0.071$. Since 0.071 and 0.007 differ in the hundredths place and $7 > 0$, $0.071 > 0.007$. Since 0.007 and 0.0069 differ in the thousandths place and $7 > 6$, $0.007 > 0.0069$. From largest to smallest, the numbers are 0.73, $\frac{7}{10}$, 0.071, 0.007, 0.0069.

Cumulative Review

67. $-15 - (-6) = -15 + 6 = -9$

68. $(356)(-28) = -9968$

69. $-45 \div 9 = -5$

70.
$$\begin{array}{r} 68 \\ 231\overline{)15708} \\ \underline{1386} \\ 1848 \\ \underline{1848} \\ 0 \end{array}$$

71. We add.
$\frac{2}{3} + \frac{1}{2} = \frac{2 \cdot 2}{3 \cdot 2} + \frac{1 \cdot 3}{2 \cdot 3} = \frac{4}{6} + \frac{3}{6} = \frac{7}{6}$ or $1\frac{1}{6}$ lb of nuts

 Copyright © 2012 Pearson Education, Inc.

72. We add.

$$\frac{3}{4}+\frac{1}{2}=\frac{3}{4}+\frac{1\cdot 2}{2\cdot 2}=\frac{3}{4}+\frac{2}{4}=\frac{5}{4} \text{ or } 1\frac{1}{4} \text{ in.}$$

Quick Quiz 8.1

1. 7.21
The last digit is in the hundredths place. Seven and twenty-one hundredths

2. a. There are four places after the decimal so there are four zeros in the denominator.

$$0.0217=\frac{217}{10{,}000}$$

b. There are three zeros in the denominator so we move the decimal in the numerator three places to the left. We need to insert a zero so that we can move the decimal three places.

$$4\frac{32}{1000}=4.032$$

3. 156.748
The round-off place digit is in the tenths place: 156.<u>7</u>48. The digit to the right is less than 5. Do not change the round-off place digit. Drop all digits to the right.
156.7

4. Answers may vary. One possible solution follows:
There are four places after the decimal so there are four zeros in the denominator.

$$8.6711=8\frac{6711}{10{,}000}$$

8.2 Exercises

1. To add numbers in decimal notation, we <u>line up</u> the decimal points.

3. When subtracting decimals, we place the decimal point in the answer <u>in line</u> with the decimal point in the problem.

5.
$$\begin{array}{r} 0.34 \\ +\ 7.21 \\ \hline 7.55 \end{array}$$

7.
$$\begin{array}{r} 1.01 \\ +\ 3.46 \\ \hline 4.47 \end{array}$$

9.
$$\begin{array}{r} 63.2000 \\ +\ 0.2348 \\ \hline 63.4348 \end{array}$$

11.
$$\begin{array}{r} 73.000 \\ 7.540 \\ +\ 0.483 \\ \hline 81.023 \end{array}$$

13.
$$\begin{array}{r} 73.1000 \\ +\ 0.3169 \\ \hline 73.4169 \end{array}$$

15.
$$\begin{array}{r} 15.000 \\ 2.730 \\ +\ 0.423 \\ \hline 18.153 \end{array}$$

17.
$$\begin{array}{r} 53.783 \\ -\ 2.460 \\ \hline 51.323 \end{array}$$

19.
$$\begin{array}{r} 16.54 \\ -\ 3.90 \\ \hline 12.64 \end{array}$$

21.
$$\begin{array}{r} 20.00 \\ -\ 0.36 \\ \hline 19.64 \end{array}$$

23. $-12.1-0.23=-12.1+(-0.23)$
To add two numbers with the same sign, add the absolute values and keep the common sign.

$$\begin{array}{r} -12.10 \\ +\ -0.23 \\ \hline -12.33 \end{array}$$

25. $-91.13-14.213=-91.13+(-14.213)$
To add two numbers with the same sign, add the absolute values and keep the common sign.

$$\begin{array}{r} -91.130 \\ +\ -14.213 \\ \hline -105.343 \end{array}$$

27. $-8.69-(-4.12)=-8.69+4.12$
To add two numbers with different signs, we keep the sign of the larger absolute and subtract.

$$\begin{array}{r} -8.69 \\ 4.12 \\ \hline -4.57 \end{array}$$

Copyright © 2012 Pearson Education, Inc.

29.
$$\begin{array}{r} 2.3x \\ +\,3.9x \\ \hline 6.2x \end{array}$$

31.
$$\begin{array}{r} 24.8y \\ -\,9.2y \\ \hline 15.6y \end{array}$$

33.
$$\begin{array}{r} 3.5x \\ +\,9.1x \\ \hline 12.6x \end{array}$$
$3.5x + 9.1x - y = 12.6x - y$

35.
$$\begin{array}{r} 1.4x \\ +\,3.5x \\ \hline 4.9x \end{array}$$
$1.4x + 6.2y + 3.5x = 4.9x + 6.2y$

37. a.
$$\begin{array}{r} -3.4 \\ +\,-2.1 \\ \hline -5.5 \end{array}$$

b. $9.7 - (-5.4) = 9.7 + 5.4$
$$\begin{array}{r} 9.7 \\ +\,5.4 \\ \hline 15.1 \end{array}$$

c. $-9.2 - 4.1 = -9.2 + (-4.1)$
$$\begin{array}{r} -9.2 \\ +\,-4.1 \\ \hline -13.3 \end{array}$$

39. a.
$$\begin{array}{r} 4.6x \\ +\,2.0x \\ \hline 6.6x \end{array}$$

b.
$$\begin{array}{r} 3.04y \\ -\,7.50y \\ \hline -4.46y \end{array}$$

c.
$$\begin{array}{r} 1.00x \\ -\,0.25x \\ \hline 0.75x \end{array}$$

41. We replace the variable with 9.
$y - 0.861 = 9 - 0.861$
$$\begin{array}{r} 9.000 \\ -\,0.861 \\ \hline 8.139 \end{array}$$

43. We replace the variable with 5.42.
$211.2 - n = 211.2 - 5.42$
$$\begin{array}{r} 211.20 \\ -\,5.42 \\ \hline 205.78 \end{array}$$

45. We replace the variable with −6.7.
$x + 2.3 = -6.7 + 2.3$
$$\begin{array}{r} -6.7 \\ +\,2.3 \\ \hline -4.4 \end{array}$$

47. To estimate the total deductions, round each deduction to the nearest dollar and then add.
162 + 61 + 48 = $271
Now round John's salary to the nearest dollar and subtract the estimated deductions.
1763 − 271 = $1492
John takes home about $1492 each month.

49. Round the closing numbers for Day 1 and Day 2 to the nearest whole number and then subtract.
2780 − 2764 = 16
The difference is about 16.

51. The bars for Day 4 and Day 5 have the greatest difference in height. Round the closing numbers for these two days to the nearest whole number and then subtract.
2799 − 2725 = 74
Day 4 and Day 5 have the greatest difference; the difference is about 74.

53. We subtract.
$$\begin{array}{r} 100.00 \\ -\,72.31 \\ \hline 27.69 \end{array}$$
Ann should get $27.69.

55. We subtract.
$$\begin{array}{r} 10.82 \\ -\,10.54 \\ \hline 0.28 \end{array}$$
The 1988 record was 0.28 second faster.

57. The fastest was 10.54 seconds in 1988 and the slowest was 11.40 seconds in 1964. We subtract.
$$\begin{array}{r} 11.40 \\ -\,10.54 \\ \hline 0.86 \end{array}$$
The difference in the fastest and slowest records is 0.86 second.

Copyright © 2012 Pearson Education, Inc.

59. $-2.3-(-0.24)+4.6-9=-2.3+0.24+4.6+(-9)$
$=-2.3+(-9)+0.24+4.6$
$=-11.3+4.84$
$=-6.46$

61. $\frac{3}{10}-1.26+(-2.3)=0.3+(-1.26)+(-2.3)$
$=0.3+(-3.56)$
$=-3.26$

63. The pattern has one 6 followed by one or more 3s. The number of 3s increases by one each time, so there should be five 3s following the last 6 shown, followed by another 6 and then some more 3s. The next seven digits are 3333363.

65. The pattern has one 8 followed by one or more 1s. The number of 1s increases by one each time, so there should be four 1s following the last 8 shown, followed by another 8 and then some more 1s. The next seven digits are 1111811.

67. The pattern has a sequence of digits starting with 9 and decreasing by 1. Each time the sequence gets 1 digit longer. Therefore, the next sequence will start with 9 and end with 4, followed by another sequence that starts with 9 and ends with 3. The next seven digits are 9876549.

Cumulative Review

69. $(231)(14)=3234$

70. $(-12)(92)=-1104$

71.
```
     245
 12)2940
    24
     54
     48
      60
      60
       0
```
$2940 \div 12 = 245$

72. $3105 \div (-3) = -1035$

Quick Quiz 8.2

1. a.
$$\begin{array}{r} 42.09 \\ +\ 3.10 \\ \hline 45.19 \end{array}$$

b.
$$\begin{array}{r} 5.03 \\ -\ 2.68 \\ \hline 2.35 \end{array}$$

c. $28.61-(-4.21)=28.61+4.21$
$$\begin{array}{r} 28.61 \\ +\ 4.21 \\ \hline 32.82 \end{array}$$

2.
$$\begin{array}{r} 5.60x \\ +\ 2.13x \\ \hline 7.73x \end{array}$$
$5.6x+2.13x+9.01=7.73x+9.01$

3. We replace the variable with −20.96.
$y-18.75=-20.96-18.75=-20.96+(-18.75)$
$$\begin{array}{r} -20.96 \\ +\ -18.75 \\ \hline -39.71 \end{array}$$

4. Answers may vary. One possible solution follows:
We replace the variable with 0.866.
$x-3.1=0.866-3.1$
We keep the sign of the larger absolute value and subtract.
$$\begin{array}{r} 0.866 \\ -\ 3.100 \\ \hline -2.144 \end{array}$$

8.3 Exercises

1. If one factor has 4 decimal places and the second factor has 2 decimal places, the product has <u>6</u> decimal places.

3. When we divide $4.62\overline{)12.7}$, we rewrite the equivalent division problem $\underline{462\overline{)1270}}$ and then divide.

5. $0.03\times0.04 \rightarrow 3\times4=12$
$0.03\times0.04=0.0012$

7. $0.05\times0.07 \rightarrow 5\times7=35$
$0.05\times0.07=0.0035$

Copyright © 2012 Pearson Education, Inc.

9. $$\begin{array}{r} 7.43 \\ \times\ \ 8.3 \\ \hline 2\ 229 \\ 59\ 44 \\ \hline 61.669 \end{array}$$

11. $$\begin{array}{r} 15.2 \\ \times 4.3 \\ \hline 456 \\ 60\ 8 \\ \hline 65.36 \end{array}$$

13. $$\begin{array}{r} 2.35 \\ \times\ (-3) \\ \hline -7.05 \end{array}$$

15. $$\begin{array}{r} -4.23 \\ \times\ \ \ 2.7 \\ \hline 2\ 961 \\ 8\ 46 \\ \hline -11.421 \end{array}$$

17. $$\begin{array}{r} -0.613 \\ \times\ (-25) \\ \hline 3\ 065 \\ 12\ 26 \\ \hline 15.325 \end{array}$$

19. $$\begin{array}{r} -2.81 \\ \times\ \ 12.1 \\ \hline 281 \\ 5\ 62 \\ 28\ 1 \\ \hline -34.001 \end{array}$$

21. Move the decimal point two places to the right.
$0.1498 \times 100 = 14.98$

23. Move the decimal point four places to the right.
$8.554 \times 10{,}000 = 85{,}540$

25. Move the decimal point to the right 4 places.
$41 \times 10^4 = 410{,}000$

27. Move the decimal point to the right 4 places.
$0.6 \times 10^4 = 6000$

29. $$\begin{array}{r} 2.16 \\ 8\overline{)17.28} \\ \underline{16} \\ 1\ 2 \\ \underline{8} \\ 48 \\ \underline{48} \\ 0 \end{array}$$

$17.28 \div 8 = 2.16$

31. $$\begin{array}{r} 0.23 \\ 14\overline{)3.22} \\ \underline{2\ 8} \\ 42 \\ \underline{42} \\ 0 \end{array}$$

$3.22 \div 14 = 0.23$

33. $$\begin{array}{r} 0.0565 \\ 64\overline{)3.6160} \\ \underline{3\ 20} \\ 416 \\ \underline{384} \\ 320 \\ \underline{320} \\ 0 \end{array}$$

$3.616 \div 64 = 0.0565$

35. $$\begin{array}{r} 3.451 \\ 24\overline{)82.824} \\ \underline{72} \\ 10\ 8 \\ \underline{9\ 6} \\ 1\ 22 \\ \underline{1\ 20} \\ 24 \\ \underline{24} \\ 0 \end{array}$$

$82.824 \div 24 = 3.451$

 Copyright © 2012 Pearson Education, Inc.

37.
$$\begin{array}{r} 0.232 \\ 14\overline{)3.250} \\ \underline{28} \\ 45 \\ \underline{42} \\ 30 \\ \underline{28} \\ 2 \end{array}$$

$3.25 \div 14 \approx 0.23$

39.
$$3.7\overline{)-0.2988} \rightarrow \begin{array}{r} -0.080 \\ 37\overline{)-2.988} \\ \underline{2\ 96} \\ 28 \\ \underline{0} \\ 28 \end{array}$$

$-0.2988 \div 3.7 \approx -0.08$

41.
$$-1.7\overline{)-20.8} \rightarrow \begin{array}{r} 12.235 \\ 17\overline{)208.000} \\ \underline{17} \\ 38 \\ \underline{34} \\ 4\ 0 \\ \underline{3\ 4} \\ 60 \\ \underline{51} \\ 90 \\ \underline{85} \\ 5 \end{array}$$

$-20.8 \div (-1.7) \approx 12.24$

43.
$$0.27\overline{)8.343} \rightarrow \begin{array}{r} 30.9 \\ 27\overline{)834.3} \\ \underline{81} \\ 24\ 3 \\ \underline{24\ 3} \\ 0 \end{array}$$

$8.343 \div 0.27 = 30.9$

45.
$$5.88\overline{)13.7592} \rightarrow \begin{array}{r} 2.34 \\ 588\overline{)1375.92} \\ \underline{1176} \\ 199\ 9 \\ \underline{176\ 4} \\ 2352 \\ \underline{2352} \\ 0 \end{array}$$

$13.7592 \div 5.88 = 2.34$

47.
$$1.8\overline{)3} \rightarrow \begin{array}{r} 1.66 \\ 18\overline{)30.00} \\ \underline{18} \\ 12\ 0 \\ \underline{10\ 8} \\ 120 \\ \underline{108} \\ 12 \end{array}$$

$3 \div 1.8 = 1.\overline{6}$

49.
$$1.1\overline{)0.6} \rightarrow \begin{array}{r} 0.5454 \\ 11\overline{)6.0000} \\ \underline{5\ 5} \\ 50 \\ \underline{44} \\ 60 \\ \underline{55} \\ 50 \\ \underline{44} \\ 6 \end{array}$$

$0.6 \div 1.1 = 0.\overline{54}$

51.
$$2.2\overline{)11.3} \rightarrow \begin{array}{r} 5.13636 \\ 22\overline{)113.00000} \\ \underline{110} \\ 3\ 0 \\ \underline{2\ 2} \\ 80 \\ \underline{66} \\ 140 \\ \underline{132} \\ 80 \\ \underline{66} \\ 140 \\ \underline{132} \\ 8 \end{array}$$

$11.3 \div 2.2 = 5.1\overline{36}$

Copyright © 2012 Pearson Education, Inc.

53. $6.6\overline{)200} \rightarrow 66\overline{)2000.0000}$

Quotient: 30.3030

198
20 0
19 8
200
198
20

$200 \div 6.6 = 30.\overline{30}$

55. $6\overline{)11.000}$

Quotient: 1.833

6
5 0
4 8
20
18
20
18
2

$\frac{11}{6} \approx 1.83$

57. $15\overline{)7.000}$

Quotient: 0.466

6 0
1 00
90
100
90
10

$\frac{7}{15} \approx 0.47$ and $12\frac{7}{15} \approx 12.47$

59. $3\overline{)1.00}$

Quotient: 0.33

9
10
9
1

$\frac{1}{3} = 0.\overline{3}$

61. $15\overline{)2.000}$

Quotient: 0.133

1 5
50
45
50
45
5

$\frac{2}{15} = 0.1\overline{3}$

63. $0.44\overline{)20.35} \rightarrow 44\overline{)2035.00}$

Quotient: 46.25

176
275
264
11 0
8 8
2 20
2 20
0

$20.35 \div 0.44 = 46.25$

65. $9\overline{)2.00}$

Quotient: 0.22

1 8
20
18
2

$\frac{2}{9} = 0.\overline{2}$

67.

$$\begin{array}{r} 4.24 \\ \times(-3.5) \\ \hline 2\ 120 \\ 12\ 72 \\ \hline -14.840 \end{array}$$

-14.840 or -14.84

69.

$$\begin{array}{r} 0.8 \\ \times 0.4 \\ \hline 0.32 \end{array}$$

Copyright © 2012 Pearson Education, Inc.

71. a.

$$\begin{array}{r} 0.0033 \\ 300\overline{)1.0000} \\ \underline{9\,00} \\ 1000 \\ \underline{900} \\ 100 \end{array}$$

$\frac{1}{300} = 0.00\overline{3}$

b.

$$\begin{array}{r} 0.033 \\ 30\overline{)1.000} \\ \underline{90} \\ 100 \\ \underline{90} \\ 10 \end{array}$$

$\frac{1}{30} = 0.0\overline{3}$

73. Use a calculator.
$-562.53 \div 13.123 = -42.8659605273...$
≈ -42.866

75. Following the pattern in the second row, the next two entries must be repeating 4s and then repeating 5s.
$\frac{4}{9} = 0.44..., \frac{5}{9} = 0.55...$

Cumulative Review

77. $12x = 96$
$\frac{12x}{12} = \frac{96}{12}$
$x = 8$
Check: $12x = 96$
$12(8) \stackrel{?}{=} 96$
$96 = 96$ ✓

78. $x - 25 = -30$
$+ \quad 25 \quad 25$
$x = -5$
Check: $x - 25 = -30$
$-5 - 25 \stackrel{?}{=} -30$
$-30 = -30$ ✓

79. $x + 45 = 17$
$+ \ -45 \quad -45$
$x = -28$
Check: $x + 45 = 17$
$-28 + 45 \stackrel{?}{=} 17$
$17 = 17$ ✓

80. $15x = 225$
$\frac{15x}{15} = \frac{225}{15}$
$x = 15$
Check: $15x = 225$
$15(15) \stackrel{?}{=} 225$
$225 = 225$ ✓

Quick Quiz 8.3

1. a.

$$\begin{array}{r} 5.07 \\ \times \ 3.1 \\ \hline 507 \\ 15\,21 \\ \hline 15.717 \end{array}$$

b. Move the decimal point to the right 3 places.
$(4.39)(10^3) = 4390$

2.

$$6.3\overline{)36.54} \rightarrow \begin{array}{r} 5.8 \\ 63\overline{)365.4} \\ \underline{315} \\ 50\,4 \\ \underline{50\,4} \\ 0 \end{array}$$

$36.54 \div 6.3 = 5.8$

3.

$$\begin{array}{r} 3.22 \\ 9\overline{)29.00} \\ \underline{27} \\ 2\,0 \\ \underline{1\,8} \\ 20 \\ \underline{18} \\ 2 \end{array}$$

$\frac{29}{9} = 3.\overline{2}$

Copyright © 2012 Pearson Education, Inc.

4. Answers may vary. One possible solution follows:
Check Marc's answer.
$0.097 \times 0.5 \stackrel{?}{=} 0.485$
$$\begin{array}{r} 0.097 \\ \times\ \ 0.5 \\ \hline 0.0485 \end{array} \stackrel{?}{=} 0.485$$
No, the answers are not the same.
$0.0485 \neq 0.485$

8.4 Exercises

1. $$\begin{array}{rl} x+3.7 = & 9.8 \\ +\ -3.7\ \ & -3.7 \\ \hline x = & 6.1 \end{array}$$

3. $$\begin{array}{rl} y-2.8 = & 6.95 \\ +\ \ 2.8\ \ & 2.8 \\ \hline y = & 9.75 \end{array}$$

5. $$\begin{array}{rl} 2.9+x = & 6 \\ +\ -2.9\ \ & -2.9 \\ \hline x = & 3.1 \end{array}$$

7. $$\begin{array}{rl} x+2.5 = & -9.6 \\ +\ -2.5\ \ & -2.5 \\ \hline x = & -12.1 \end{array}$$

9. $4x = 11.24$
$$\frac{4x}{4} = \frac{11.24}{4}$$
$x = 2.81$

11. $5.1x = 25.5$
$$\frac{5.1x}{5.1} = \frac{25.5}{5.1}$$
$x = 5$

13. $-5.6x = -19.04$
$$\frac{-5.6x}{-5.6} = \frac{-19.04}{-5.6}$$
$x = 3.4$

15. $$\begin{array}{rl} -5.2x-3.3 = & 22.7 \\ +\ \ \ \ 3.3\ \ & 3.3 \\ \hline -5.2x = & 26 \end{array}$$
$$\frac{-5.2x}{-5.2} = \frac{26}{-5.2}$$
$x = -5$

17. $$\begin{array}{rl} -3x-5.3 = & 11.23 \\ +\ \ \ \ 5.3\ \ & 5.3 \\ \hline -3x = & 16.53 \end{array}$$
$$\frac{-3x}{-3} = \frac{16.53}{-3}$$
$x = -5.51$

19. $$\begin{array}{rl} 0.9x+8.7 = & 15.9 \\ +\ \ \ -8.7\ \ & -8.7 \\ \hline 0.9x = & 7.2 \end{array}$$
$$\frac{0.9x}{0.9} = \frac{7.2}{0.9}$$
$x = 8$

21. $$\begin{array}{rl} 2(x-1) = & 26.4 \\ 2(x)-2(1) = & 26.4 \\ 2x-2 = & 26.4 \\ +\ \ \ \ 2\ \ & 2 \\ \hline 2x = & 28.4 \end{array}$$
$$\frac{2x}{2} = \frac{28.4}{2}$$
$x = 14.2$

23. $$\begin{array}{rl} 2(x+3.2) = & x+9.9 \\ 2(x)+2(3.2) = & x+9.9 \\ 2x+6.4 = & x+9.9 \\ +\ \ -x\ \ & -x \\ \hline x+6.4 = & 9.9 \\ +\ -6.4\ \ & -6.4 \\ \hline x = & 3.5 \end{array}$$

25. We multiply by 10.
$$\begin{aligned} 0.3x+0.2 &= 1.7 \\ 10(0.3x+0.2) &= 10(1.7) \\ 10(0.3x)+10(0.2) &= 10(1.7) \\ 3x+2 &= 17 \\ 3x &= 15 \\ \frac{3x}{3} &= \frac{15}{3} \\ x &= 5 \end{aligned}$$

27. We multiply by 10.
$$\begin{aligned} 0.3x-0.6 &= 5.4 \\ 10(0.3x-0.6) &= 10(5.4) \\ 10(0.3x)-10(0.6) &= 10(5.4) \\ 3x-6 &= 54 \\ 3x &= 60 \\ \frac{3x}{3} &= \frac{60}{3} \\ x &= 20 \end{aligned}$$

Copyright © 2012 Pearson Education, Inc.

29. We multiply by 100.

$$\begin{aligned} 0.08x - 1.1 &= 1.22 \\ 100(0.08x - 1.1) &= 100(1.22) \\ 100(0.08x) - 100(1.1) &= 100(1.22) \\ 8x - 110 &= 122 \\ 8x &= 232 \\ \frac{8x}{8} &= \frac{232}{8} \\ x &= 29 \end{aligned}$$

31. We multiply by 100.

$$\begin{aligned} 0.15x + 0.23 &= 1.43 \\ 100(0.15x + 0.23) &= 100(1.43) \\ 100(0.15x) + 100(0.23) &= 100(1.43) \\ 15x + 23 &= 143 \\ 15x &= 120 \\ \frac{15x}{15} &= \frac{120}{15} \\ x &= 8 \end{aligned}$$

33.

$$\begin{aligned} 5.6 + x &= -4.8 \\ + -5.6 \quad &\quad -5.6 \\ \hline x &= -10.4 \end{aligned}$$

35.

$$\begin{aligned} 4.7x &= 14.1 \\ \frac{4.7x}{4.7} &= \frac{14.1}{4.7} \\ x &= 3 \end{aligned}$$

37.

$$\begin{aligned} 3(x + 1.4) &= 6.9 \\ 3(x) + 3(1.4) &= 6.9 \\ 3x + 4.2 &= 6.9 \\ 3x &= 2.7 \\ \frac{3x}{3} &= \frac{2.7}{3} \\ x &= 0.9 \end{aligned}$$

39.

$$\begin{aligned} 6x + 10.5 &= x + 21 \\ + -x \quad &\quad -x \\ \hline 5x + 10.5 &= 21 \\ + \quad -10.5 &\quad -10.5 \\ \hline 5x &= 10.5 \\ \frac{5x}{5} &= \frac{10.5}{5} \\ x &= 2.1 \end{aligned}$$

41. a. $\frac{243.2 \text{ mi}}{16 \text{ gal}} = 15.2 \text{ mi/gal}$

b.

$$\begin{aligned} \frac{243.2 \text{ mi}}{16 \text{ gal}} &= \frac{380 \text{ mi}}{x \text{ gal}} \\ \frac{243.2}{16} &= \frac{380}{x} \\ x \cdot 243.2 &= 16 \cdot 380 \\ 243.2x &= 6080 \\ \frac{243.2x}{243.2} &= \frac{6080}{243.2} \\ x &= 25 \end{aligned}$$

Sam would use 25 gallons of gas if he drove 380 miles.

43. a. We multiply.
17(0.24) = $4.08

b. We multiply.
17(0.14) = $2.38

c. We subtract.
4.08 − 2.38 = $1.70

45. a. We divide.
65 ÷ 2.5 = 26 containers

b. We multiply.
2.5(5.70) = $14.25

47. We multiply and then add.

$$\begin{aligned} &2(22.50) + 43.97 + 4(8.88) \\ &= 45.00 + 43.97 + 35.52 \\ &= \$124.49 \end{aligned}$$

49. We subtract to find the number of miles over 200 miles: 423 − 200 = 223 miles.

$$\begin{aligned} \boxed{18.95} \times \boxed{3} + \boxed{0.12} \times \boxed{223} &= 56.85 + 26.7 \\ &= \boxed{\$83.61} \end{aligned}$$

51. We find the cost for each plan. No fee plan:

$0 + 0.10(75) + 0.15(20) = 7.50 + 3.00 = \$10.50$

Single rate plan:

$$\begin{aligned} 4.95 + 0.07(75) + 0.10(20) &= 4.95 + 5.25 + 2.00 \\ &= \$12.20 \end{aligned}$$

The plan with no fee has the lower cost, so it is the better deal.

53. a. Since we are comparing the number of dimes to the number of quarters, we let the variable represent the number of quarters.
Q = number of quarters
The number of dimes is 5 times the number of quarters.
$5Q$ = number of dimes

Copyright © 2012 Pearson Education, Inc.

b. The value of the quarters is $(0.25)Q$ and the value of the dimes is $(0.10)(5Q)$. The total value of the coins is \$4.50.
$(0.25)Q + (0.10)(5Q) = 4.50$

c.
$$\begin{aligned}(0.25)Q+(0.10)(5Q)&=4.50\\ 0.25Q+0.50Q&=4.50\\ 25Q+50Q&=450\\ 75Q&=450\\ \frac{75Q}{75}&=\frac{450}{75}\\ Q&=6\end{aligned}$$
$5Q = 5(6) = 30$
Cody needs 6 quarters and 30 dimes.

d. Do we have five times as many dimes as quarters? Yes, 30 dimes is 5×6, or five times the number of quarters.
Is the value of the coins equal to \$4.50? Yes, 30 dimes = \$3.00 and
6 quarters = \$1.50; \$3.00 + \$1.50 = \$4.50.

55. Since we are comparing the number of nickels to the number of quarters, we let the variable represent the number of quarters.
Q = number of quarters
The number of nickels is 5 times the number of quarters.
$5Q$ = number of nickels
The value of the quarters is $(0.25)Q$ and the value of the nickels is $(0.05)(5Q)$. The total value of the coins is \$3.50.
$$\begin{aligned}(0.25)Q+(0.05)(5Q)&=3.50\\ 0.25Q+0.25Q&=3.50\\ 25Q+25Q&=350\\ 50Q&=350\\ \frac{50Q}{50}&=\frac{350}{50}\\ Q&=7\end{aligned}$$
$5Q = 5(7) = 35$
Janet has 7 quarters and 35 nickels.
Check:
Do we have five times as many nickels as quarters? Yes, 35 nickels is 5×7, or five times the number of quarters.
Is the value of the coins equal to \$3.50? Yes, 35 nickels = \$1.75 and 7 quarters = \$1.75;
\$1.75 + \$1.75 = \$3.50.

57. Use a calculator to perform the division.
$$\begin{aligned}23.098x&=103.941\\ \frac{23.098x}{23.098}&=\frac{103.941}{23.098}\\ x&=4.5\end{aligned}$$

59. Use a calculator to subtract.
$$\begin{aligned}x+3.0012&=21.566\\ +\,-3.0012\quad&\;\;-3.0012\\ \hline x&=18.5648\end{aligned}$$

61. Replace t with 3.5 and a with 15.
$$\begin{aligned}a&=\frac{v}{t}\\ 15&=\frac{v}{3.5}\\ 3.5(15)&=3.5\left(\frac{v}{3.5}\right)\\ 52.5&=v\end{aligned}$$
The velocity is 52.5 feet per second.

Cumulative Review

63. We find the cross products and set them equal.
$$\begin{aligned}\frac{x}{500}&=\frac{54}{100}\\ 100\cdot x&=500\cdot 54\\ 100x&=27{,}000\\ \frac{100x}{100}&=\frac{27{,}000}{100}\\ x&=270\end{aligned}$$

64. We find the cross products and set them equal.
$$\begin{aligned}\frac{a}{40}&=\frac{13}{5}\\ 5\cdot a&=40\cdot 13\\ 5a&=520\\ \frac{5a}{5}&=\frac{520}{5}\\ a&=104\end{aligned}$$

65.
$$\begin{array}{r}0.33\\ 3\overline{)1.00}\\ \underline{9}\\ 10\\ \underline{9}\\ 1\end{array}$$
$\frac{1}{3}=0.\overline{3}$

66.
$$\begin{array}{r}0.833\\ 6\overline{)5.000}\\ \underline{4\,8}\\ 20\\ \underline{18}\\ 20\\ \underline{18}\\ 2\end{array}$$
$\frac{5}{6}=0.8\overline{3}$

 Copyright © 2012 Pearson Education, Inc.

67. We subtract.

$$\begin{array}{rcr} 25 & \rightarrow & 24\frac{3}{3} \\ -\ 12\frac{1}{3} & \rightarrow & -\ 12\frac{1}{3} \\ \hline & & 12\frac{2}{3}\text{ lb} \end{array}$$

68. We subtract.

$$\begin{array}{rcrcr} 11\frac{7}{16} & \rightarrow & 11\frac{7}{16} & \rightarrow & 10\frac{23}{16} \\ -\ 10\frac{7}{8} & \rightarrow & -\ 10\frac{14}{16} & \rightarrow & -\ 10\frac{14}{16} \\ \hline & & & & \frac{9}{16}\text{ in.} \end{array}$$

Quick Quiz 8.4

1. $0.3x + 0.2x = 2.5$
$0.5x = 2.5$
$\frac{0.5x}{0.5} = \frac{2.5}{0.5}$
$x = 5$

2. $3(x - 1.4) = 3.3$
$3(x) - 3(1.4) = 3.3$
$3x - 4.2 = 3.3$
$+ \quad 4.2 \quad 4.2$
$3x = 7.5$
$\frac{3x}{3} = \frac{7.5}{3}$
$x = 2.5$

3. a. Since we are comparing the number of nickels to the number of quarters, we let the variable represent the number of quarters.
Q = number of quarters
The number of nickels is twice the number of quarters.
$2Q$ = number of nickels

b. The value of the quarters is $(0.25)Q$ and the value of the nickels is $(0.05)(2Q)$. The total value of the coins is $1.05.
$(0.25)Q + (0.05)(2Q) = 1.05$

c. $(0.25)Q + (0.05)(2Q) = 1.05$
$0.25Q + 0.10Q = 1.05$
$25Q + 10Q = 105$
$35Q = 105$
$\frac{35Q}{35} = \frac{105}{35}$
$Q = 3$
$2Q = 2(3) = 6$
Zach has 3 quarters and 6 nickels.

d. Do we have twice as many nickels as quarters? Yes, 6 nickels is 2×3, or twice the number of quarters.
Is the value of the coins equal to $1.05?
Yes, 6 nickels = $0.30 and
3 quarters = $0.75; $0.30 + $0.75 = $1.05.

4. Answers may vary. One possible solution follows:
To solve the equation $2.2 + 2.4 = 5(x - 1) + 3x$:
1. Add 2.2 + 2.4. $4.6 = 5(x - 1) + 3x$
2. Remove parentheses. $4.6 = 5x - 5 + 3x$
3. Combine like terms. $4.6 = 8x - 5$
4. Add 5 to both sides. $4.6 = 8x - 5$
$+5 \quad +5$
$9.6 = 8x$
5. Divide both sides by 8. $\frac{9.6}{8} = \frac{8x}{8}$
$1.2 = x$

How Am I Doing? Sections 8.1–8.4

1. Move the decimal point in the numerator 2 places to the left.
$\frac{2}{100} = 0.02$

2. The last digit is in the thousandths place.
$0.027 = \frac{27}{1000}$

3. 0.56 ? 0.566
0.560 ? 0.566
The numbers differ in the thousandths place.
Since $0 < 6$, $0.560 < 0.566$.

4. 4212.65133
The round-off place digit is in the thousandths place: 4212.65<u>1</u>33. The digit to the right is less than 5. Do not change the round-off place digit.
4212.651

5.
$$\begin{array}{r} 35.0 \\ 4.73 \\ +\ 0.623 \\ \hline 40.353 \end{array}$$

6. $-81.14 - 15.313 = -81.14 + (-15.313)$
$$\begin{array}{r} -81.14 \\ -15.313 \\ \hline -96.453 \end{array}$$

7.
$$\begin{array}{r} 2.3x \\ +\ 4.4x \\ \hline 6.7x \end{array}$$
$2.3x + 3.1y + 4.4x = 6.7x + 3.1y$

8. We replace the variable with 5.8.
$y - 0.921 = 5.8 - 0.921$
$$\begin{array}{r} 5.800 \\ -\ 0.921 \\ \hline 4.879 \end{array}$$

9.
$$\begin{array}{r} -3.23 \\ \times\ \ 1.61 \\ \hline 323 \\ 1\,938 \\ 3\,23 \\ \hline -5.2003 \end{array}$$

10. Move the decimal point to the right 3 places.
$0.2783 \times 10^3 = 278.3$

11. $2.6\overline{)13.806} \rightarrow 26\overline{)138.06}$
$$\begin{array}{r} 5.31 \\ 26\overline{)138.06} \\ \underline{130} \\ 8\,0 \\ \underline{7\,8} \\ 26 \\ \underline{26} \\ 0 \end{array}$$
$13.806 \div 2.6 = 5.31$

12.
$$\begin{array}{r} 3.833 \\ 6\overline{)23.000} \\ \underline{18} \\ 5\,0 \\ \underline{4\,8} \\ 20 \\ \underline{18} \\ 20 \\ \underline{18} \\ 2 \end{array}$$
$\frac{23}{6} = 3.8\overline{3}$

13.
$$\begin{aligned} -2.1x - 4.4 &= 3.16 \\ +\qquad 4.4 &\quad 4.4 \\ \hline -2.1x &= 7.56 \\ \frac{-2.1x}{-2.1} &= \frac{7.56}{-2.1} \\ x &= -3.6 \end{aligned}$$

14.
$$\begin{aligned} 2(x+4.5) &= x+9.8 \\ 2(x)+2(4.5) &= x+9.8 \\ 2x+9 &= x+9.8 \\ +\ -x \quad & \quad -x \\ \hline x+9 &= 9.8 \\ +\ -9 \quad & \quad -9 \\ \hline x &= 0.8 \end{aligned}$$

15. We subtract to find the number of hours of overtime: 53 − 40 = 13 hours.
We multiply and then add.
40(9.25) + 13(13.88) = 370 + 180.44 = $550.44
Lester will earn $550.44.

8.5 Exercises

1. To estimate 10% of a whole number, we can delete the last digit.

3. We can find 15% by adding 5% and 10%.

5. We can find 6% by adding 5% and 1%.

7. 10% of 701
We first round to the nearest hundred.
700
Then we drop the last digit.
70

9. $5\%\text{ of }701 \approx \frac{1}{2} \times (10\%\text{ of }700) = \frac{1}{2} \times 70 = 35$

11. $20\%\text{ of }701 \approx 2 \times (10\%\text{ of }700) = 2 \times 70 = 140$

Copyright © 2012 Pearson Education, Inc.

13. 1% of 205
We first round to the nearest hundred.
200
Then we drop the last two digits.
2

15. $$\begin{aligned}7\% \text{ of } 205 &\approx 7\% \text{ of } 200\\ &= 7\times(1\% \text{ of } 200)\\ &= 7\times 2\\ &= 14\end{aligned}$$

17. $$\begin{aligned}30\% \text{ of } 205 &\approx 30\% \text{ of } 200\\ &= 3\times(10\% \text{ of } 200)\\ &= 3\times 20\\ &= 60\end{aligned}$$

19. 10% of 1020
We first round to the nearest thousand.
1000
Then we drop the last digit.
100

21. $$\begin{aligned}15\% \text{ of } 1020 &\approx 15\% \text{ of } 1000\\ &= 10\% \text{ of } 1000 + 5\% \text{ of } 1000\\ &= 100 + \frac{1}{2}\times(10\% \text{ of } 1000)\\ &= 100 + \frac{1}{2}\times 100\\ &= 100 + 50\\ &= 150\end{aligned}$$

23. $$\begin{aligned}20\% \text{ of } 1020 &\approx 20\% \text{ of } 1000\\ &= 10\% \text{ of } 1000 + 10\% \text{ of } 1000\\ &= 100 + 100\\ &= 200\end{aligned}$$

25. 1% of 3015
We first round to the nearest thousand.
3000
Then we drop the last two digits.
30

27. $$\begin{aligned}2\% \text{ of } 3015 &\approx 2\% \text{ of } 3000\\ &= 1\% \text{ of } 3000 + 1\% \text{ of } 3000\\ &= 30 + 30\\ &= 60\end{aligned}$$

29. $$\begin{aligned}&3\% \text{ of } 3015\\ &\approx 3\% \text{ of } 3000\\ &= 1\% \text{ of } 3000 + 1\% \text{ of } 3000 + 1\% \text{ of } 3000\\ &= 30 + 30 + 30\\ &= 90\end{aligned}$$

31. 10% of 320,050
First we round to the nearest ten thousand.
320,000
Then we drop the last digit.
32,000

33. $$\begin{aligned}5\% \text{ of } 320{,}050 &\approx \frac{1}{2}\times(10\% \text{ of } 320{,}000)\\ &= \frac{1}{2}\times 32{,}000\\ &= 16{,}000\end{aligned}$$

35. $$\begin{aligned}&15\% \text{ of } 320{,}050\\ &\approx (10\% \text{ of } 320{,}000) + (5\% \text{ of } 320{,}000)\\ &= 32{,}000 + 16{,}000\\ &= 48{,}000\end{aligned}$$

37. 1% of 250,030
We first round to the nearest ten thousand.
250,000
Then we drop the last two digits.
2500

39. $$\begin{aligned}4\% \text{ of } 250{,}030 &\approx 4\times(1\% \text{ of } 250{,}000)\\ &= 4\times 2500\\ &= 10{,}000\end{aligned}$$

41. $$\begin{aligned}8\% \text{ of } 250{,}030 &\approx 2\times(4\% \text{ of } 250{,}000)\\ &= 2\times 10{,}000\\ &= 20{,}000\end{aligned}$$

43. $$\begin{aligned}&7\% \text{ of } \$22{,}000\\ &= 5\% \text{ of } \$22{,}000 + 2\% \text{ of } \$22{,}000\\ &= \frac{1}{2}\times(10\% \text{ of } \$22{,}000) + 2\times(1\% \text{ of } \$22{,}000)\\ &= \frac{1}{2}\times \$2200 + 2\times \$220\\ &= \$1100 + \$440\\ &= \$1540\end{aligned}$$

45. We round \$329,500 to the nearest ten thousand dollars: \$330,000.

$$\begin{aligned}5\% \text{ of } \$329{,}500 &\approx \frac{1}{2}\times(10\% \text{ of } \$330{,}000)\\ &= \frac{1}{2}\times \$33{,}000\\ &= \$16{,}500\end{aligned}$$

Nico paid the realtor about \$16,500.

Copyright © 2012 Pearson Education, Inc.

47. a. Ski gloves: \$29.99 → \$30
Jacket: \$199 → \$200
Sweater: \$49.99 → \$50
\$30 + \$200 + \$50 + \$50 = \$330
The total cost before the discount was about \$330.

b. 10% of \$330 = \$33
$40\% \text{ of } \$330 = \$33+\$33+\$33+\$33 = \$132$
The estimated discount rounded to the nearest ten is \$130.

c. We subtract: \$330 – \$130 = \$200.

Cumulative Review

49. We find the cross products and set them equal.

$$\frac{5}{18}=\frac{20}{n}$$
$$n\cdot 5=18\cdot 20$$
$$5n=360$$
$$\frac{5n}{5}=\frac{360}{5}$$
$$n=72$$

50. We find the cross products and set them equal.

$$\frac{n}{36}=\frac{7}{3}$$
$$3\cdot n=36\cdot 7$$
$$3n=252$$
$$\frac{3n}{3}=\frac{252}{3}$$
$$n=84$$

51. First we find the total filled.

$$\frac{1}{4}+\frac{3}{8}=\frac{1\cdot 2}{4\cdot 2}+\frac{3}{8}=\frac{2}{8}+\frac{3}{8}=\frac{5}{8}$$

The tank is $\frac{5}{8}$ full.

Subtract from 1 to find the amount unfilled.

$$1-\frac{5}{8}=\frac{8}{8}-\frac{5}{8}=\frac{3}{8}$$

Thus $\frac{3}{8}$ of the tank remains unfilled.

52. First we find the total miles after jogging and walking.

$$5\frac{3}{4} \rightarrow 1\frac{3}{4}$$
$$-4\frac{1\cdot 2}{2\cdot 2} \rightarrow +2\frac{2}{4}$$
$$3\frac{5}{4}=4\frac{1}{4}\text{ mi}$$

Now we subtract to find the distance left.

$$5 \rightarrow 4\frac{4}{4}$$
$$-4\frac{1}{4} \rightarrow -4\frac{1}{4}$$
$$\frac{3}{4}\text{mi}$$

Mary Beth is $\frac{3}{4}$ mile short of completing her workout.

Quick Quiz 8.5

1. $5\% \text{ of } 2995 \approx \frac{1}{2}\times(10\% \text{ of } 3000) = \frac{1}{2}\times 300 = 150$

2. $20\% \text{ of } 60{,}015 \approx 10\% \text{ of } 60{,}000+10\% \text{ of } 6{,}000 = 6000+6000 = 12{,}000$

3. $15\% \text{ of } \$14{,}050$
$\approx 10\% \text{ of } \$14{,}000+5\% \text{ of } \$14{,}000$
$= \$1400+\frac{1}{2}\times\$1400$
$= \$1400+\$700$
$= \$2100$
Jan earned about \$2100.

4. Answers may vary. Two possible solutions follow:
Find 35% of 200.
1. 35% of 200
$= 3x(10\% \text{ of } 200)+5x(1\% \text{ of } 200)$
$= 3\times 20+5\times 2$
$= 60+10$
$= 70$

2. $35\% \text{ of } 200 = 35\times(1\% \text{ of } 200) = 35\times 2 = 70$

 Copyright © 2012 Pearson Education, Inc.

8.6 Exercises

1. To change a percent to a decimal, move the decimal point 2 places to the <u>left</u> and drop the <u>% sign</u>.

3. $\frac{42}{100} = 42\%$
42% of the students voted.

5. $\frac{63}{100} = 63\%$
63% of the powerboats had a radar navigation system.

7. $\frac{28}{100} = 28\%$
28% of people use electric toothbrushes.

9. $\frac{113}{100} = 113\%$
This year's attendance was 113% of last year's attendance.

11. $\frac{0.9}{100} = 0.9\%$
0.9% is saturated fat.

13.

Decimal	Percent
0.576	57.6%
0.249	24.9%
0.003	0.3%
1.546	154.6%

15.

Decimal	Percent
3.7	370%
0.238	23.8%
0.006	0.6%
12.882	1288.2%

17. Move the decimal point two places to the left.
36% = 0.36

19. Move the decimal point two places to the left.
53.8% = 0.538

21. Move the decimal point two places to the right.
0.075 = 7.5%

23. Move the decimal point two places to the left.
2.33% = 0.0233

25. Move the decimal point two places to the right.
0.03413 = 3.413%
Thus 3.413% of Alaska is covered by water.

27.

Fraction Form	Decimal Form	Percent Form
$\frac{4}{5}$	0.8	80%
$\frac{27}{100}$	0.27	27%
$\frac{7}{1000}$	0.007	0.7%
$4\frac{1}{3}$	$4.\overline{3}$	$433.\overline{3}\%$

29.

Fraction Form	Decimal Form	Percent Form
$\frac{5}{16}$	0.3125	31.25%
$2\frac{3}{5}$	2.6	260%
$\frac{1}{1000}$	0.001	$\frac{1}{10}\%$
$6\frac{1}{2}$	6.5	650%

31. $4 \div 32 = 0.125$
$\frac{4}{32} = 0.125 = 12.5\%$

33. $1 \div 5 = 0.2$
$\frac{1}{5}\% = 0.2\% = 0.002 = \frac{2}{1000} = \frac{1}{500}$

35. a. $14 \div 40 = 0.35$
$\frac{14}{40} = 0.35 = 35\%$

b. $22.3\% = 0.223 = \frac{223}{1000}$

37. $1\frac{1}{4}\% = 1.25\% = 0.0125 = \frac{125}{10,000} = \frac{1}{80}$

39. $1 \div 40 = 0.025$
$\frac{1}{40} = 0.025 = 2.5\%$

Copyright © 2012 Pearson Education, Inc.

41. $\frac{9}{14} = 9 \div 14 = 0.642857...$
$= 64.2857...\%$
$\approx 64.29\%$

43. $\frac{7}{9} = 7 \div 9 = 0.77777... = 77.777...\% \approx 77.78\%$

45. $5.5\% = 0.055 = \frac{55}{1000} = \frac{11}{200}$

47. $11.5\% = 0.115 = \frac{115}{1000} = \frac{23}{200}$

49. $15.9\% = 0.159 = \frac{159}{1000}$

51. The percents in the second row increase by 20 each time.

Fraction	$\frac{3}{5}$	$\frac{4}{5}$
Percent	60%	80%

53. The percents in the second row increase by 12.5 each time.

Fraction	$4\frac{2}{8}$	$4\frac{3}{8}$
Percent	425%	437.5%

Cumulative Review

54. Three times what number is equal to forty-eight?
$3n = 48$
$\frac{3n}{3} = \frac{48}{3}$
$n = 16$
The number is 16.

55. Twice what number is equal to three hundred thirty?
$2x = 330$
$\frac{2x}{2} = \frac{330}{2}$
$x = 165$
The number is 165.

56. One-fourth of what number is equal to 60?
$\frac{1}{4}x = 60$
$4\left(\frac{1}{4}x\right) = 4(60)$
$x = 240$
The number is 240.

57. What is one-third of sixty-nine?
$x = \frac{1}{3}(69)$
$x = 23$
Twenty-three is one-third of sixty-nine.

58. First we find the total of the expenses.
\$64.55 + \$34.50 + \$55.90 = \$154.95
Next we divide by 3.
\$154.95 ÷ 3 = \$51.65
Each roommate must contribute \$51.65.

59. We subtract.
5.1 − 3.7 = 1.4 million square miles

Quick Quiz 8.6

1. a. $\frac{15}{24} = 15 \div 24 = 0.625$

b. $\frac{15}{24} = 0.625 = 62.5\%$

2. a. $0.14 = \frac{14}{100} = \frac{7}{50}$

b. $0.14 = 14\%$

3. a. $0.05\% = 0.0005 = \frac{5}{10,000} = \frac{1}{2000}$

b. $0.05\% = 0.0005$

4. Answers may vary. One possible solution follows:
Change 0.43% to a decimal. Move the decimal two places to the left.
0.43% = 0.0043
Then change it to a fraction. There are 4 decimal places, so there are 4 zeros in the denominator.
$0.43\% = 0.0043 = \frac{43}{10,000}$

Copyright © 2012 Pearson Education, Inc.

8.7 Exercises

1. 35 is much more than $\frac{1}{2}$ of 40.

3. 9 is very close to 10 so it cannot be $\frac{1}{4}$ of 10.

5. What is 32% of 90?
$x = 32\% \times 90$
$x = 0.32 \times 90$
$x = 28.8$
28.8 is 32% of 90.

7. Find 26% of 145.
$n = 26\% \times 145$
$n = 0.26 \times 145$
$n = 37.7$
26% of 145 is 37.7.

9. What is 52% of 60?
$x = 52\% \times 60$
$x = 0.52 \times 60$
$x = 31.2$
31.2 is 52% of 60.

11. What is 150% of 40?
$x = 150\% \times 40$
$x = 1.50 \times 40$
$x = 60$
60 is 150% of 40.

13. What is 15% of $18.45?
$x = 15\% \times 18.45$
$x = 0.15 \times 18.45$
$x = 2.7675 \approx 2.77$
Bui should leave $2.77.

15. Find 60% of 650.
$n = 60\% \times 650$
$n = 0.60 \times 650$
$n = 390$
390 students are transferring to a four-year college.

17. 54 is what percent of 30?
$54 = n\% \times 30$
$\frac{54}{30} = n\%$
$1.8 = n\%$
$180 = n$
54 is 180% of 30.

19. What percent of 650 is 70?
$n\% \times 650 = 70$
$n\% = \frac{70}{650}$
$n\% = 0.10769...$
$n = 10.769...\% \approx 10.77\%$
70 is about 10.77% of 650.

21. What percent of 60 is 18?
$n\% \times 60 = 18$
$n\% = \frac{18}{60}$
$n\% = 0.3$
$n = 30\%$
30% of 60 is 18.

23. 15 is what percent of 80?
$15 = n\% \times 80$
$\frac{15}{80} = n\%$
$0.1875 = n\%$
$18.75 = n$
18.75% of the calories are from fat.

25. 2 is what percent of 5?
$2 = n\% \times 5$
$\frac{2}{5} = n\%$
$0.4 = n\%$
$40 = n$
Tasha made 40% of the shots.

27. 56 is 70% of what number?
$56 = 70\% \times n$
$56 = 0.70 \times n$
$\frac{56}{0.70} = n$
$80 = n$
56 is 70% of 80.

29. 24 is 40% of what number?
$24 = 40\% \times n$
$24 = 0.40 \times n$
$\frac{24}{0.40} = n$
$60 = n$
24 is 40% of 60.

Copyright © 2012 Pearson Education, Inc.

31. 70 is 20% of what number?
$70 = 20\% \times n$
$70 = 0.20 \times n$
$\frac{70}{0.20} = n$
$350 = n$
70 is 20% of 350.

33. 12 is 24% of what number?
$12 = 24\% \times n$
$12 = 0.24 \times n$
$\frac{12}{0.24} = n$
$50 = n$
There are 50 employees at the company.

35. 8 is 125% of what number?
$8 = 125\% \times n$
$8 = 1.25 \times n$
$\frac{8}{1.25} = n$
$6.4 = n$
The original trail was 6.4 miles.

37. Let x = the charge for members.
$128 is 60% more than the charge for members.
$100\%x + 60\%x = \$128$
$160\%x = 128$
$1.60x = 128$
$x = \frac{128}{1.60}$
$x = 80$
Members pay $80 to play a round of golf.

39. Let x = amount of winter bill.
$95 is 90% more than the amount of a winter bill.
$100\%x + 90\%x = \$95$
$190\%x = 95$
$1.90x = 95$
$x = \frac{95}{1.90}$
$x = 50$
Mary Beth can expect her winter bill to be $50.

41. 44 is 50% of what number?
$44 = 50\% \times n$
$44 = 0.50 \times n$
$\frac{44}{0.50} = n$
$88 = n$
44 is 50% of 88.

43. 125% of 60 is what number?
$125\% \times 60 = n$
$1.25 \times 60 = n$
$75 = n$
125% of 60 is 75.

45. What is 27% of 78?
$x = 27\% \times 78$
$x = 0.27 \times 78$
$x = 21.06$
21.06 is 27% of 78.

47. 15.66 is what percent of 87?
$15.66 = n\% \times 87$
$\frac{15.66}{87} = n\%$
$0.18 = n\%$
$18 = n$
15.66 is 18% of 87.

49. 135 is 45% of what number?
$135 = 45\% \times n$
$135 = 0.45 \times n$
$\frac{135}{0.45} = n$
$300 = n$
135 is 45% of 300.

51. 110 is what percent of 440?
$110 = n\% \times 440$
$\frac{110}{440} = n\%$
$0.25 = n\%$
$25 = n$
110 is 25% of 440.

53. What is 73% of 130 million?
$x = 73\% \times 130$ mil
$x = 0.73 \times 130$ mil
$x = 94.9$ mil
94.9 million red roses were sold.

55. 86 is what percent of 92?
$86 = n\% \times 92$
$\frac{86}{92} = n\%$
$0.934... = n\%$
$93.4... = n$
$93 \approx n$
Robert answered about 93% of the questions correctly.

Copyright © 2012 Pearson Education, Inc.

57. We subtract.
34.5% − 24.5% = 10%
The percentage of aluminum recycled is 10% greater than the percentage of glass recycled.

59. Find the percentage not spent on recreation.
13% + 10% + 27% + 33% + 10% = 93%
Subtract this result from 100%.
100% − 93% = 7%
Jeremy spends 7% of his income on recreation.

61. Jeremy saves 13% of $1250. What is 13% of $1250?
$x = 13\% \times 1250$
$x = 0.13 \times 1250$
$x = 162.50$
Jeremy saves $162.50 each month.

63. What is 67.3% of 348.9?
$x = 67.3\% \times 348.9$
$x = 0.673 \times 348.9$
$x = 234.8097$
Use a calculator for the multiplication.
234.8097 is 67.3% of 348.9.

65. 368 is 20% of what number?
$368 = 20\% \times n$
$368 = 0.20 \times n$
$\frac{368}{0.20} = n$
$1840 = n$
Use a calculator for the division. 368 is 20% of 1840.

67. Find 34.6% of 1,400,000.
$n = 34.6\% \times 1{,}400{,}000$
$n = 0.346 \times 1{,}400{,}000$
$n = 484{,}400$
484,400 U.S. surfers are women.

Cumulative Review

69.
$$\begin{aligned} 2x+3 &= 13 \\ + \quad -3 &\quad -3 \\ \hline 2x &= 10 \\ \frac{2x}{2} &= \frac{10}{2} \\ x &= 5 \end{aligned}$$

70.
$$\begin{aligned} 3x-1 &= 2x+4 \\ + \; -2x &\quad -2x \\ \hline x-1 &= 4 \\ + \quad 1 &\quad 1 \\ \hline x &= 5 \end{aligned}$$

71.
$$\begin{aligned} 5x-3 &= 3x+9 \\ + \; -3x &\quad -3x \\ \hline 2x-3 &= 9 \\ + \quad 3 &\quad 3 \\ \hline 2x &= 12 \\ \frac{2x}{2} &= \frac{12}{2} \\ x &= 6 \end{aligned}$$

72.
$$\begin{aligned} 2(3x+1) &= 2x-6 \\ 6x+2 &= 2x-6 \\ + \; -2x &\quad -2x \\ \hline 4x+2 &= -6 \\ + \quad -2 &\quad -2 \\ \hline 4x &= -8 \\ \frac{4x}{4} &= \frac{-8}{4} \\ x &= -2 \end{aligned}$$

Quick Quiz 8.7

1. What is 35% of 60?
$x = 35\% \times 60$
$x = 0.35 \times 60$
$x = 21$
21 is 35% of 60.

2. 15 is what percent of 50?
$15 = n\% \times 50$
$\frac{15}{50} = n\%$
$0.3 = n\%$
$30 = n$
15 is 30% of 50.

3. 8 is 20% of what number?
$8 = 20\% \times n$
$8 = 0.20 \times n$
$\frac{8}{0.20} = n$
$40 = n$
8 is 20% of 40.

Copyright © 2012 Pearson Education, Inc.

4. Answers may vary. One possible solution follows:
0.8% of windows are defective. There are 375 windows in a shipment. How many windows in the shipment are defective?
Let x = the number of defective windows in the shipment.
$x = 0.8\%$ of 375 windows
Write the equation.
$x = 0.8\% \times 375 = 0.008 \times 375 = 3$
3 of the windows in the shipment should be defective.

8.8 Exercises

1. Since 100% of 80 is 80, it is obvious that 150% of 80 is greater than 80.

3. 16% of 250 is 40.
The value of p is 16.
The base is the entire quantity. It follows the word *of*: $b = 250$.
The amount is the part compared to the whole: $a = 40$.

5. What is 95% of 420?
The value of p is 95.
The base follows the word *of*: $b = 420$.
The amount is unknown. We let a = the amount.

7. 69% of what is 8230?
The value of p is 69.
The base is unknown. We represent the base by the variable b.
The amount 8230 is the part of the base: $a = 8230$.

9. 63 is what percent of 90?
We let p represent the unknown percent.
The base follows the word *of*: $b = 90$.
The amount is the part compared to the whole: $a = 63$.

11. What percent of 47 is 10?
We let p represent the unknown percent.
The base follows the word *of*: $b = 47$.
The amount is the part compared to the whole: $a = 10$.

13. 400 is 160% of what?
The value of p is 160.
The base is unknown. We represent the base by the variable b.
The amount 400 is the part of the base: $a = 400$.

15. 24% of 200 is what?
$p = 24$, $b = 200$.
$\frac{a}{b} = \frac{p}{100}$ becomes $\frac{a}{200} = \frac{24}{100}$.

$$\frac{a}{200} = \frac{6}{25}$$
$$25a = (200)(6)$$
$$25a = 1200$$
$$\frac{25a}{25} = \frac{1200}{25}$$
$$a = 48$$

24% of 200 is 48.

17. Find 250% of 30.
$p = 250$, $b = 30$
$\frac{a}{b} = \frac{p}{100}$ becomes $\frac{a}{30} = \frac{250}{100}$.

$$\frac{a}{30} = \frac{5}{2}$$
$$2a = (30)(5)$$
$$2a = 150$$
$$\frac{2a}{2} = \frac{150}{2}$$
$$a = 75$$

250% of 30 is 75.

19. 0.6% of 4000 is what?
$p = 0.6$, $b = 4000$
$\frac{a}{b} = \frac{p}{100}$ becomes $\frac{a}{4000} = \frac{0.6}{100}$.

$$\frac{a}{4000} = \frac{0.6}{100}$$
$$100a = (4000)(0.6)$$
$$100a = 2400$$
$$\frac{100a}{100} = \frac{2400}{100}$$
$$a = 24$$

0.6% of 4000 is 24.

21. 82 is 50% of what?
$p = 50$, $a = 82$
$\frac{a}{b} = \frac{p}{100}$ becomes $\frac{82}{b} = \frac{50}{100}$.

$$\frac{82}{b} = \frac{1}{2}$$
$$2(82) = b$$
$$164 = b$$

82 is 50% of 164.

Copyright © 2012 Pearson Education, Inc.

23. 150% of what is 75?
$p = 150$, $a = 75$
$\frac{a}{b} = \frac{p}{100}$ becomes $\frac{75}{b} = \frac{150}{100}$.
$$\frac{75}{b} = \frac{3}{2}$$
$$2(75) = 3b$$
$$150 = 3b$$
$$\frac{150}{3} = \frac{3b}{3}$$
$$50 = b$$
150% of 50 is 75.

25. 4000 is 0.8% of what?
$p = 0.8$, $a = 4000$
$\frac{a}{b} = \frac{p}{100}$ becomes $\frac{4000}{b} = \frac{0.8}{100}$.
$$\frac{4000}{b} = \frac{0.8}{100}$$
$$(100)(4000) = 0.8b$$
$$400,000 = 0.8b$$
$$\frac{400,000}{0.8} = \frac{0.8b}{0.8}$$
$$500,000 = b$$
4000 is 0.8% of 500,000.

27. 70 is what percent of 280?
$b = 280$, $a = 70$
$\frac{a}{b} = \frac{p}{100}$ becomes $\frac{70}{280} = \frac{p}{100}$.
$$\frac{1}{4} = \frac{p}{100}$$
$$100 = 4p$$
$$\frac{100}{4} = \frac{4p}{4}$$
$$25 = p$$
70 is 25% of 280.

29. What percent of 140 is 11.2?
$b = 140$, $a = 11.2$
$\frac{a}{b} = \frac{p}{100}$ becomes $\frac{11.2}{140} = \frac{p}{100}$.
$$\frac{11.2}{140} = \frac{p}{100}$$
$$(100)(11.2) = 140p$$
$$1120 = 140p$$
$$\frac{1120}{140} = \frac{140p}{140}$$
$$8 = p$$
8% of 140 is 11.2.

31. What percent of \$5000 is \$90?
$b = 5000$, $a = 90$
$\frac{a}{b} = \frac{p}{100}$ becomes $\frac{90}{5000} = \frac{p}{100}$.
$$\frac{9}{500} = \frac{p}{100}$$
$$(100)(9) = 500p$$
$$900 = 500p$$
$$\frac{900}{500} = \frac{500p}{500}$$
$$1.8 = p$$
1.8% of \$5000 is \$90.

33. 26% of 350 is what?
$p = 26$, $b = 350$
$\frac{a}{b} = \frac{p}{100}$ becomes $\frac{a}{350} = \frac{26}{100}$.
$$\frac{a}{350} = \frac{13}{50}$$
$$50a = (350)(13)$$
$$50a = 4550$$
$$\frac{50a}{50} = \frac{4550}{50}$$
$$a = 91$$
26% of 350 is 91.

35. 180% of what is 720?
$p = 180$, $a = 720$
$\frac{a}{b} = \frac{p}{100}$ becomes $\frac{720}{b} = \frac{180}{100}$.
$$\frac{720}{b} = \frac{9}{5}$$
$$5(720) = 9b$$
$$3600 = 9b$$
$$\frac{3600}{9} = \frac{9b}{9}$$
$$400 = b$$
180% of 400 is 720.

37. 75 is what percent of 400?
$b = 400$, $a = 75$
$\frac{a}{b} = \frac{p}{100}$ becomes $\frac{75}{400} = \frac{p}{100}$.
$$\frac{3}{16} = \frac{p}{100}$$
$$(100)(3) = 16p$$
$$300 = 16p$$
$$\frac{300}{16} = \frac{16p}{16}$$
$$18.75 = p$$
75 is 18.75% of 400.

Copyright © 2012 Pearson Education, Inc.

39. Find 0.2% of 650.
$p = 0.2$, $b = 650$
$\frac{a}{b} = \frac{p}{100}$ becomes $\frac{a}{650} = \frac{0.2}{100}$.

$$\frac{a}{650} = \frac{0.2}{100}$$
$$100a = (650)(0.2)$$
$$100a = 130$$
$$\frac{100a}{100} = \frac{130}{100}$$
$$a = 1.3$$

0.2% of 650 is 1.3.

41. What percent of 25 is 15.2?
$b = 25$, $a = 15.2$
$\frac{a}{b} = \frac{p}{100}$ becomes $\frac{15.2}{25} = \frac{p}{100}$.

$$\frac{15.2}{25} = \frac{p}{100}$$
$$(100)(15.2) = 25p$$
$$1520 = 25p$$
$$\frac{1520}{25} = \frac{25p}{25}$$
$$60.8 = p$$

60.8% of 25 is 15.2.

43. 68 is 40% of what?
$p = 40$, $a = 68$
$\frac{a}{b} = \frac{p}{100}$ becomes $\frac{68}{b} = \frac{40}{100}$.

$$\frac{68}{b} = \frac{2}{5}$$
$$(5)(68) = 2b$$
$$340 = 2b$$
$$\frac{340}{2} = \frac{2b}{2}$$
$$170 = b$$

60 is 40% of 170.

45. 94.6 is what percent of 220?
$b = 220$, $a = 94.6$
$\frac{a}{b} = \frac{p}{100}$ becomes $\frac{94.6}{220} = \frac{p}{100}$.

$$\frac{94.6}{220} = \frac{p}{100}$$
$$(100)(94.6) = 220p$$
$$9460 = 220p$$
$$\frac{9460}{220} = \frac{220p}{220}$$
$$43 = p$$

94.6 is 43% of 220.

47. What is 12.5% of 380?
$p = 12.5$, $b = 380$
$\frac{a}{b} = \frac{p}{100}$ becomes $\frac{a}{380} = \frac{12.5}{100}$.

$$\frac{a}{380} = \frac{12.5}{100}$$
$$100a = (380)(12.5)$$
$$100a = 4750$$
$$\frac{100a}{100} = \frac{4750}{100}$$
$$a = 47.5$$

47.5 is 12.5% of 380.

49. Find 0.05% of 5600.
$p = 0.05$, $b = 5600$
$\frac{a}{b} = \frac{p}{100}$ becomes $\frac{a}{5600} = \frac{0.05}{100}$.

$$\frac{a}{5600} = \frac{0.05}{100}$$
$$100a = (5600)(0.05)$$
$$100a = 280$$
$$\frac{100a}{100} = \frac{280}{100}$$
$$a = 2.8$$

0.05% of 5600 is 2.8.

51. What is 80% of 250?
$p = 80$, $b = 250$
$\frac{a}{b} = \frac{p}{100}$ becomes $\frac{a}{250} = \frac{80}{100}$.

$$\frac{a}{250} = \frac{4}{5}$$
$$5a = (250)(4)$$
$$5a = 1000$$
$$\frac{5a}{5} = \frac{1000}{5}$$
$$a = 200$$

200 apartments must be rented to cover expenses.

53. a. Subtract 60% from 100%.
100% − 60% = 40%
40% of the openings should be available after the first week.

 Copyright © 2012 Pearson Education, Inc.

b. Find 40% of 150?
$p = 40$, $b = 150$

$\frac{a}{b} = \frac{p}{100}$ becomes $\frac{a}{150} = \frac{40}{100}$.

$$\frac{a}{150} = \frac{2}{5}$$
$$5a = (150)(2)$$
$$5a = 300$$
$$\frac{5a}{5} = \frac{300}{5}$$
$$a = 60$$

60 openings should be available after the first week.

55. a. What percent of \$1950 is \$42.90?
$b = 1950$, $a = 42.90$

$\frac{a}{b} = \frac{p}{100o}$ becomes $\frac{42.90}{1950} = \frac{p}{100}$.

$$\frac{42.90}{1950} = \frac{p}{100}$$
$$(100)(42.90) = 1950p$$
$$4290 = 1950p$$
$$\frac{4290}{1950} = \frac{1950p}{1950}$$
$$2.2 = p$$

2.2% of Owen's salary is withheld for the vacation savings plan.

b. Subtract 2.2% from 100%.
100% − 2.2% = 97.8%
97.8% of Owen's salary is not withheld for the vacation savings plan.

57. 90 is 75% off what?
$p = 75$, $a = 90$

$\frac{a}{b} = \frac{p}{100}$ becomes $\frac{90}{b} = \frac{75}{100}$.

$$\frac{90}{b} = \frac{3}{4}$$
$$(4)(90) = 3b$$
$$360 = 3b$$
$$\frac{360}{3} = \frac{3b}{3}$$
$$120 = b$$

There are 120 offices in the building.

59. What percent of 9 is 6?
$b = 9$, $a = 6$

$\frac{a}{b} = \frac{p}{100}$ becomes $\frac{6}{9} = \frac{p}{100}$.

$$\frac{2}{3} = \frac{p}{100}$$
$$(100)(2) = 3p$$
$$200 = 3p$$
$$\frac{200}{3} = \frac{3p}{3}$$
$$66.66... = p$$
$$67 \approx p$$

Delroy answered about 67% correctly.

61. What is $19\frac{1}{4}\%$ of 798?

$p = 19\frac{1}{4} = 19.25$, $b = 798$

$\frac{a}{b} = \frac{p}{100}$ becomes $\frac{a}{798} = \frac{19.25}{100}$.

$$\frac{a}{798} = \frac{19.25}{100}$$
$$100a = (798)(19.25)$$
$$100a = 15,361.5$$
$$\frac{100a}{100} = \frac{15,361.5}{100}$$
$$a = 153.615 \approx 153.62$$

Use a calculator for the multiplication. $19\frac{1}{4}\%$ of 798 is about 153.62.

63. Find 18% of 20% of \$3300.
First find 20% of \$3300.
$p = 20$, $b = 3300$

$\frac{a}{b} = \frac{p}{100}$ becomes $\frac{a}{3300} = \frac{20}{100}$.

$$\frac{a}{3300} = \frac{1}{5}$$
$$5a = 3300$$
$$\frac{5a}{5} = \frac{3300}{5}$$
$$a = 660$$

20% of \$3300 is \$660.
Now find 18% of \$660.

$\frac{a}{b} = \frac{p}{100}$ becomes $\frac{a}{660} = \frac{18}{100}$.

Copyright © 2012 Pearson Education, Inc.

$\frac{a}{660}=\frac{9}{50}$
$50a=(660)(9)$
$50a=5940$
$\frac{50a}{50}=\frac{5940}{50}$
$a=118.8$
Use a calculator for the multiplication and division.
18% of \$660 is \$118.80, so 18% of 20% of \$3300 is \$118.80.

Cumulative Review

65. $A=LW=(7\text{ in.})(4\text{ in.})=28\text{ in}^2$

66. $V=LWH=(8\text{ cm})(4\text{ cm})(6\text{ cm})=192\text{ cm}^3$

67. $A=s^2=(2\text{ ft})^2=4\text{ ft}^2$

68. $P=4s=4(3\text{ in.})=12\text{ in.}$

Quick Quiz 8.8

1. What is 18% of 90?
$p=18, b=90$
$\frac{a}{b}=\frac{p}{100}$ becomes $\frac{a}{90}=\frac{18}{100}$.
$\frac{a}{90}=\frac{9}{50}$
$50a=(90)(9)$
$50a=810$
$\frac{50a}{50}=\frac{810}{50}$
$a=16.2$
16.2 is 18% of 90.

2. 45 is what percent of 125?
$b=125, a=45$
$\frac{a}{b}=\frac{p}{100}$ becomes $\frac{45}{125}=\frac{p}{100}$.
$\frac{9}{25}=\frac{p}{100}$
$(100)(9)=25p$
$900=25p$
$\frac{900}{25}=\frac{25p}{25}$
$36=p$
45 is 36% of 125.

3. 56 is 80% of what number?
$p=80, a=56$
$\frac{a}{b}=\frac{p}{100}$ becomes $\frac{56}{b}=\frac{80}{100}$.
$\frac{56}{b}=\frac{4}{5}$
$(5)(56)=4b$
$280=4b$
$\frac{280}{4}=\frac{4b}{4}$
$70=b$
56 is 80% of 70.

4. Answers may vary. One possible solution follows:
In the percent proportion, what can you say about the *percent number* if the value of the *amount* is larger than the *base*?
$\frac{\text{amount}}{\text{base}}=\frac{\text{percent number}}{100}$
If a is larger than b, we can write $a=b(1+n)$ where n is a positive number. Substitute for a,
$\frac{b(1+n)}{b}=\frac{p}{100}$
$1+n=\frac{p}{100}$
$100(1+n)=p$
Since n is a positive number, $(1+n)$ is larger than 1, and p is larger than 100.

8.9 Exercises

1. Principal: the amount deposited or borrowed

3. Time: the period of time interest is calculated

5. Commission = commission rate × total sales

7. $n=18\%\times\$4200$
$n=0.18\times\$4200$
$n=\$756$
Becca earned \$756 in commission.

9. $n=25\%\times\$13{,}500$
$n=0.25\times\$13{,}500$
$n=\$3375$
Brandon's commission was \$3375.

 Copyright © 2012 Pearson Education, Inc.

11. $75 = n\% \times 250$

$\frac{75}{250} = n\%$

$0.3 = n\%$

$30 = n$

The price decreased by 30%.

13. $360 = n\% \times 1800$

$\frac{360}{1800} = n\%$

$0.2 = n\%$

$20 = n$

The price is reduced by 20%.

15. First we find the amount of the reduction.
$10\% \times 4500 = 0.10 \times 4500 = \$450$
Now we find this year's cost.
$\$4500 - \$450 = \$4050$
This year's cost is $4050.

17. First we find the amount of the raise.
$7\% \times 42{,}000 = 0.07 \times 42{,}000 = \$2940$
Now we find his new salary.
$\$42{,}000 + \$2940 = \$44{,}940$
Jerome's new salary is $44,940.

19. First we find the increase in tickets sold.
$15\% \times 760 = 0.15 \times 760 = 114$
Now we find the total.
$760 + 114 = 874$ tickets
874 people will attend this year.

21. a. $P = \$2500$, $R = 6\%$, $T = 1$ year
$I = P \times R \times T$
$I = \$2500 \times 0.06 \times 1 = \$150$
The interest for one year is $150.

b. $\$2500 + \$150 = \$2650$
Pam will have $2650 at the end of the year.

23. $P = \$500,\ R = 8\%,\ T = 6 \text{ months} = \frac{1}{2} \text{ year}$

$I = P \times R \times T$

$I = \$500 \times 0.08 \times \frac{1}{2} = \$20$

The interest is $20.

25. a. $n = 20\% \times \$4150$
$n = 0.20 \times \$4150$
$n = \$830$
Meleena's commission was $830.

b. $\$1500 + \$830 = \$2330$
Meleena's total earnings were $2330.

27. First find the interest.
$P = \$6600$, $R = 9\%$, $T = 2$ years
$I = P \times R \times T = \$6600 \times 0.09 \times 2 = \$1188$
Now find the total to be repaid.
$\$6600 + \$1188 = \$7788$
Divide by 24 to find the payment.
$\$7788 \div 24 = \$324.50$
Each payment is $324.50.

29. First find the amount Sam lost.
$15\% \times 265 = 0.15 \times 265 = 39.75$ lb
Now find his new weight.
$265 - 39.75 = 225.25$ lb
Sam weighed 225.25 pounds after his weight loss.

31. a. We add.
$\$289 + \$289 + \$421 = \$999$
The total price is $999.

b. Find the discount.
$30\% \times \$999 = 0.30 \times \$999 = \$299.70$
We subtract.
$\$999 - \$299.70 = \$699.30$
Kristen paid $699.30.

33. $35 is 5% of what?

$35 = 5\% \times n$

$35 = 0.05n$

$\frac{35}{0.05} = \frac{0.05n}{0.05n}$

$700 = n$

Marcia will need to purchase $700 of merchandise to save $35.

35. a. The rent in 2009 plus a 2.2% increase equals the rent in 2010:

$n + 0.022n = 1232$

$1.022n = 1232$

$\frac{1.022n}{1.022} = \frac{1232}{1.022}$

$n \approx 1205.48$

The average rent was $1205.48 in 2009.

b. The 2010 rent plus a 2.2% increase equals the 2011 rent.

$1232 + 0.022 \times 1232 = n$

$1259.10 \approx n$

If the same percent increase occurred, the 2011 rent was $1259.10.

Copyright © 2012 Pearson Education, Inc.

37. a. Find 3% of 35.
$3\% \times 35 = 0.03 \times 35 = 1.05$
Jerry can expect to sell one vacuum cleaner, so his total sales will be \$320.
Find 11% of \$320.
$11\% \times 320 = 0.11 \times 320 = \$35.20$

b. Find 9% of 35.
$9\% \times 35 = 0.09 \times 35 = 3.15$
Jerry can expect to sell three vacuum cleaners, so his total sales will be
$3 \times \$320 = \$960.$
Find 11% of \$960.
$11\% \times 960 = 0.11 \times 960 = \$105.60$
If Jerry wants to make at least \$100, he should pay \$5.60 or less for the list.

Cumulative Review

39. $A = bh = (6 \text{ cm})(4 \text{ cm}) = 24 \text{ cm}^2$

40. $V = LWH = (9 \text{ cm})(4 \text{ cm})(7 \text{ cm}) = 252 \text{ cm}^3$

Quick Quiz 8.9

1. $40\% \times \$120 = 0.40 \times \$120 = \$48$
The price is reduced \$48.

2. $P = \$12{,}000,\ R = 14\%,\ T = 3$ years
$I = P \times R \times T$
$I = \$12{,}000 \times 0.14 \times 3 = \$5040$
The interest is \$5040.

3. First find the raise.
$12\% \times \$28{,}000 = 0.12 \times \$28{,}000 = \$3360$
Now find the new salary.
$\$28{,}000 + \$3360 = \$31{,}360$
Joe will earn \$31,360 his second year.

4. Answers may vary. One possible solution follows:
Use the simple interest formula.
$I = P \times R \times T$
$P = \$6500,\ R = 9\%$ per year,
$T = 4 \text{ months} = \frac{4}{12} \text{ year}$

$$
\begin{aligned}
I &= P \times R \times T \\
&= \$6500 \times 9\% \times \frac{1}{3} \\
&= 6500 \times 0.09 \times \frac{1}{3} \\
&= 195
\end{aligned}
$$

The interest is \$195.

Use Math to Save Money

1. Job 1: Divide the annual salary by 52 since there are 52 weeks in one year.
$\$50{,}310 \div 52 = \$967.50$
The weekly gross income is \$967.50.
Job 2: We multiply to find the gross weekly earnings.
$6 \times 5 \times \$20.50 = \$615$
The gross pay per week is \$615.

2. Job 1:
15% of $\$967.50 = 0.15 \times \$967.50 \approx \$145.13$
The weekly payroll deductions are \$145.13.
Job 2: 12% of $\$615 = 0.12 \times \$615 = \$73.80$
The weekly payroll deductions are \$73.80.

3. Job 1:

$$
\begin{aligned}
&\$175 + \$75 + \$25 + \$50 + (\$0.35 \times 45 \times 5) \\
&= \$175 + \$75 + \$25 + \$50 + \$78.75 \\
&= \$403.75
\end{aligned}
$$

Job-related expenses will be \$403.75.
Job 2:

$$
\begin{aligned}
&\$150 + \$50 + \$25 + \$25 + (\$0.35 \times 5 \times 5) \\
&= \$150 + \$50 + \$25 + \$25 + \$8.75 \\
&= \$258.75
\end{aligned}
$$

Job-related expenses will be \$258.75.

4. Job 1:

$$
\begin{aligned}
&\$967.50 - (\$145.13 + \$403.75) \\
&= \$967.50 - \$548.88 \\
&= \$418.62
\end{aligned}
$$

Sharon will have \$418.62 left to spend.
Job 2:

$$
\begin{aligned}
\$615 - (\$73.80 + \$258.75) &= \$615 - \$332.55 \\
&= \$282.45
\end{aligned}
$$

Sharon will have \$282.45 left to spend.

5. Job 1:
Total time: $45 + 5 = 50$ hours
$\frac{\$418.62}{50 \text{ hr}} \approx \$8.37$ per hour
The real hourly wage is \$8.37.
Job 2:
Total time: $30 + 1 = 31$ hours
$\frac{\$282.45}{31 \text{ hr}} \approx \$9.11$
The real hourly wage is \$9.11.

6. She should take job 2, since it has the greater real hourly wage.

7. She should take Job 1, since she will have more left to spend each week.

Copyright © 2012 Pearson Education, Inc.

You Try It

1. $20.48 = 20\frac{48}{100} = 20\frac{12}{25}$

2. 1.380 ? 1.308
The numbers differ in the hundredths place.
Since 8 > 0, 1.38 > 1.308.

3. a. 27.1048
The round-off place digit is in the hundredths place: 27.1<u>0</u>48. The digit to the right is less than 5. Do not change the round-off place digit. Drop all digits to the right.
27.10

b. 1453.1286
The round-off place digit is in the thousandths place: 1453.12<u>8</u>6. The digit to the right is greater than or equal to 5. Increase the round-off place digit by 1. Drop all digits to the right.
1453.129

4. a.
$$\begin{array}{r} 59.000 \\ 2.730 \\ +\ 0.345 \\ \hline 62.075 \end{array}$$

b. $-13.6 - 0.82 = -13.6 + (-0.82)$
Add the absolute values and keep the common sign.
$$\begin{array}{r} -13.60 \\ +\ -0.82 \\ \hline -14.42 \end{array}$$

5.
$$\begin{array}{r} 2.6y \\ -\ 1.4y \\ \hline 1.2y \end{array}$$
$2.6y + 9.1x - 1.4y = 9.1x + 1.2y$

6. a.
$$\begin{array}{r} 7.34 \\ \times\ \ 4.3 \\ \hline 2\ 02 \\ 29\ 36\ \ \\ \hline 31.562 \end{array}$$

b.
$$\begin{array}{r} -3.16 \\ \times\ \ 2.7 \\ \hline 2\ 212 \\ 6\ 32\ \ \\ \hline -8.532 \end{array}$$

7. a. Move the decimal point three places to the right.
$85.26 \times 10^3 = 85,260$

b. Move the decimal point two places to the right.
$0.00012 \times 100 = 0.012$

8. $0.05\overline{)2.715} \rightarrow 5\overline{)271.5}$
$$\begin{array}{r} 54.3 \\ 5\overline{)271.5} \\ \underline{25} \\ 21 \\ \underline{20} \\ 1\ 5 \\ \underline{1\ 5} \\ 0 \end{array}$$
$2.715 \div 0.05 = 54.3$

9.
$$\begin{array}{r} 0.4166 \\ 12\overline{)5.0000} \\ \underline{4\ 8} \\ 20 \\ \underline{12} \\ 80 \\ \underline{72} \\ 80 \\ \underline{72} \\ 8 \end{array}$$
$\frac{5}{12} = 0.41\overline{6}$ or 0.4166...

10.
$$\begin{aligned} 3(x-2) &= -12.69 \\ 3x-6 &= -12.69 \\ +\quad 6 &\quad\ \ 6 \\ \hline 3x &= -6.69 \\ \frac{3x}{3} &= \frac{-6.69}{3} \\ x &= -2.23 \end{aligned}$$

11. 11% of \$35,050
≈ 10% of \$35,000 + 1% of 35,000
= \$3500 + \$350
= \$3850
Mark will earn about \$3850 in commission.

12. Move the decimal point two places to the right.
1.7 = 170%

13. Move the decimal point two places to the left.
0.13% = 0.0013

Copyright © 2012 Pearson Education, Inc.

14. $1 \div 5 = 0.2$

$4\frac{1}{5} = 4.2 = 420\%$

15. $64\% = \frac{64}{100} = \frac{16}{25}$

16. What percent of 75 is 15?

$n\% \times 75 = 15$

$75n\% = 15$

$\frac{75n\%}{75} = \frac{15}{75}$

$n\% = 0.20$

15 is 20% of 75.

17. What is 18% of 60? The amount a is unknown. The base $b = 60$ and percent $p = 18$.

$\frac{a}{b} = \frac{p}{100}$ becomes $\frac{a}{60} = \frac{18}{100}$

$\frac{a}{60} = \frac{9}{50}$

$50a = 60(9)$

$50a = 540$

$\frac{50a}{50} = \frac{540}{50}$

$a = 10.8$

10.8 is 18% of 60.

18. commission = (0.14)($3250) = $455

Tim earned $455 in commission.

19. raise = (0.07)($42,000) = $2940

new salary = $42,000 + $2940 = $44,940

Danny's new salary is $44,940.

20. $I = P \times R \times T$

$I = (12{,}000)(0.14)(3) = 5040$

The interest is $5040.

Chapter 8 Review Problems

1. 0.679

The last digit is in the thousandths place.

Six hundred seventy-nine thousandths

2. 7.0083

The last digit is in the ten thousandths place.

Seven and eighty-three ten thousandths

3. Forty-six and 85/100

4. There are three places after the decimal, so there are three zeros in the denominator.

$4.267 = 4\frac{267}{1000}$

5. There are two places after the decimal, so there are two zeros in the denominator.

$43.91 = 43\frac{91}{100}$

6. There are three zeros in the denominator so we move the decimal in the numerator three places to the left.

$32\frac{761}{1000} = 32.761$

7. There are three zeros in the denominator so we move the decimal in the numerator three places to the left. We need to insert a zero so that we can move the decimal three places.

$54\frac{26}{1000} = 54.026$

8. 0.523 ? 0.524

The thousandths digits differ. Since $3 < 4$, $0.523 < 0.524$.

9. 0.16 ? 0.168

0.160 ? 0.168

The thousandths digits differ. Since $0 < 8$, $0.160 < 0.168$.

10. 842.8569

The round-off place digit is in the hundredths place: $842.8\underline{5}69$. The digit to the right is greater than or equal to 5. Increase the round-off place digit by 1. Drop all digits to the right.

842.86

11. 406.7809

The round-off place digit is in the thousandths place: $406.78\underline{0}9$. The digit to the right is greater than or equal to 5. Increase the round-off place digit by 1. Drop all digits to the right.

406.781

12.
$$\begin{array}{r} -5.200 \\ 0.317 \\ \hline -4.883 \end{array}$$

Copyright © 2012 Pearson Education, Inc.

13.
$$\begin{array}{r} 0.588 \\ 36.000 \\ +\ 8.430 \\ \hline 45.018 \end{array}$$

14.
$$\begin{array}{r} 25.98 \\ -\ 2.33 \\ \hline 23.65 \end{array}$$

15. $-9.355 - 2.48 = -9.355 + (-2.48)$
$$\begin{array}{r} -9.355 \\ +\ (-2.48) \\ \hline -11.835 \end{array}$$

16.
$$\begin{array}{r} -9.2 \\ +\ (-5.4) \\ \hline -14.6 \end{array}$$

17. $-5.3 - (-6.67) = -5.3 + 6.67 = 6.67 - 5.3$
$$\begin{array}{r} 6.67 \\ -\ 5.3 \\ \hline 1.37 \end{array}$$

18. To subtract numbers in decimal notation, we <u>line up</u> the decimal points.

19. When adding 85 + 36.5, we rewrite 85 as <u>85.0</u> so that we can line up <u>the decimal points</u>.

20. We replace the variable with 2.4.
$x - 9.3 = 2.4 - 9.3 = -9.3 + 2.4$
$$\begin{array}{r} -9.3 \\ +\ 2.4 \\ \hline -6.9 \end{array}$$

21. We replace the variable with −2.3.
$y + 17.2 = -2.3 + 17.2 = 17.2 - 2.3$
$$\begin{array}{r} 17.2 \\ -\ 2.3 \\ \hline 14.9 \end{array}$$

22.
$$\begin{array}{r} 4.6x \\ +\ 7.2x \\ \hline 11.8x \end{array}$$

23.
$$\begin{array}{r} 8.6x \\ +\ 3.9x \\ \hline 12.5x \end{array}$$

24. Round each amount to the nearest dollar and then add.
45 + 50 + 44 + 47 = $186
Jamahl's gasoline was about $186 for the month.

25.
$$\begin{array}{r} 0.082 \\ \times\ 0.02 \\ \hline 0.00164 \end{array}$$

26.
$$\begin{array}{r} 7.21 \\ \times\ 5.68 \\ \hline 5768 \\ 4\ 326\ \ \\ 36\ 05\ \ \ \ \\ \hline 40.9528 \end{array}$$

27.
$$\begin{array}{r} -41.25 \\ \times\ \ 3.01 \\ \hline 4125 \\ 0\ 000\ \ \\ 123\ 75\ \ \ \ \\ \hline -124.1625 \end{array}$$

28.
$$\begin{array}{r} -9.01 \\ \times\ (-5.6) \\ \hline 5\ 406 \\ 45\ 05\ \ \\ \hline 50.456 \end{array}$$

29. If one factor has 2 decimal places and the second factor has 3 decimal places, the product has <u>five</u> decimal places.

30. If one factor has 3 decimal places and the second factor has 4 decimal places, the product has <u>seven</u> decimal places.

31. Move the decimal to the right 2 places.
$0.1249 \times 100 = 12.49$

32. Move the decimal to the right 3 places.
$1.6 \times 1000 = 1600$

33. Move the decimal to the right 5 places.
$41 \times 10^5 = 4{,}100{,}000$

34. Subtract to find the number of overtime hours.
52 − 40 = 12 overtime hours
$$\begin{aligned} 40(\$9.10) + 12(\$13.65) &= \$364 + \$163.8 \\ &= \$527.80 \end{aligned}$$
Mark will earn $527.80.

Copyright © 2012 Pearson Education, Inc.

35. Subtract to find the number of miles over 36,000 miles.
$44{,}322.6 - 36{,}000 = 8{,}322.6$ mi
$\$4500 + 8322.6(\$0.12) \approx \$4500 + \$998.71$
$= \$5498.71$
It will cost Mark \$5498.71 to buy the truck.

36. $12\overline{)8.660}$ (quotient 0.721)
8 4
26
24
20
12
8

$8.66 \div 12 \approx 0.72$

37. $-7.2\overline{)-8.52} \rightarrow 72\overline{)85.200}$ (quotient 1.183)
72
13 2
7 2
600
576
240
216
24

$-8.52 \div -7.2 \approx 1.18$

38. $0.44\overline{)21.2} \rightarrow 44\overline{)2120.0000}$ (quotient 48.1818)
176
360
352
8 0
4 4
3 60
3 52
80
44
360
352
8

$21.2 \div 0.44 = 48.\overline{18}$

39. When we divide $7.21\overline{)25.9}$, we rewrite the equivalent division problem $\underline{721\overline{)2590}}$ and then divide.

40. $12\overline{)4.349}$

41. $8\overline{)86.00}$ (quotient 10.75)
8
06 0
5 6
40
40
0

$\frac{86}{8} = 10.75$

42. $9\frac{1}{5} = 9 + \frac{1}{5}$

$5\overline{)1.0}$ (quotient 0.2)
1 0
0

$9 + \frac{1}{5} = 9 + 0.2 = 9.2$

$9\frac{1}{5} = 9.2$

43. **a.** We divide to find the number of containers.
$38.5 \div 3.5 = 11$ containers

b. We multiply to find the cost.
$3.5(6.20) = \$21.70$

44.
$$\begin{aligned} x - 2.68 &= 8.23 \\ +\quad 2.68 &\quad\; 2.68 \\ \hline x &= 10.91 \end{aligned}$$

45.
$$\begin{aligned} -1.6x &= 5.44 \\ \frac{-1.6x}{-1.6} &= \frac{5.44}{-1.6} \\ x &= -3.4 \end{aligned}$$

46.
$$\begin{aligned} 2x + 2.4 &= 8.7 \\ +\quad -2.4 &\quad -2.4 \\ \hline 2x &= 6.3 \\ \frac{2x}{2} &= \frac{6.3}{2} \\ x &= 3.15 \end{aligned}$$

Copyright © 2012 Pearson Education, Inc.

47.
$$\begin{aligned} 5(x+1.4) &= x+2.6 \\ 5x+7 &= x+2.6 \\ +\quad -x \quad & \quad -x \\ 4x+7 &= 2.6 \\ +\quad -7 & \quad -7 \\ 4x &= -4.4 \\ \frac{4x}{4} &= \frac{-4.4}{4} \\ x &= -1.1 \end{aligned}$$

48. Since we are comparing the number of nickels to the number of dimes, we let the variable represent the number of dimes.
D = number of dimes
The number of nickels is 6 times the number of dimes.
$6D$ = number of nickels
The value of the dimes is $(0.10)D$ and the value of the nickels is $(0.05)(6D)$. The total value of the coins is \$12.
$$\begin{aligned} (0.10)D+(0.05)(6D) &= 12 \\ D+3D &= 120 \\ 4D &= 120 \\ \frac{4D}{4} &= \frac{120}{4} \\ D &= 30 \end{aligned}$$
$6D = 6(30) = 180$
Ernie needs 30 dimes and 180 nickels.
Check:
Do we have six times as many nickels as dimes? Yes, 180 nickels is 6×30, or six times the number of dimes.
Is the value of the coins equal to \$12? Yes, 180 nickels = \$9 and 30 dimes = \$3;
\$9 + \$3 = \$12.

49. 10% of 200,022
We first round to the nearest hundred thousand.
200,000
Then we drop the last digit.
20,000

50. 1% of 200,022
We first round to the nearest hundred thousand.
200,000
Then we drop the last two digits.
2000

51.
$$\begin{aligned} 7\% \text{ of } 200{,}022 &\approx 7\times(1\% \text{ of } 200{,}000) \\ &= 7\times 2000 \\ &= 14{,}000 \end{aligned}$$

52. 20% of 200,022
$\approx$ (10% of 200,000) + (10% of 200,000)
= 20,000 + 20,000
= 40,000

53. $\frac{85}{100} = 85\%$
85% of the citizens voted.

54. $\frac{132}{110} = 1.2 = 120\%$
This year's enrollment is 120% of last year's enrollment.

55. Move the decimal point to the right two places.
0.016 = 1.6%

56. 124% = 1.24

57.

Decimal Form	Percent Form
0.379	37.9%
0.428	42.8%
0.005	0.5%
3.47	347%

58.

Decimal Form	Percent Form
1.2	120%
0.035	3.5%
0.0025	0.25%
0.567	56.7%

59. $\frac{56}{168} = 0.\overline{3} = 33.\overline{3}\%$

60. $81.3\% = 0.813 = \frac{813}{1000}$

Copyright © 2012 Pearson Education, Inc.

61.

Fraction Form	Decimal Form	Percent Form
$\frac{3}{8}$	0.375	37.5%
$\frac{18}{25}$	0.72	72%
$\frac{1}{200}$	0.005	$0.5\% = \frac{1}{2}\%$
$7\frac{2}{5}$	7.4	740%

62.

Fraction Form	Decimal Form	Percent Form
$\frac{3}{4}$	0.75	75%
$\frac{14}{25}$	0.56	56%
$\frac{1}{500}$	0.002	$0.2\% = \frac{1}{5}\%$
$3\frac{6}{15}$	3.4	340%

63. 82 is 25% of what number?

$$82 = 25\% \times n$$
$$82 = 0.25 \times n$$
$$\frac{82}{0.25} = n$$
$$328 = n$$

82 is 25% of 328.

64. 15 is what percent of 300?

$$15 = n\% \times 300$$
$$\frac{15}{300} = n\%$$
$$0.05 = n\%$$
$$5 = n$$

15 is 5% of 300.

65. Find 45% of 120.

$n = 45\% \times 120$
$n = 0.45 \times 120$
$n = 54$
54 is 45% of 120.

66. Let x = the amount for food and drinks.
Then $15\% \times x$ = the amount of the tip.

$$100\%x + 15\%x = \$18$$
$$115\%x = 18$$
$$1.15x = 18$$
$$x = \frac{18}{1.15}$$
$$x \approx 15.65$$

José can spend at most $15.65 on food and drinks.

67. 26 is 10% of what number?

$$26 = 10\% \times n$$
$$26 = 0.10 \times n$$
$$\frac{26}{0.10} = n$$
$$260 = n$$

There are 260 employees at A&R Accounting.

68. What is 8% of $22,000?

$x = 8\% \times 22{,}000$
$x = 0.08 \times 22{,}000$
$x = 1760$
Danny paid $1760 sales tax.

69. $4.25 is what percent of $32.40?

$$4.25 = n\% \times 32.40$$
$$\frac{4.25}{32.40} = n\%$$
$$0.13117... = n\%$$
$$13.117... = n$$
$$13.12 \approx n$$

Dean's tip was about 13.12% of the total.

70. What is 85% of 400?
The value of p is 85.
The base follows the word *of*: $b = 400$.
The amount is unknown. We let
a = the amount.

71. 20 is what percent of 100?
We let p represent the unknown percent.
The base follows the word *of*: $b = 100$.
The amount is the part compared to the whole:
$a = 20$.

Copyright © 2012 Pearson Education, Inc.

72. 9 is 45% of what?
$p = 45, a = 9$

$\frac{a}{b} = \frac{p}{100}$ becomes $\frac{9}{b} = \frac{45}{100}$.

$\frac{9}{b} = \frac{9}{20}$
$(20)(9) = 9b$
$180 = 9b$
$\frac{180}{9} = \frac{9b}{9}$
$20 = b$

9 is 45% of 20.

73. 18 is what percent of 50?
$b = 50, a = 18$

$\frac{a}{b} = \frac{p}{100}$ becomes $\frac{18}{50} = \frac{p}{100}$.

$\frac{9}{25} = \frac{p}{100}$
$(100)(9) = 25p$
$900 = 25p$
$\frac{900}{25} = \frac{25p}{25}$
$36 = p$

18 is 36% of 50.

74. a. The original price of the dishwasher in the advertisement is 225 + 75 = $300.

b. What percent of 300 is 75?
$b = 300, a = 75$

$\frac{75}{300} = \frac{p}{100}$
$\frac{1}{4} = \frac{p}{100}$
$100 = 4p$
$\frac{100}{4} = \frac{4p}{p}$
$25 = p$

The dishwasher is marked down 25%.

75. a. Subtract 70% from 100%.
100% − 70% = 30%
30% of the spaces should be available after the first week.

b. Find 30% of 250?
$p = 30, b = 250$

$\frac{a}{b} = \frac{p}{100}$ becomes $\frac{a}{250} = \frac{30}{100}$.

$\frac{a}{250} = \frac{3}{10}$
$10a = (250)(3)$
$10a = 750$
$\frac{10a}{10} = \frac{750}{10}$
$a = 75$

75 spaces should be available after the first week.

76. $n = 18\% \times \$13,250$
$n = 0.18 \times \$13,250$
$n = \$2385$
Jason's commission is $2385.

77. $P = \$8500, R = 11\%, T = 1$ yr
$I = P \times R \times T = \$8500 \times 0.11 \times 1 = \$935$
The interest is $935.

How Am I Doing? Chapter 8 Test

1. 207.402
The last digit is in the thousandths place.
Two hundred seven and four hundred two thousandths

2. There are three places after the decimal so there are three zeros in the denominator.

$0.013 = \frac{13}{1000}$

3. There are two zeros in the denominator so we move the decimal in the numerator two places to the left.

$\frac{51}{100} = 0.51$

4. 0.45 ? 0.412
The hundredths digits differ. Since 5 > 1,
0.45 > 0.412

5. 746.136
The round-off place digit is in the hundredths place: 746.1<u>3</u>6. The digit to the right is greater than or equal to 5. Increase the round-off place digit by 1. Drop all digits to the right.
746.14

Copyright © 2012 Pearson Education, Inc.

6. We replace the variable with −2.07.
$x = 0.12 = -2.07 + 0.12$
$$\begin{array}{r} -2.07 \\ +\ 0.12 \\ \hline -1.95 \end{array}$$

7.
$$\begin{array}{r} 3.1x \\ +\ 1.06x \\ \hline 4.16x \end{array}$$
$3.1x + 2.01y + 1.06x = 4.16x + 2.01y$

8.
$$\begin{array}{r} 12.93 \\ +\ 0.21 \\ \hline 13.14 \end{array}$$

9.
$$\begin{array}{r} 18.80 \\ -\ 6.23 \\ \hline 12.57 \end{array}$$

10. $(-13.2) - (-7.1) = -13.2 + 7.1$
$$\begin{array}{r} -13.2 \\ +\ 7.1 \\ \hline -6.1 \end{array}$$

11.
$$\begin{array}{r} 8.24 \\ \times\ 1.2 \\ \hline 1\ 648 \\ 8\ 24 \\ \hline 9.888 \end{array}$$

12. Move the decimal point to the right 3 places.
$(4.72)(10^3) = 4720$

13. $3.5\overline{)15.75} \to 35\overline{)157.5}$
$$\begin{array}{r} 4.5 \\ 35\overline{)157.5} \\ \underline{140} \\ 17\ 5 \\ \underline{17\ 5} \\ 0 \end{array}$$
$15.75 \div 3.5 = 4.5$

14.
$$\begin{array}{r} 6.33 \\ 3\overline{)19.00} \\ \underline{18} \\ 1\ 0 \\ \underline{9} \\ 10 \\ \underline{9} \\ 1 \end{array}$$
$\frac{19}{3} = 6.\overline{3}$

15.
$$\begin{aligned} 0.5x + 0.2x &= 2.8 \\ 0.7x &= 2.8 \\ \frac{0.7x}{0.7} &= \frac{2.8}{0.7} \\ x &= 4 \end{aligned}$$

16.
$$\begin{aligned} 2(y + 1.3) &= 7.8 \\ 2y + 2.6 &= 7.8 \\ +\quad -2.6 &\ -2.6 \\ \hline 2y &= 5.2 \\ \frac{2y}{2} &= \frac{5.2}{2} \\ y &= 2.6 \end{aligned}$$

17. $\frac{7}{100} = 7\%$
7% of the computer chips are defective.

18. a. $10\% \text{ of } 504 \approx 10\% \text{ of } 500 = 50$

b.
$$\begin{aligned} 5\% \text{ of } 504 &\approx \frac{1}{2} \times (10\% \text{ of } 500) \\ &= \frac{1}{2} \times 50 \\ &= 25 \end{aligned}$$

c. $1\% \text{ of } 504 \approx 1\% \text{ of } 500 = 5$

d. $6\% \text{ of } 504 \approx 6 \times (1\% \text{ of } 500) = 6 \times 5 = 30$

e.
$$\begin{aligned} 15\% \text{ of } 504 &\approx 10\% \text{ of } 500 + 5\% \text{ of } 500 \\ &= 50 + 25 \\ &= 75 \end{aligned}$$

19. a. $\frac{76}{95} = 76 \div 95 = 0.8$

b. $\frac{76}{95} = 0.8 = 80\%$

Copyright © 2012 Pearson Education, Inc.

20. a. $0.10 = \frac{10}{100} = \frac{1}{10}$

b. $0.10 = 10\%$

21. a. $5\% = \frac{5}{100} = \frac{1}{20}$

b. $5\% = 0.05$

22. 12 is 30% of what number?

$12 = 30\% \times n$

$12 = 0.30 \times n$

$\frac{12}{0.30} = n$

$40 = n$

12 is 30% of 40.

23. 5 is what percent of 250?

$5 = n\% \times 250$

$\frac{5}{250} = n\%$

$0.02 = n\%$

$2 = n$

5 is 2% of 250.

24. What is 20% of 48?

$n = 20\% \times 48$

$n = 0.20 \times 48$

$n = 9.6$

9.6 is 20% of 48.

25.

$$\begin{array}{r} 150.00 \\ -\ 112.33 \\ \hline 37.67 \end{array}$$

After the first purchase, Jerald had \$37.67 in points.

$$\begin{array}{r} 69.99 \\ -\ 37.67 \\ \hline 32.32 \end{array}$$

Jerald had to pay \$32.32 in cash.

26. Subtract to find the number of miles over 30,000 miles.

$34{,}100.5 - 30{,}000 = 4100.5$ miles

$$\begin{aligned} \$5500 + 4100.5(\$0.12) &= \$5500 + \$492.06 \\ &= \$5992.06 \end{aligned}$$

27. 35% of what number is 49?

$35\% \times n = 49$

$0.35 \times n = 49$

$n = \frac{49}{0.35}$

$n = 140$

There are 140 employees.

28. a. Find 24% of \$3300.

$n = 24\% \times \$3300$

$n = 0.24 \times \$3300$

$n = \$792$

The price is reduced \$792.

b. Subtract \$792 from \$3300.

$\$3300 - \$792 = \$2508$

The sale price is \$2508.

29. Let x = the weeknight rate.

The weeknight rate plus a 20% increase equals the Saturday rate.

$100\%x + 20\%x = \$249$

$120\%x = \$249$

$1.20x = \$249$

$x = \frac{\$249}{1.20}$

$x = \$207.50$

Juan will pay \$207.50 for Sunday night.

30. What is 15% of \$10,500?

$x = 15\% \times 10{,}500$

$x = 0.15 \times 10{,}500$

$x = 1575$

Sylvia's commission was \$1575.

31. $P = \$8200$, $R = 11\%$, $T = 2$ yr

$I = P \times R \times T = \$8200 \times 0.11 \times 2 = \$1804$

The interest is \$1804.

32. Calculate the price reduction for each camera. Then subtract the reduction and the manufacturer's discount from the price of each.

One-Step:

Find 20% of \$350.

$n = 20\% \times \$350 = 0.20 \times 350 = \$70$

Subtract to find the reduced price.

$\$350 - \$70 = \$280$

Now subtract the manufacturer's discount.

$\$280 - \$25 = \$255$

The final price of the One-Step is \$255.

One-Touch:

Find 15% of \$350.

$n = 15\% \times \$350 = 0.15 \times \$350 = \$52.50$

Copyright © 2012 Pearson Education, Inc.

Subtract to find the reduced price.
$350 – $52.50 = $297.50
Now subtract the manufacturer's discount.
$297.50 – $50 = $247.50
The final price of the One-Step is $247.50.
The final price of the One-Touch is lower, so it is the better deal.

Copyright © 2012 Pearson Education, Inc.

Chapter 9

9.1 Exercises

1. Computer Support Specialist has the most symbols on the pictograph and thus will require the most people.

3. Since there are ten symbols beside Computer Support Specialist, and each one equals 50,000 jobs, we have $10 \times 50{,}000 = 500{,}000$ Computer Support Specialists.
Since there is one-half symbol beside Desktop Publisher and one symbol equals 50,000 jobs, we have $\frac{1}{2} \times 50{,}000 = 25{,}000$ Desktop Publishers.
$500{,}000 + 25{,}000 = 525{,}000$
525,000 Computer Support Specialists and Desktop Publishers will be needed.

5. There are eight more symbols beside Computer Support Specialist than beside Network Systems and Data Communication Analyst.
$8 \times 50{,}000 = 400{,}000$
400,000 more Computer Support Specialists will be needed than Network Systems and Data Communication Analysts.

7. Since there are 16 symbols beside Asia, and each one equals 20 people, we have
$16 \times 20 = 320$ people per square mile.

9. Europe has 20 more people per square mile than South America because there is one extra symbol beside Europe. (Each symbol equals 20 people.)

11. 32.2% of the household expenses went for housing.

13. a. Housing and transportation are the two largest expenses.

b. If we add 32.20 and 21.56, we get 53.76. Thus, housing and transportation together account for 53.76% of the household expenses.

15. $32.2\% \text{ of } \$27{,}776 = (0.322)(27{,}776) \approx \$8943.87$

17. 17.1% of sports-related ER visits were caused by basketball injuries.

19. If we add 10.4 + 16.1, we get 26.5. Thus 26.5% of the injuries were caused by football or cycling.

21. The entire circle represents 100% and basketball accounts for 17.1% of the injuries. Thus, we subtract 100 − 17.1 = 82.9% to get the percent of the injuries not caused by basketball.

23. The bar graph rises to 42.7 for Boston and the average yearly total. Thus, the average yearly snowfall for Boston is 42.7 inches.

25. The bar for the average yearly total is shorter than the bar for the season snowfall as of March for Anchorage, AK and Buffalo, NY.

27. From the double-bar graph, we see that average yearly snowfall for Great Falls, Montana is 63.3 inches and the season snowfall as of March was 31.9 inches.
63.3 − 31.9 = 31.4 in.
The difference was 31.4 inches.

29. By observation, the greatest difference in the heights of the two bars for any city occurs for Flagstaff, AZ. This difference is
109.1 − 66.5 = 42.6 inches.
There is also a large difference between the height of the bars for Great Falls, MT, but this difference is only 31.4 inches. Thus, the largest difference is for Flagstaff, AZ.

31. The bars for snowfalls as of March for Flagstaff, AZ, Anchorage, AK, and Buffalo, NY are taller than the corresponding bar for Denver, CO.

33. The lowest point on the line graph for the Lilly Cafe corresponds to June, so the Lilly Cafe had the fewest number of customers in June.

35. a. Since the dot corresponding to June on the line graph for the Bay Shore Restaurant is opposite 4.5, we know that the Bay Shore Restaurant had approximately 4500 customers in June.

b. Since the dot corresponding to June on the line graph for the Bay Shore Restaurant is higher than the dot for May, the number of customers increased from May to June.

Copyright © 2012 Pearson Education, Inc.

37. There are increases between April and May, between May and June, and between June and July, where the dot for the latter month in each pair is higher on the graph. The biggest jump occurs between June and July, so the largest increase is between June and July.

39. The highest dot on the graph for Sea-Tac corresponds to January, so the precipitation at Sea-Tac was the greatest in January.
The highest dot on the graph for the Kansas City airport corresponds to May, so the precipitation at the Kansas City airport was the greatest in May.

41. The dots for Sea-Tac are lower on the graph than the dots for Kansas City airport for the months of April, May, and June.

43. Since the dot on the Sea-Tac graph for June is opposite 1.5, we know that the average precipitation at Sea-Tac in June was 1.5 inches.

45. Draw a vertical and a horizontal line. Since we are comparing the high and low temperatures, we place the label *Temperature* on the vertical line. We label the horizontal line with the names of the cities.

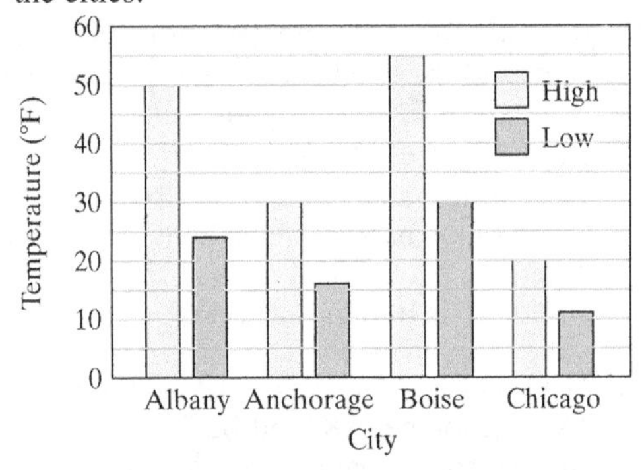

47. Draw a vertical and a horizontal line. Since we are comparing the costs of life insurance premiums for men and women, we place the label *Dollars* on the vertical line. We place the label *Age* on the horizontal.

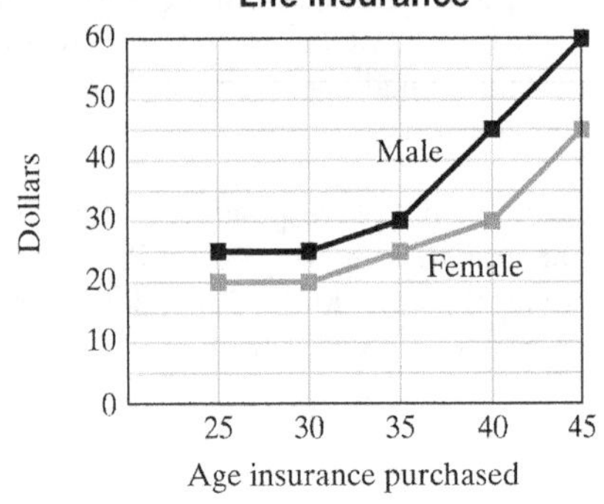

49. Draw a vertical and a horizontal line. Since we are comparing the quarterly profits for the two different years, we place the label *Profit* on the vertical line. We place the label *Quarter* on the horizontal line.

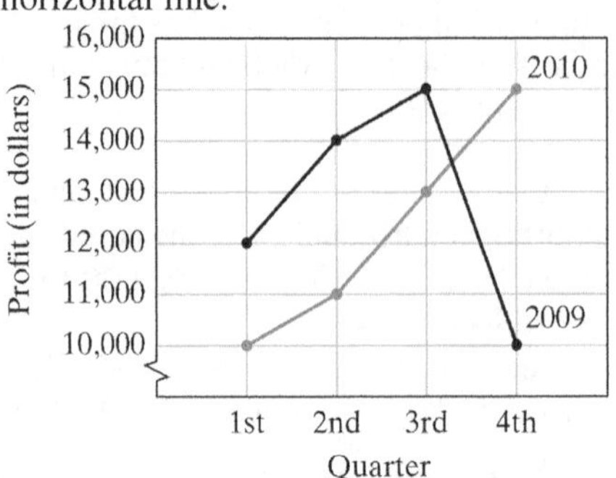

51. a. $8.9 + 17.5 + 17.5 + 30.8 = 74.7\%$
74.7% of the tickets were allocated to individual teams.

b. The host team receives 8.9% of the 100,000 available tickets.
$n = 8.9\%(100,000) = 0.089(100,000) = 8900$
The host team receives 8900 tickets.

c. The AFC and NFC teams each receive 17.5% of the available tickets.
$n = 17.5\%$ of 100,000 tickets
$= 0.175 \times 100,000$
$= 17,500$ tickets
The revenue from those tickets is
$r = 17,500 \times \$102 = \$1,785,000$
Each team receives $1,785,000.

d. The league office gives away 25% of its tickets. The number of tickets given away is
$n =$ number of league office tickets $\times 25\%$
$= 25\% \times 25.3\% \times 100,000$
$= 0.25 \times 0.253 \times 100,000$
$= 6325$ tickets
The value is $v = 6325 \times \$102 = \$645,150$.
The promotion is valued at $645,150.

Cumulative Review

52. $5x - 2(3x^2 + 1) - (-4x^2 + 6x + 3)$
$= 5x - 6x^2 - 2 + 4x^2 - 6x - 3$
$= -6x^2 + 4x^2 + 5x - 6x - 2 - 3$
$= -2x^2 - x - 5$

53. $(x - 2)(x + 4) = x^2 + 4x - 2x - 8 = x^2 + 2x - 8$

 Copyright © 2012 Pearson Education, Inc.

54. The GCF is $2x$.

$$8x^2y + 2xy + 4x = 2x \cdot 4xy + 2x \cdot y + 2x \cdot 2$$
$$= 2x(4xy + y + 2)$$

Quick Quiz 9.1

1. Draw a vertical and a horizontal line. Since we are comparing the quarterly profits for the two different years, we place the label *Profit* on the vertical line. We place the label *Quarter* on the horizontal line.

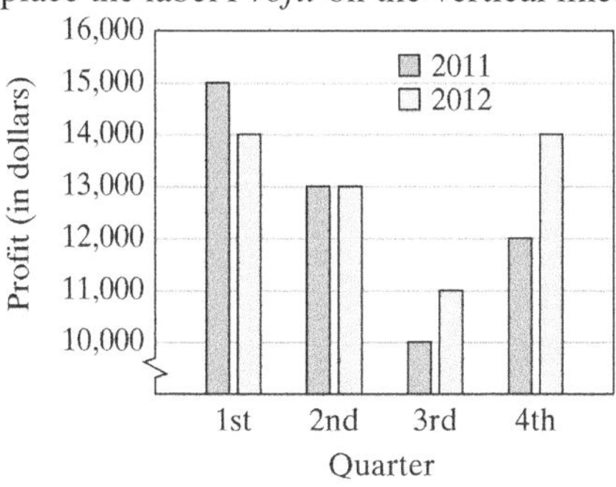

2. $8.9 + 17.5 + 17.5 + 30.8 = 74.7\%$
74.7% of the tickets were allocated to individual teams.
$100 - 74.7 = 25.3\%$
25.3% of the tickets were not allocated to individual teams. Or, reading directly from the graph, the only one that is not a team is "League Office" which is labeled 25.3%.

3. Draw a vertical and a horizontal line. Since we are comparing the numbers of male and female customers at different times of the day, we place the label *Number of Customers* on the vertical line. We place the label *Time of Day* on the horizontal line.

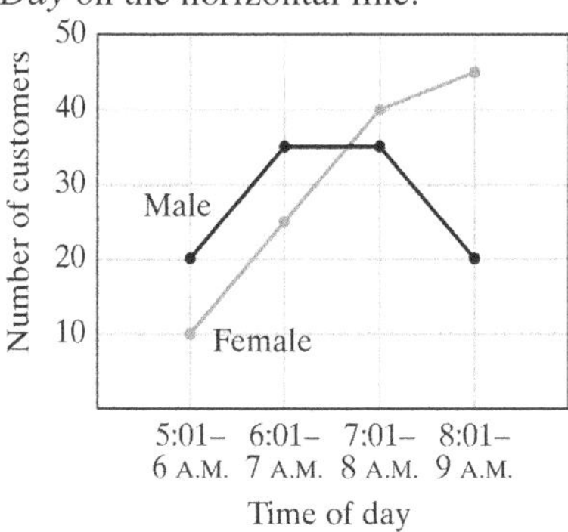

4. Answers may vary.

a. The student enrollment is
Morning (M) 45%
Weekend (W) 10%
Afternoon (A) $2 \times$ weekend $= 2 \times 10\% = 20\%$
Evening (E) The rest of the students
Total enrollment $= 100\% = M + W + A + E$

$$\begin{aligned} 100\% &= 45\% + 10\% + 20\% + E \\ 100\% &= 75\% + E \\ \underline{-75\%} & \quad \underline{-75\%} \\ 25\% &= E \end{aligned}$$

Evening enrollment is 25% of the students.

Copyright © 2012 Pearson Education, Inc.

b. 25% is $\frac{1}{4}$ of the circle.

9.2 Exercises

1. The mean (average) of a set of values is the sum of the values divided by the number of values.

3. $\frac{84+91+86+95+98}{5} = \frac{454}{5} = 90.8$

5. $\frac{2+3+3+4+2.5+2.5+4}{7} = \frac{21}{7} = 3$ hr

7. $\frac{23+45+63+34+21+42}{6} = \frac{228}{6} = 38$

9. $\frac{189,000+185,000+162,000+145,000+162,000}{5} = \frac{848,000}{5} = \$168,600$

11. $\frac{6+5+4+5+8}{5} = \frac{28}{5} = 5.6$ sales

13. $\frac{0+2+3+2+2}{5+4+6+5+4} = \frac{9}{24} = 0.375$

15. $\frac{276+350+391+336}{12+14+17+14} = \frac{1353}{57} \approx 23.7$ mi/gal

17. 22, 42, 45, 47, 50, 51, 58
↑
median = 47

19. 865, 968, 999, 1023, 1052, 1152
↑
median $= \frac{999+1023}{2} = \frac{2022}{2} = 1011$

21. 0.34, 0.52, 0.58, 0.69, 0.71
↑
median = 0.58

23. 1.1, 2.8, 2.9, 3.4, 3.7, 3.9
↑
median $= \frac{2.9+3.4}{2} = \frac{6.3}{2} = 3.15$

Copyright © 2012 Pearson Education, Inc.

25. \$11,600
\$15,700
\$17,000 ←
\$23,500
\$26,700
\$31,500

$$\text{median} = \frac{17,000+23,500}{2} = \frac{40,500}{2} = \$20,250$$

27. 12, 20, 24, 26, 31, 40, 62, 108
↑

$$\text{median} = \frac{26+31}{2} = \frac{57}{2} = 28.5 \text{ min}$$

29. \$5.99
\$7.99
\$9.99 ←
\$11.99
\$13.99
\$17.99

$$\text{median} = \frac{9.99+11.99}{2} = \frac{21.98}{2} = \$10.99$$

31. 22, 36, 36, 37, 44, 48, 53, 60, 64, 71
↑

$$\text{median} = \frac{44+48}{2} = \frac{92}{2} = 46 \text{ actors}$$

33. The mode of 60, 65, 68, 60, 72, 59, and 80 is 60 since it occurs twice in the set of data.

35. The data 121, 150, 117, 150, 121, 180, 127, and 123 are bimodal since both 121 and 150 occur twice.

37. The mode of \$249, \$649, \$439, \$259, \$269, and \$249 is \$249 since it occurs twice in the set of data.

39. 21, 82, 42, 55, 42, 45, 49

a. $$\frac{21+82+42+55+42+45+49}{7} = \frac{336}{7} = 48$$

b. 21, 42, 42, 45, 49, 55, 82
↑
median = 45

c. The mode is 42 since it occurs twice in the set of data.

41. 2.7, 7.1, 6.9, 7.5, 6.1

a. $$\frac{2.7+7.1+6.9+7.5+6.1}{5} = \frac{30.3}{5} = 6.06$$

b. 2.7, 6.1, 6.9, 7.1, 7.5
↑
median = 6.9

c. none

43. 97, 81, 92, 73, 86, 81

a. $$\frac{97+81+92+73+86+81}{6} = \frac{510}{6} = 85$$

b. 73, 81, 81, 86, 92, 97
↑

$$\text{median} = \frac{81+86}{2} = \frac{167}{2} = 83.5$$

c. The mode is 81 since it occurs twice in the set of data.

45. a. $$\frac{1350+1600+2400+2800+3200}{5} = \frac{11,350}{5} = \$2270$$

b. 1350, 1600, 2400, 2800, 3200
↑
median = \$2400

c. There is no mode since no salary occurs more than once.

47. a. 1500 + 1700 + 1650 + 1300 + 1440 + 1580 + 1820 + 1380 + 2900 + 6300 = 21,570
21,570 ÷ 10 = \$2157

b. 1300
1380
1440
1500
1580 ←
1650
1700
1820
2900
6300

$$\text{median} = \frac{1580+1650}{2} = \frac{3230}{2} = \$1615$$

Copyright © 2012 Pearson Education, Inc.

c. Median, because mean is affected by the high salary of \$6300.

49. 1987, 2576, 3700, 4700, 5000, 7200, 8764, 9365
↑

$$\text{median} = \frac{4700+5000}{2} = \frac{9700}{2} = 4850$$

Cumulative Review

51. Replace x with 26.
$\frac{x}{2}+4 = \frac{26}{2}+4 = 13+4 = 17$

52. Replace x with 7.
$\frac{35}{x}-9 = \frac{35}{7}-9 = 5-9 = -4$

53. Replace x with 5.
$2x + 1 = 2(5) + 1 = 10 + 1 = 11$

54. Replace x with 0.
$3x - 7 = 3(0) - 7 = 0 - 7 = -7$

55. $\frac{3}{8}\cdot 9200 = \$3450$

56. $20,320-(-131.2) = 20,320+131.2$
$= 20,451.2 \text{ ft}$

Quick Quiz 9.2

1. $\frac{15+13+14+11+13+12}{6} = \frac{78}{6} = 13$

2. a. 3, 5, 9, 11, 14, 17, 19
↑
median = 11

b. 20, 35, 40, 50, 65, 75
↑

$$\text{median} = \frac{40+50}{2} = \frac{90}{2} = 45$$

3. a. The mode of 6, 9, 5, 7, 6, and 4 is 6 since it occurs twice.

b. The data 5, 6, 4, 3, 5, 7, and 4 are bimodal since both 4 and 5 occur twice.

4. Answers may vary. One possible solution follows:
First, list the number of inquiries in order.
297, 778, 801, 887, 887, 926, 926
The values fall into three "groups." One value is approximately 300, two values near 800, and four near 900. Based on these "groups," we can expect values in the mid to high 800's.
The mean may not be the best estimate because there is a very low value (297) that will affect the mean more than the other estimates, median or mode.
The median is less sensitive to high and low values. In this case, the median is 887, which agrees with the expectation of values in the mid to high 800's.
There are two modes, 887 and 926. The mode of 887 agrees with the median and our expectation of mid to high 800's. However, the mode at 926 does not. Thus, the best estimate of the number of inquiries is the median.

How Am I Doing? Sections 9.1–9.2

1. 26% of the city is used for business.

2. Homes take up the largest percentage of the land.

3. Together homes and condominiums take up 35 + 19 = 54% of the land.
$54\% \text{ of } 70 = 0.54(70) = 37.8 \text{ mi}^2$
37.8 square miles have homes and condominiums.

4. The bar graph rises to 4 inches for February and 2011, so the monthly rainfall was 4 inches in February 2011.

5. 5 − 3 = 2 in.
The difference was 2 inches.

6. The shortest bar corresponds to January and 2011, so January 2011 had the least rainfall.

7. $\frac{4+3+3+5}{4} = \frac{15}{4} = 3.75 \text{ in.}$

8. 7, 9, 14, 19, 25, 28, 32
↑
median = 19

9. 2, 5, 9, 13, 18, 23
↑

$$\text{median} = \frac{9+13}{2} = \frac{22}{2} = 11$$

Copyright © 2012 Pearson Education, Inc.

10. The mode of 79, 85, 81, 83, and 85 is 85 since it occurs twice.

11. The data 8, 5, 7, 8, 4, 9, and 7 are bimodal since both 7 and 8 occur twice.

12. 2.1, 9.2, 9.6, 8.7

Mean: $\frac{2.1+9.2+9.6+8.7}{4}=\frac{29.6}{4}=7.4$

Median: 2.1, 8.7, 9.2, 9.6
↑

$\text{median}=\frac{8.7+9.2}{2}=\frac{17.9}{2}=8.95$

Mode: There is no mode since no item occurs more than once in the set of data.

9.3 Exercises

1. Starting at the origin, move 2 units right followed by 1 unit down.

3. (year, number of students)
↓ ↓
(2003, 4700)
(2004, 4800)
(2005, 4950)
(2006, 4850)

5. (shoe size, height)
↓ ↓
Lena: (4, 60)
Janie: (7, 62)
Mark: (9, 66)

7. To plot (−2, 2), or $x = -2$, $y = 2$, we start at the origin and move 2 units in the negative x-direction followed by 2 units in the positive y-direction.

9. To plot (−1, 4), or $x = -1$, $y = 4$, we start at the origin and move 1 unit in the negative x-direction followed by 4 units in the positive y-direction.

11. To plot (3, −2), or $x = 3$, $y = -2$, we start at the origin and move 3 units in the positive x-direction followed by 2 units in the negative y-direction.

13. To plot (−2, −4), or $x = -2$, $y = -4$, we start at the origin and move 2 units in the negative x-direction followed by 4 units in the negative y-direction.

7–13.

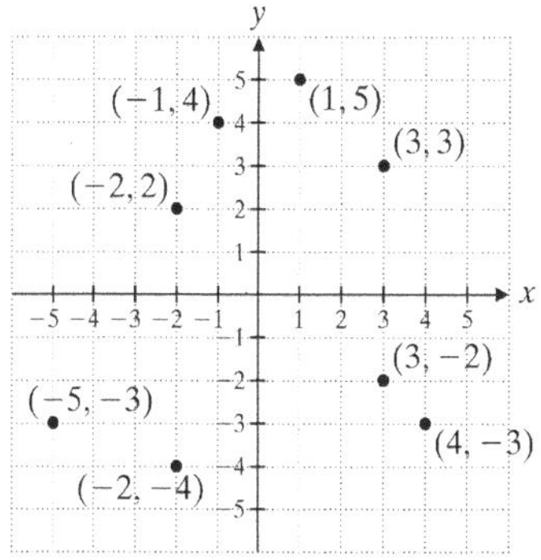

15. To plot $\left(4\frac{1}{2}, 3\right)$, or $x=4\frac{1}{2}$, $y = 3$, we start at the origin and move $4\frac{1}{2}$ units in the positive x-direction followed by 3 units in the positive y-direction.

17. To plot $\left(-1\frac{1}{2}, -4\right)$, or $x=-1\frac{1}{2}$, $y = -4$, we start at the origin and move $1\frac{1}{2}$ units in the negative x-direction followed by 4 units in the negative y-direction.

19. To plot $\left(2\frac{1}{2}, -1\right)$, or $x=2\frac{1}{2}$, $y = -1$, we start at the origin and move $2\frac{1}{2}$ units in the positive x-direction followed by 1 unit in the negative y-direction.

21. To plot $\left(-3\frac{1}{2}, 2\right)$, or $x=-3\frac{1}{2}$, $y = 2$, we start at the origin and move $3\frac{1}{2}$ units in the negative x-direction followed by 2 units in the positive y-direction.

15–21.

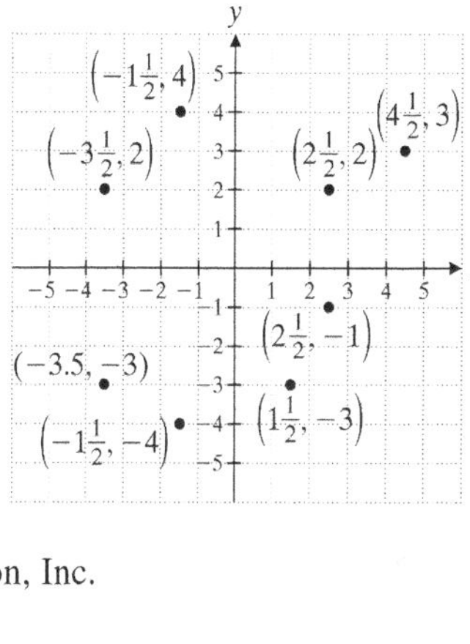

Copyright © 2012 Pearson Education, Inc.

23. $K = (-3, -3)$

25. $M = (3, 3)$

27. $O = (-1, 4)$

29. $Q = \left(-3\frac{1}{2}, 4\right)$

31. $S = \left(1\frac{1}{2}, -3\right)$

33. $U = (5, -1)$

35. a. $(-4, -2)$

b.

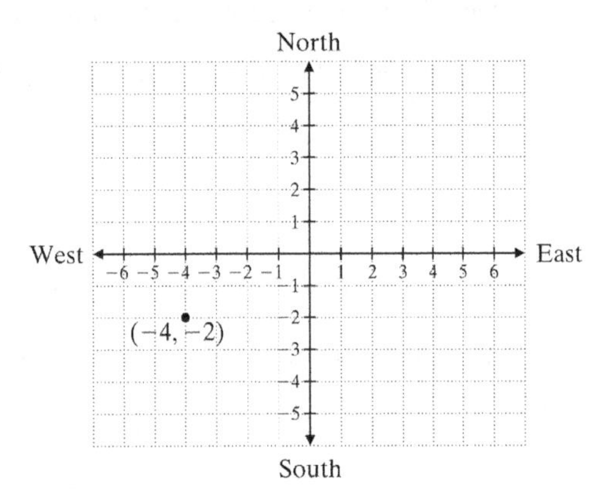

37. a. $(2, 1)$

b.

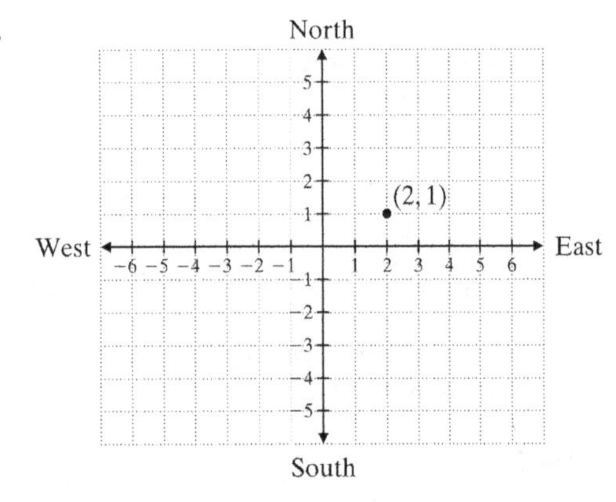

39. $(2, 4), (2, -1), (2, -3), (2, 0)$

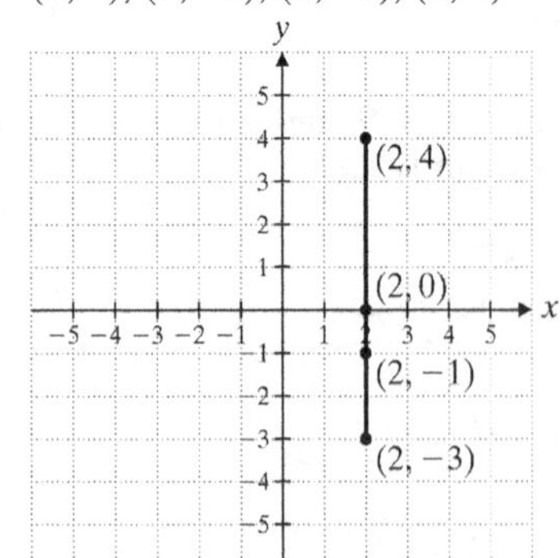

41. $(1, 3), (-5, 3), (0, 3), (-2, 3)$

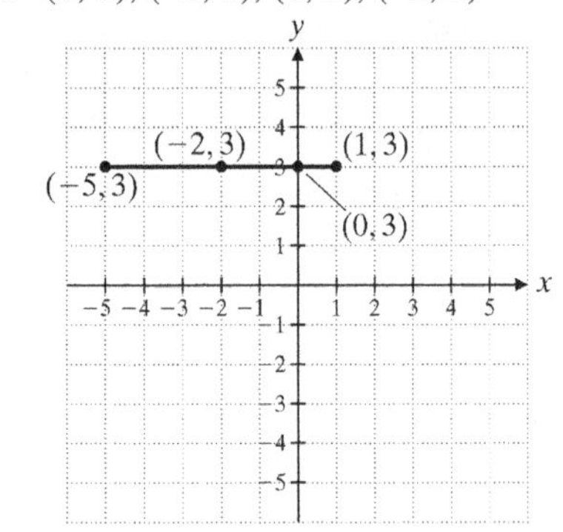

43. $(0, -1), (-2, -1), (-4, -1), (4, -1)$

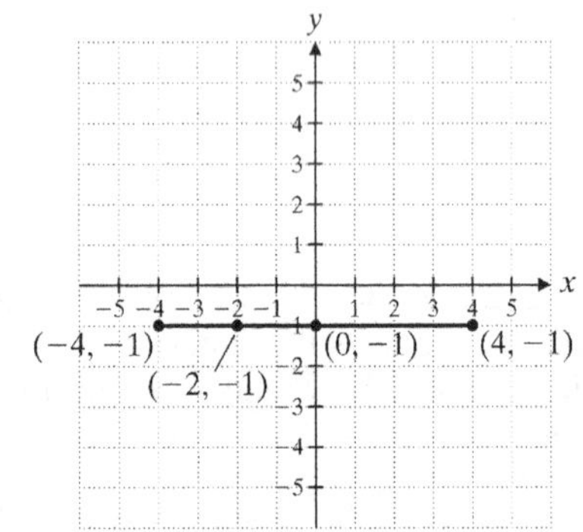

45. The figure is a rectangle.

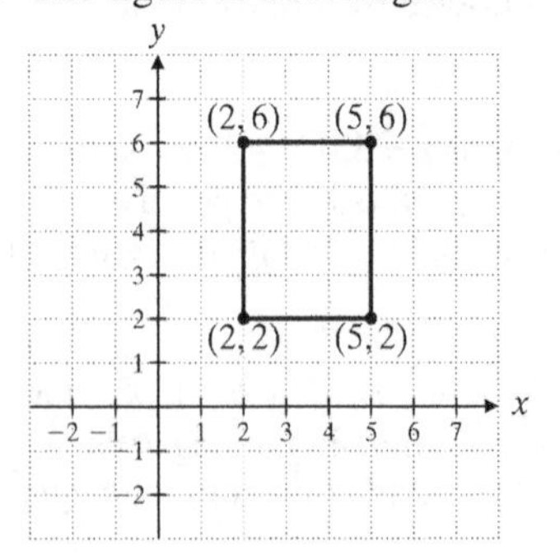

Copyright © 2012 Pearson Education, Inc.

Cumulative Review

47. We multiply.
$39(150 \text{ million km}) = 5850 \text{ million km}$

48. $\frac{20x^6}{55x^{10}} = \frac{20}{55}x^{6-10} = \frac{4}{11}x^{-4} = \frac{4}{11x^4}$

49. $(2^3)^2 = 2^{3\cdot 2} = 2^6$

50. $(2x)(y^4)(3x^2)(2y^6) = 2\cdot 3\cdot 2\cdot x^{1+2}\cdot y^{4+6}$
$= 12x^3y^{10}$

Quick Quiz 9.3

1. a. To plot $(-1, 4)$, or $x = -1$, $y = 4$, we start at the origin and move 1 unit in the negative x-direction followed by 4 units in the positive y-direction.

b. To plot $\left(2\frac{1}{2}, -3\right)$, or $x = 2\frac{1}{2}$, $y = -3$, we start at the origin and move $2\frac{1}{2}$ units in the positive x-direction followed by 3 units in the negative y-direction.

c. To plot $(-2, -4)$, or $x = -2$, $y = -4$, we start at the origin and move 2 units in the negative x-direction followed by 4 units in the negative y-direction.

d. To plot $(0, 1)$, or $x = 0$, $y = 1$, we start at the origin and move 0 units in the x-direction followed by 1 unit in the positive y-direction.

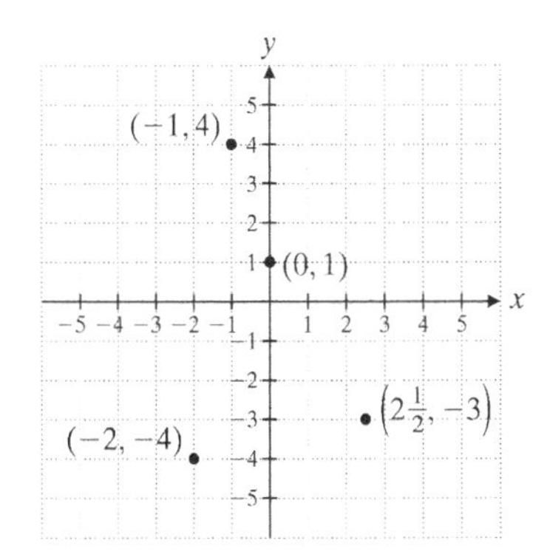

2. a. $S = (-5, 2)$

b. $T = (1, 1)$

c. $U = (-3, -3)$

d. $V = (0, -4)$

3.

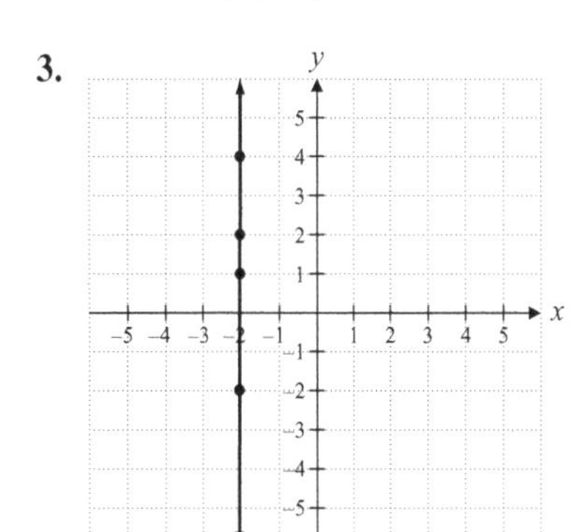

4. Answers may vary. One possible solution follows:
Line M passes through the points (k, b), (n, b) and (a, b). Since all of the y-values are the same, yet the line, M, is parallel to the x-axis b units away from it.

9.4 Exercises

1. No, because substituting $x = 2$ and $y = 5$ into $x + 2y = 4$ does not yield a true statement.

3. $x + y = 4$
For each ordered pair, we can choose any two numbers whose sum is 4.
$0 + 4 = 4$: $(0, 4)$
$4 + 0 = 4$: $(4, 0)$
$1 + 3 = 4$: $(1, 3)$
$-1 + 5 = 4$: $(-1, 5)$
Answers may vary.

5. $x + y = 12$
For each ordered pair, we can choose any two numbers whose sum is 12.
$0 + 12 = 12$: $(0, 12)$
$12 + 0 = 12$: $(12, 0)$
$1 + 11 = 12$: $(1, 11)$
$-1 + 13 = 12$: $(-1, 13)$
Answers may vary.

7. a. $y = 35x$
We replace y with 140 and solve for x.
$140 = 35x$
$\frac{140}{35} = \frac{35x}{35}$
$4 = x$
It takes 4 minutes to label 140 bottles. The ordered pair is $(4, 140)$.

b. $y = 35x$
We replace y with 280 and solve for x.
$$280 = 35x$$
$$\frac{280}{35} = \frac{35x}{35}$$
$$8 = x$$
It takes 8 minutes to label 280 bottles. The ordered pair is (8, 280).

9. a. $y = 80x$
We replace y with 240 and solve for x.
$$240 = 80x$$
$$\frac{240}{80} = \frac{80x}{80}$$
$$3 = x$$
It takes 3 minutes to type 240 words. The ordered pair is (3, 240).

b. $y = 80x$
We replace y with 400 and solve for x.
$$400 = 80x$$
$$\frac{400}{80} = \frac{80x}{80}$$
$$5 = x$$
It takes 5 minutes to type 400 words. The ordered pair is (5, 400).

11. $x + 2y = 16$

(0, _):
$$0 + 2y = 16$$
$$2y = 16$$
$$y = 8$$

(_, 0)
$$x + 2(0) = 16$$
$$x + 0 = 16$$
$$x = 16$$

(_, 4)
$$x + 2(4) = 16$$
$$x + 8 = 16$$
$$x = 8$$

(x, y)
(0, 8)
(16, 0)
(8, 4)

13. $x + y = 5$

(_, 2):
$$x + 2 = 5$$
$$x = 3$$

(0, _):
$$0 + y = 5$$
$$y = 5$$

(1, _):
$$1 + y = 5$$
$$y = 4$$

(x, y)
(3, 2)
(0, 5)
(1, 4)

15. $y = x + 2$

(−1, _):
$$y = -1 + 2$$
$$y = 1$$

(_, 3):
$$3 = x + 2$$
$$1 = x$$

(_, 0):
$$0 = x + 2$$
$$-2 = x$$

(x, y)
(−1, 1)
(1, 3)
(−2, 0)

17. $y = 5x + 3$

(0, _):
$$y = 5(0) + 3$$
$$y = 0 + 3$$
$$y = 3$$

(−1, _):
$$y = 5(-1) + 3$$
$$y = -5 + 3$$
$$y = -2$$

(1, _)
$$y = 5(1) + 3$$
$$y = 5 + 3$$
$$y = 8$$

(x, y)
(0, 3)
(−1, −2)
(1, 8)

19. $y = 5x - 3$

We choose three values for x: 0, $\frac{3}{5}$, and 1.

$$y = 5(0) - 3$$
$$y = 0 - 3$$
$$y = -3$$

$$y = 5\left(\frac{3}{5}\right) - 3$$
$$y = 3 - 3$$
$$y = 0$$

$$y = 5(1) - 3$$
$$y = 5 - 3$$
$$y = 2$$

(x, y)
(0, −3)
$\left(\frac{3}{5}, 0\right)$
(1, 2)

Answers may vary.

21. $y = x + 6$
We choose three values for x: 0, −6, and 1.

$$y = 0 + 6$$
$$y = 6$$

$$y = -6 + 6$$
$$y = 0$$

$$y = 1 + 6$$
$$y = 7$$

(x, y)
(0, 6)
(−6, 0)
(1, 7)

Answers may vary.

Copyright © 2012 Pearson Education, Inc.

23. $y = 2x + 2$

We choose three values for x: −2, −1, and 0.

$y = 2(-2) + 2$ $y = -4 + 2$ $y = -2$

$y = 2(-1) + 2$ $y = -2 + 2$ $y = 0$

$y = 2(0) + 2$ $y = 0 + 2$ $y = 2$

(x, y)
(−2, −2)
(−1, 0)
(0, 2)

Plot these ordered pairs and draw a straight line through the points.

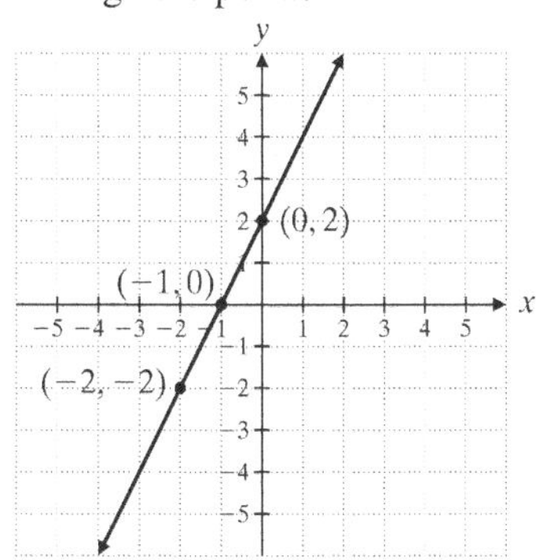

25. $y = -3x + 1$

We choose three values for x: −1, 0, and 1.

$y = -3(-1) + 1$ $y = 3 + 1$ $y = 4$

$y = -3(0) + 1$ $y = 0 + 1$ $y = 1$

$y = -3(1) + 1$ $y = -3 + 1$ $y = -2$

(x, y)
(−1, 4)
(0, 1)
(1, −2)

Plot these ordered pairs and draw a straight line through the points.

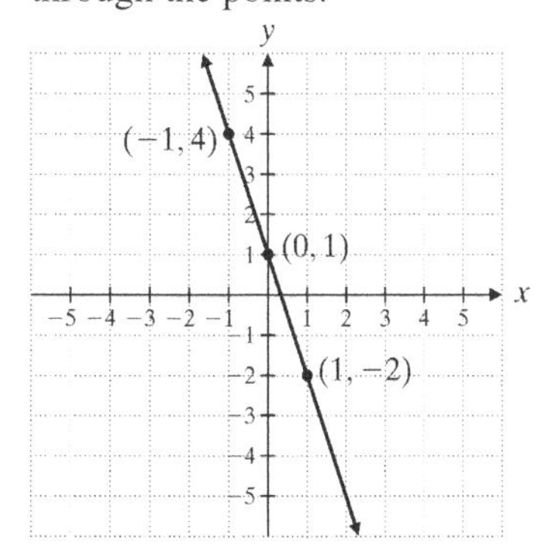

27. $y = 5x - 4$

We choose three values for x: 0, 1, and 2.

$y = 5(0) - 4$ $y = 0 - 4$ $y = -4$

$y = 5(1) - 4$ $y = 5 - 4$ $y = 1$

$y = 5(2) - 4$ $y = 10 - 4$ $y = 6$

(x, y)
(0, −4)
(1, 1)
(2, 6)

Plot these ordered pairs and draw a straight line through the points.

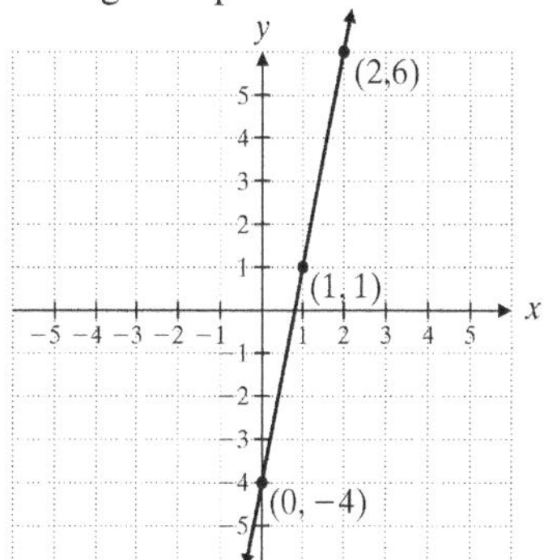

29. $y = 3x - 2$

We choose three values for x: −1, 0, and 1.

$y = 3(-1) - 2$ $y = -3 - 2$ $y = -5$

$y = 3(0) - 2$ $y = 0 - 2$ $y = -2$

$y = 3(1) - 2$ $y = 3 - 2$ $y = 1$

(x, y)
(−1, −5)
(0, −2)
(1, 1)

Plot these ordered pairs and draw a straight line through the points.

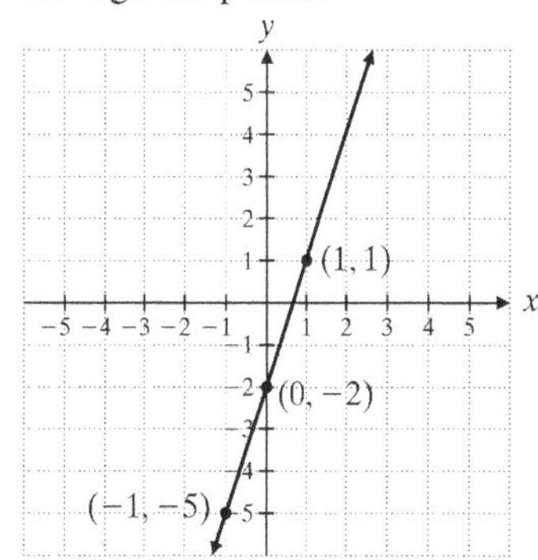

Copyright © 2012 Pearson Education, Inc.

31. $y = -5x - 7$

We choose three values for x: -2, -1, and $-\frac{3}{2}$.

$y = -5(-2) - 7$
$y = 10 - 7$
$y = 3$

$y = -5(-1) - 7$
$y = 5 - 7$
$y = -2$

$y = -5\left(-\frac{3}{2}\right) - 7$
$y = \frac{15}{2} - \frac{14}{2}$
$y = \frac{1}{2}$

(x, y)
$(-2, 3)$
$(-1, -2)$
$\left(-\frac{3}{2}, \frac{1}{2}\right)$

Plot these ordered pairs and draw a straight line through the points.

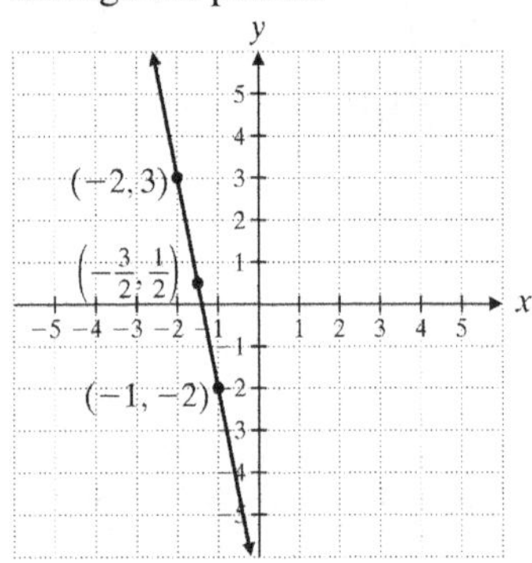

33. $y = 3$
The x-coordinate can be any number as long as y is 3.
$(-4, 3)$, $(0, 3)$, $(5, 3)$

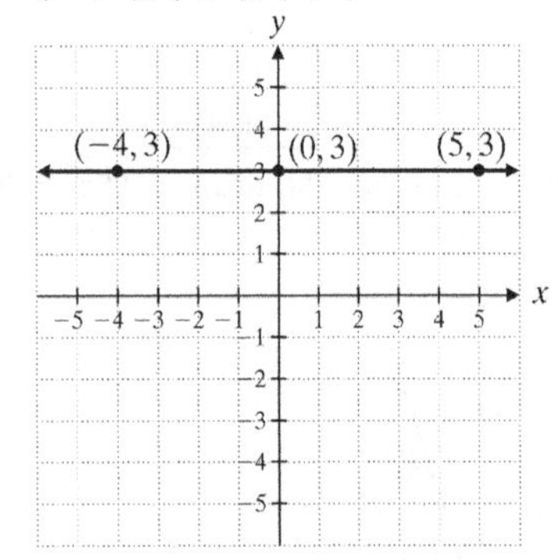

35. $x = -2$
The y-coordinate can be any number as long as x is -2.
$(-2, -3)$, $(-2, 0)$, $(-2, 4)$

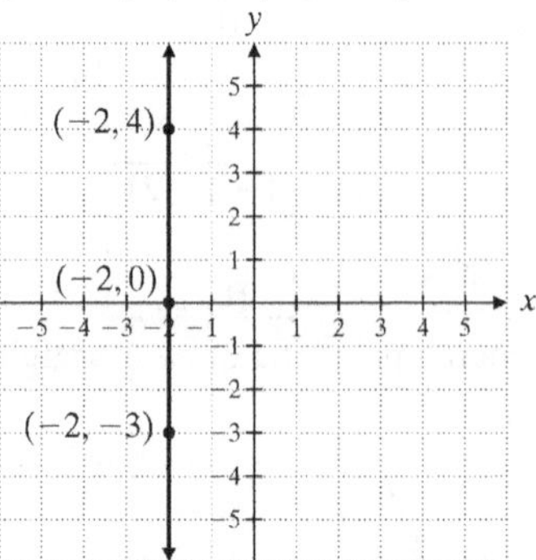

37. $(6, 0)$ is not a solution because it is not on the line formed by the other three points.

39. Any ordered pair that is on the line formed by the points on the graph is a solution. For example, $(2, 2)$ and $(3, 1)$ are solutions.

Cumulative Review

41.
$$\begin{aligned} 2x + 3 &= 10 \\ + \quad -3 &\quad -3 \\ \hline 2x &= 7 \\ \frac{2x}{2} &= \frac{7}{2} \\ x &= \frac{7}{2} \end{aligned}$$

42.
$$\begin{aligned} 5x + 2 &= 11 \\ + \quad -2 &\quad -2 \\ \hline 5x &= 9 \\ \frac{5x}{5} &= \frac{9}{5} \\ x &= \frac{9}{5} \end{aligned}$$

43.
$$\begin{aligned} -3x - 5 &= 14 \\ + \quad 5 &\quad 5 \\ \hline -3x &= 19 \\ \frac{-3x}{-3} &= \frac{19}{-3} \\ x &= -\frac{19}{3} \end{aligned}$$

Copyright © 2012 Pearson Education, Inc.

44.
$$\begin{array}{r} -2x-3=12 \\ +\quad 3\quad 3 \\ \hline -2x=15 \\ \frac{-2x}{-2}=\frac{15}{-2} \\ x=-\frac{15}{2} \end{array}$$

Quick Quiz 9.4

1. $3x + y = 15$

$(-1, \_)$:
$$3(-1)+y=15$$
$$-3+y=15$$
$$y=18$$

$(\_, 0)$
$$3x+0=15$$
$$3x=15$$
$$x=5$$

$(\_, 3)$:
$$3x+3=15$$
$$3x=12$$
$$x=4$$

(x, y)
(−1, 18)
(5, 0)
(4, 3)

2. $y = -2x + 4$

We choose three values for x: 0, 1, and 2.

$y = -2(0) + 4$
$y = 0 + 4$
$y = 4$

$y = -2(1) + 4$
$y = -2 + 4$
$y = 2$

$y = -2(2)+4$
$y = -4+4$
$y = 0$

(x, y)
(0, 4)
(1, 2)
(2, 0)

Plot these ordered pairs and draw a straight line through the points.

3. $y = 4$

The x-coordinate can be any number as long as y is 4.

(−2, 4), (0, 4), (1, 4)

2–3.

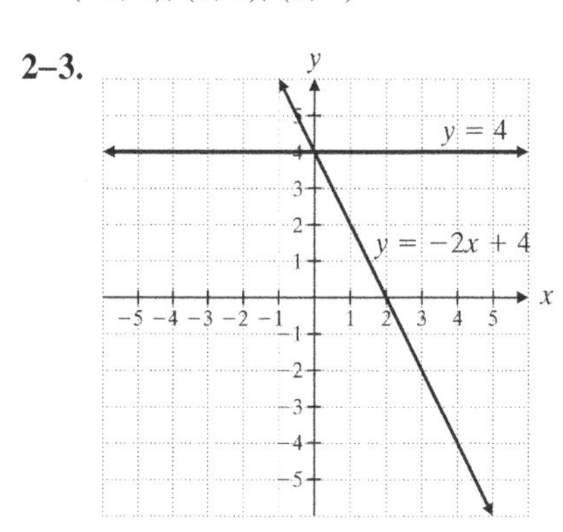

4. Answers may vary. One possible solution follows:

If a point lies on a graphed line, it will be a solution to the equation of that line. Are Mark's answers solutions to the equation $y = 2x - 1$?

(−3, 4)
$$y \stackrel{?}{=} 2x-1$$
$$4 \stackrel{?}{=} 2(-3)-1$$
$$4 \stackrel{?}{=} -6-1$$
$$4 \neq -7$$

No, the point (−3, 4) does not lie on the line.

(0, −1)
$$y \stackrel{?}{=} 2x-1$$
$$-1 \stackrel{?}{=} 2(0)-1$$
$$-1 \stackrel{?}{=} 0-1$$
$$-1 = -1$$

Yes, the point (0, −1) lies on the line.

So, Mark's answer of the two points is not correct.

Use Math to Save Money

1. Since each point costs \$2000, multiply \$2000 by 2 to find the cost of two points.

$2 \times \$2000 = \$4000$

It will cost \$4000 to pay two points.

2. $\$1135.58 - \$1073.64 = \$61.94$

They will save \$61.94 each month if they get the 5% loan instead of the 5.5% loan.

3. $\frac{\$4000}{\$61.94} \approx 64.58$

It will take about 65 months to reach the break-even point.

4. Since six years is 72 months and 65 months is the break-even point, they should pay the two points if they expect to stay in the house for six years.

5. Since three years is 36 months and they won't break even until 65 months, they should not pay the points if they expect to stay in the house for three years.

6. There are 360 months in 30 years. Multiply the monthly savings by 360.

$360 \times \$61.94 \approx \$22,300$

Now subtract the cost of the points.

$\$22,300 - \$4000 = \$18,300$

Mike and Sue would save \$18,300 over 30 years.

Copyright © 2012 Pearson Education, Inc.

You Try It

1. There are six barrel symbols corresponding to 2009 and each symbol represents 500 barrels, so 3000 barrels were produced in 2009. There are four barrel symbols corresponding to 2010 and each symbol represents 500 barrels, so 2000 barrels were produced in 2010. There are two barrel symbols corresponding to 2011 and each symbol represents 500 barrels, so 1000 barrels were produced in 2011.

2. a. The sector for age 32–50 is labeled 48%, so 48% of the police force is between 32 and 50 years old.

b. The sector for under age 23 is labeled 10%.
10% of 200 = (0.10)(200) = 20
Twenty men and women in the police force are under 23 years old.

3. a. The bar for West Coast and 2010 extends to 2 on the vertical axis, so 2000 flat-screen television sets were sold on the West Coast in 2010.

b. 7000 sets were sold in 2010; 5000 sets were sold in 2011.
7000 − 5000 = 2000
2000 more flat-screen television sets were sold in 2010 than were sold in 2011 in the Midwest.

4. a. The point on the graph for 2010 above September corresponds to 4 on the vertical axis, so 4000 visitors came to the park in September 2010.

b. The points on the graph for 2009 are lower than the points for 2010 in July and August, so there were fewer visitors in 2009 than in 2010 during July and August.

c. There are three increases shown on the graph, July 2009 to August 2009 (an increase of 2000 visitors), August 2009 to September 2009 (an increase of 1000 visitors), and July 2010 to August 2010 (an increase of 1000 visitors). The sharpest increase in attendance took place between July 2009 and August 2009.

5. Draw a vertical and a horizontal line. Since we are comparing souvenir sales for the two different weeks, we graph *Sales* on the vertical line. We graph *Souvenir* on the horizontal line.

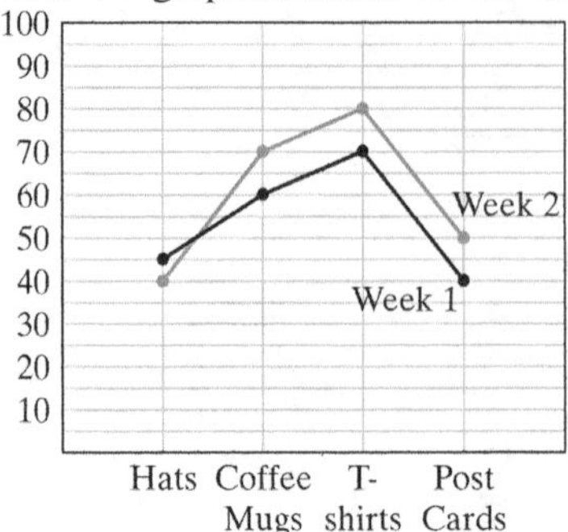

6. Draw a vertical and a horizontal line. Since we are comparing souvenir sales for the two different weeks, we graph *Sales* on the vertical line. We graph *Souvenir* on the horizontal line.

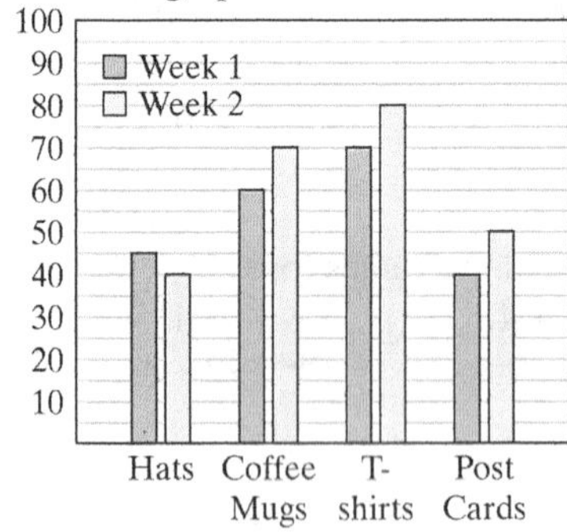

7. $\frac{12+7+9+13+11+8}{6} = \frac{60}{6} = 10$
The mean is 10.

8. a. 5, 8, 12, 14, 16, 23, 25
↑ median (14)
The median is 14.

b. 5, 8, 12, 14, 16, 23
↑ median (between 12 and 14)
$\text{median} = \frac{12+14}{2} = 13$

9. The data is bimodal since both 2 and 6 occur twice.

Copyright © 2012 Pearson Education, Inc.

10. To plot (3, −2), or $x = 3$, $y = -2$, we start at the origin and move 3 units in the positive x-direction followed by 2 units in the negative y-direction.

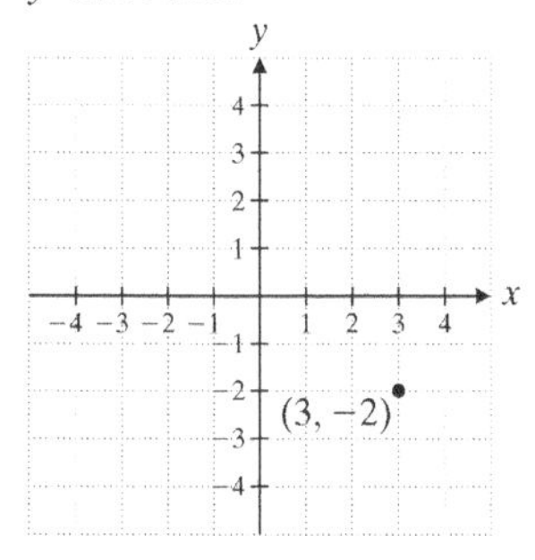

11. $2x - 3y = 6$

We choose three values for x: −3, 0, 3.

$$\begin{aligned} 2(-3) - 3y &= 6 \\ -6 - 3y &= 6 \\ -3y &= 12 \\ y &= -4 \end{aligned} \qquad \begin{aligned} 2(0) - 3y &= 6 \\ 0 - 3y &= 6 \\ -3y &= 6 \\ y &= -2 \end{aligned}$$

$$\begin{aligned} 2(3) - 3y &= 6 \\ 6 - 3y &= 6 \\ -3y &= 0 \\ y &= 0 \end{aligned}$$

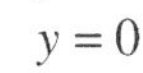
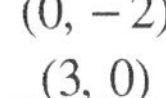

(x, y)
(−3, −4)
(0, −2)
(3, 0)

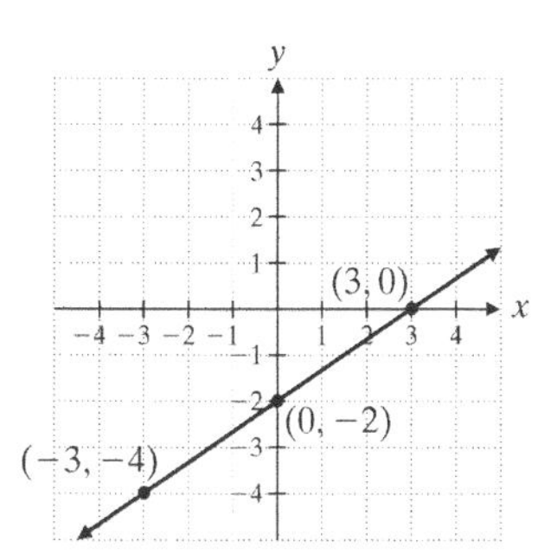

Chapter 9 Review Problems

1. Since there are seven symbols beside prealgebra and each symbol represents 50 students, $7 \times 50 = 350$ students enrolled in prealgebra.

2. Since there are six symbols beside calculus and each symbol represents 50 students, $6 \times 50 = 300$ students enrolled in calculus.

3. There are three more symbols beside algebra than there are beside calculus.
$3 \times 50 = 150$
There are 150 more students enrolled in algebra than calculus.

4. There are 16 total symbols beside prealgebra and algebra.
$16 \times 50 = 800$
The combined total enrollment in prealgebra and algebra is 800.

5. 15% of the budget is allotted for transportation.

6. 4% of the budget is allotted for savings.

7. If we add 27 and 30, we get 57. Thus, 57% of the budget is used up by food and rent.

8. If we add 15, 8, and 4, we get 27. Thus, 27% of the budget is used up by transportation, utilities/Internet, and savings.

9. 8% of \$4400 = $0.08 \times \$4400 = \$352$
\$352 per month is budgeted for utilities/Internet.

10. 15% of \$4400 = $0.15 \times \$4400 = \$660$
\$660 per month is budgeted for transportation.

11. 30% + 4% = 34%
34% of \$4400 = $0.34 \times \$4400 = \$1496$
\$1496 per month is budgeted for rent and savings.

12. 15% + 27% = 42%
42% of \$4400 = $0.42 \times \$4400 = \$1848$
\$1848 per month is budgeted for transportation and food.

13. a. Of the bars for 2006, the bar for China is the tallest. Thus, China had the largest population in 2006.

b. Of the bars for 2050, the bar for India is the tallest. Thus, India is predicted to have the largest population in 2050.

14. a. Of the bars for 2006, the bar for Brazil is the shortest. Thus, Brazil had the smallest population in 2006 among the countries represented.

b. Of the bars for 2050, the bar for Brazil is the shortest. Thus, Brazil is predicted to have the smallest population in 2050 among the countries represented.

15. Among the bars for 2050, only the bars for China and India are taller than the bar for the United States, so the United States is predicted to have the third largest population in 2050.

Copyright © 2012 Pearson Education, Inc.

16. Among the bars for 2050, the bars for China, India, and the United States are taller than the bar for Indonesia, so Indonesia is predicted to have the fourth largest population in 2050.

17. 1.4 billion − 1.3 billion = 0.1 billion
The population of China is predicted to be 0.1 billion more in 2050 than in 2006.

18. 0.420 billion − 0.299 billion = 0.121 billion
The population of the United States is predicted to be 0.121 billion more in 2050 than in 2006.

19. By observation, the greatest difference in the heights of the bars for 2006 and 2050 for any country occurs for India. Thus, India is expected to have the largest increase in population.

20. 0.285 billion − 0.225 billion = 0.06 billion
The population of Indonesia is predicted to be 0.06 billion more in 2050 than in 2006.

21. a. Since the dot corresponding to air conditioning on the private sale line graph is opposite 400, we know that the value of an air conditioner on a private sale was \$400.

b. Since the dot corresponding to air conditioning on the trade-in line graph is opposite 350, we know that the value of an air conditioner on a trade-in was \$350.

22. a. Since the dot corresponding to automatic transmission on the private sale line graph is opposite 350, we know that the value of an automatic transmission on a private sale was \$350.

b. Since the dot corresponding to automatic transmission on the trade-in line graph is opposite 300, we know that the value of an automatic transmission on a trade-in was \$300.

23. The distances between the dots on the private sale line and the trade-in line are the greatest for antilock brakes, automatic transmission, and air conditioning.

24. The distances between the dots on the private sale line and the trade-in line are the least for CD changers, alarm systems and aluminum wheels.

25. $\frac{300+350+250+175+400+175}{6}=\frac{1650}{6}$
$= \$275$

26. $\frac{250+300+225+150+350+150}{6}=\frac{1425}{6}$
$= \$237.50$

27. 175, 175, 250, 300, 350, 400
↑

$\text{median} = \frac{250+300}{2}=\frac{550}{2}=\$275$

28. 150, 150, 225, 250, 300, 350
↑

$\text{median} = \frac{225+250}{2}=\frac{475}{2}=\$237.50$

29. Draw a vertical and a horizontal line. Since we are comparing the sales of the various souvenirs for Weeks 1 and 3, we place the label *Number of Sales* on the vertical line. We label the horizontal line with the names of the souvenirs.

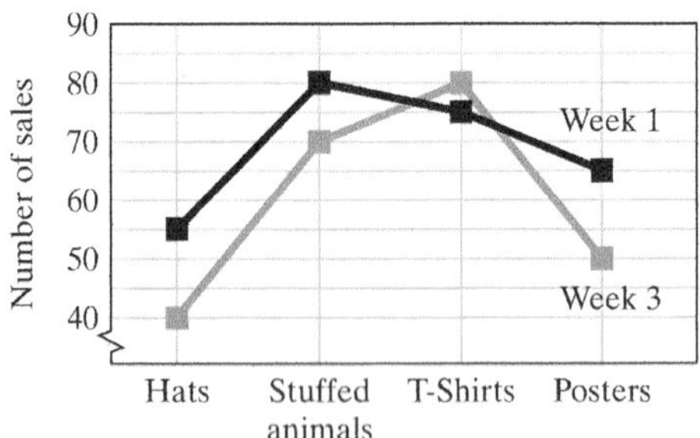

30. Draw a vertical and a horizontal line. Since we are comparing the sales of the various souvenirs for Weeks 1 and 2, we place the label *Numbers of Sales* on the vertical line. We label the horizontal line with the names of the souvenirs.

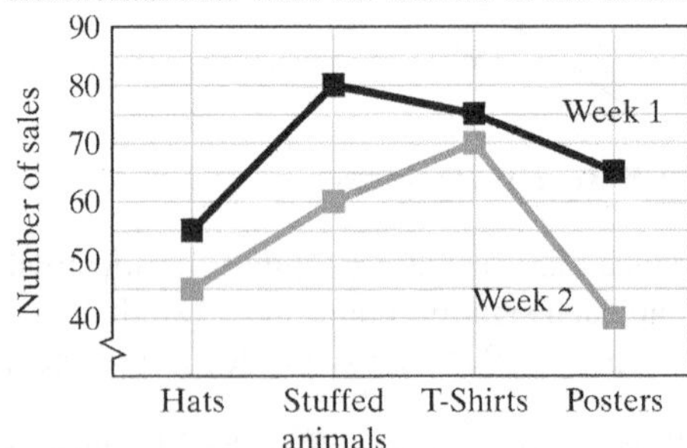

31. Draw a vertical and a horizontal line. Since we are comparing the numbers of male and female bus riders for five days, we place the label *Number of Bus Riders* on the vertical line. We place the label *Day* on the horizontal line.

Copyright © 2012 Pearson Education, Inc.

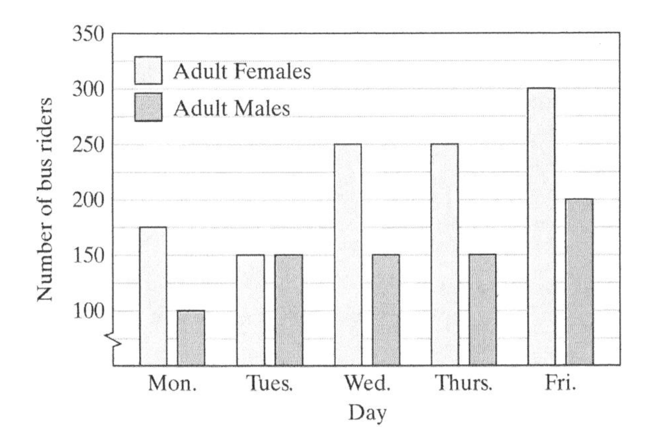

32. Draw a vertical and a horizontal line. Since we are comparing the numbers of children and female bus riders for five days, we place the label *Number of Bus Riders* on the vertical line. We place the label *Day* on the horizontal line.

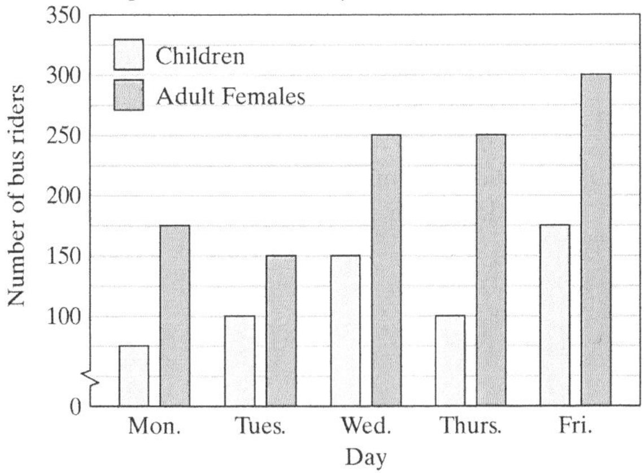

33. 57, 65, 69, 77, 82, 87, 88, 93, 100
↑
median = 82

34. 58, 77, 79, 81, 83, 87, 88, 91, 104
↑
median = 83

35. 0, 1, 4, 5, 9, 18, 19, 19, 20, 21, 22, 25, 27, 36, 38, 43
↑

$$\text{median} = \frac{19+20}{2} = \frac{39}{2} = 19.5$$

36. 0, 3, 9, 13, 14, 15, 16, 18, 19, 21, 24, 25, 26, 28, 31, 36
↑

$$\text{median} = \frac{18+19}{2} = \frac{37}{2} = 18.5$$

37. $\frac{86+83+88+95+97+100+81}{7} = \frac{630}{7} = 90°\text{F}$

38. $\frac{87+105+89+120+139+160+98}{7} = \frac{798}{7}$
$= \$114$

39. $\frac{76+20+91+57+42+21+75+82}{8} = \frac{464}{8} = 58$

40. 151 + 140 + 148 + 156 + 183 + 201 + 205 + 228 + 231 + 237 = 1180

$\frac{1880}{10} = 188$

41. The mode is 13 since it appears twice in the set of data.

42. The mode is 18 since it appears three times in the set of data.

43. (year, number sold)
↓ ↓
(2002, 5000)
(2003, 6000)
(2004, 5500)
(2005, 6250)

44. (year, number sold)
↓ ↓
(2006, 7000)
(2007, 6500)
(2008, 7500)
(2009, 7250)

45. To plot (3, 2), or $x = 3$, $y = 2$, we start at the origin and move 3 units in the positive x-direction followed by 2 units in the positive y-direction.

46. To plot $\left(2, 3\frac{1}{2}\right)$, or $x = 2$, $y = 3\frac{1}{2}$, we start at the origin and move 2 units in the positive x-direction followed by $3\frac{1}{2}$ units in the positive y-direction.

47. To plot (−2, 0), or $x = -2$, $y = 0$, we start at the origin and move 2 units in the negative x-direction followed by 0 units in the y-direction.

48. To plot (−3, −1), or $x = -3$, $y = -1$, we start at the origin and move 3 units in the negative x-direction followed by 1 unit in the negative y-direction.

Copyright © 2012 Pearson Education, Inc.

45–48.

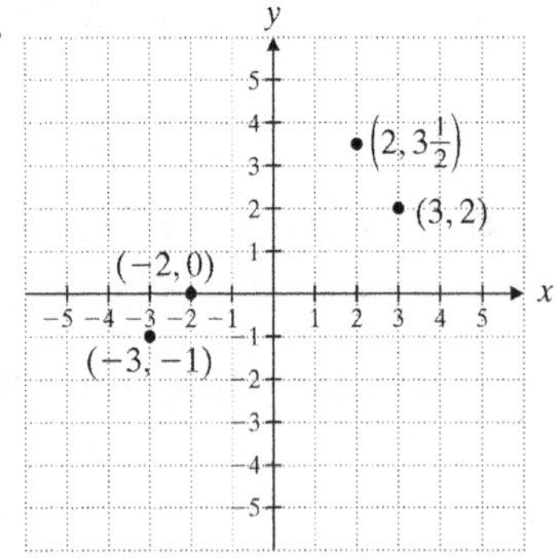

49. $R = (-5, 3)$

50. $S = (-1, -1)$

51. $T = \left(0, 2\frac{1}{2}\right)$

52. $U = (4, 1)$

53. (2, 0), (2, 3), (2, −1)

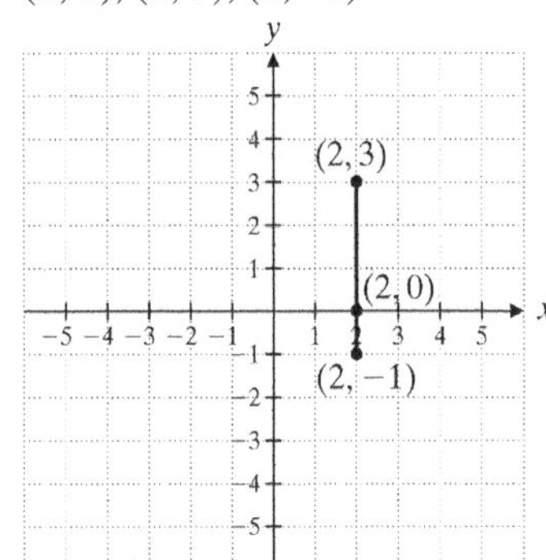

54. (3, 1), (−4, 1), (0, 1)

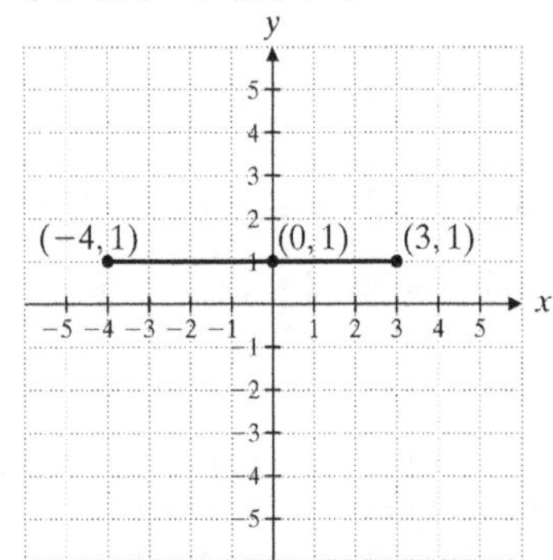

55. $x + y = 8$
For each ordered pair, we can choose any two numbers whose sum is 8.
$6 + 2 = 8$: (6, 2)
$8 + 0 = 8$: (8, 0)
$7 + 1 = 8$: (7, 1)
Answers may vary.

56. $x + y = 3$
For each ordered pair, we can choose any two numbers whose sum is 3.
$0 + 3 = 3$: (0, 3)
$2 + 1 = 3$: (2, 1)
$3 + 0 = 3$: (3, 0)
Answers may vary.

57. **a.** $y = 70x$
We replace y with 2800 and solve for x.

$$280 = 70x$$
$$\frac{280}{70} = \frac{70x}{70}$$
$$4 = x$$

It takes 4 minutes to type 280 words. The ordered pair is (4, 280).

b. $y = 70x$
We replace y with 350 and solve for x.

$$350 = 70x$$
$$\frac{350}{70} = \frac{70x}{70}$$
$$5 = x$$

It takes 5 minutes to type 350 words. The ordered pair is (5, 350).

58. $y = 2x - 6$

(_, −4):
$$-4 = 2x - 6$$
$$2 = 2x$$
$$1 = x$$

(_, −6)
$$-6 = 2x - 6$$
$$0 = 2x$$
$$0 = x$$

(_, −8)
$$-8 = 2x - 6$$
$$-2 = 2x$$
$$-1 = x$$

(x, y)
(1, −4)
(0, −6)
(−1, −8)

59. $y = -6x + 2$

(_, 2):
$$2 = -6x + 2$$
$$0 = -6x$$
$$0 = x$$

(_, 8):
$$8 = -6x + 2$$
$$6 = -6x$$
$$-1 = x$$

(_, −4)
$$-4 = -6x + 2$$
$$-6 = -6x$$
$$1 = x$$

(x, y)
(0, 2)
(−1, 8)
(1, −4)

Copyright © 2012 Pearson Education, Inc.

60. $y = 3x - 1$

We choose three values for x: -1, 0, and 1.

$y = 3(-1) - 1$ $\quad$ $y = 3(0) - 1$
$y = -3 - 1$ $\quad$ $y = 0 - 1$
$y = -4$ $\quad$ $y = -1$

$y = 3(1) - 1$
$y = 3 - 1$
$y = 2$

(x, y)
$(-1, -4)$
$(0, -1)$
$(1, 2)$

Plot these ordered pairs and draw a straight line through the points.

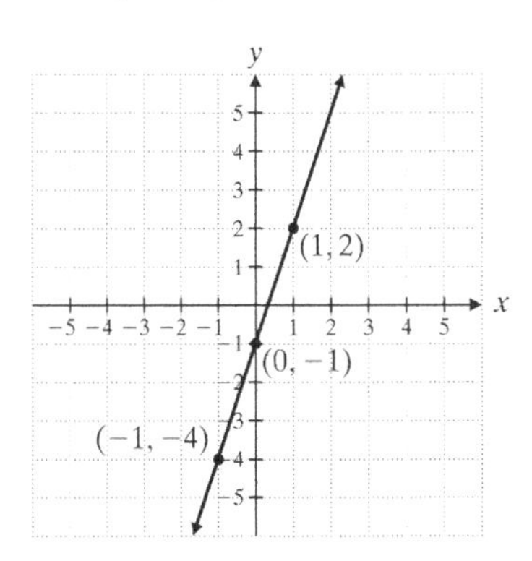

61. $y = -5x - 4$

We choose three values for x: -1, 0, and $-\frac{3}{2}$.

$y = -5(-1) - 4$ $\quad$ $y = -5(0) - 4$
$y = 5 - 4$ $\quad$ $y = 0 - 4$
$y = 1$ $\quad$ $y = -4$

$y = -5\left(-\frac{3}{2}\right) - 4$
$y = \frac{15}{2} - \frac{8}{2}$
$y = \frac{7}{2}$

(x, y)
$(-1, 1)$
$(0, -4)$
$\left(-\frac{3}{2}, \frac{7}{2}\right)$

Plot these ordered pairs and draw a straight line through the points.

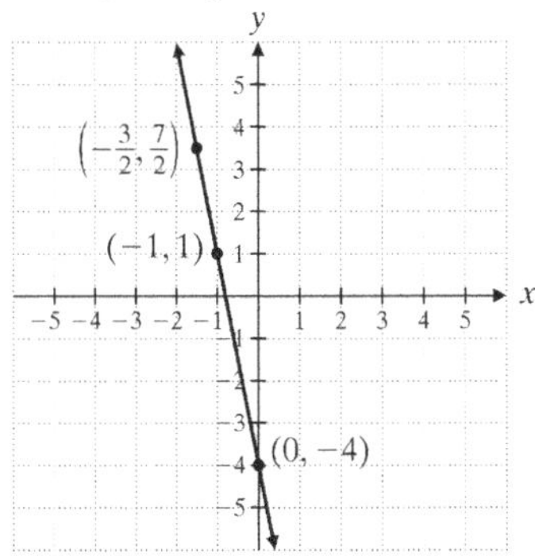

62. $y = 4x - 6$

We choose three values for x: 1, 2, and $\frac{3}{2}$.

$y = 4(1) - 6$ $\quad$ $y = 4(2) - 6$
$y = 4 - 6$ $\quad$ $y = 8 - 6$
$y = -2$ $\quad$ $y = 2$

$y = 4\left(\frac{3}{2}\right) - 6$
$y = 6 - 6$
$y = 0$

(x, y)
$(1, -2)$
$(2, 2)$
$\left(\frac{3}{2}, 0\right)$

Plot these ordered pairs and draw a straight line through the points.

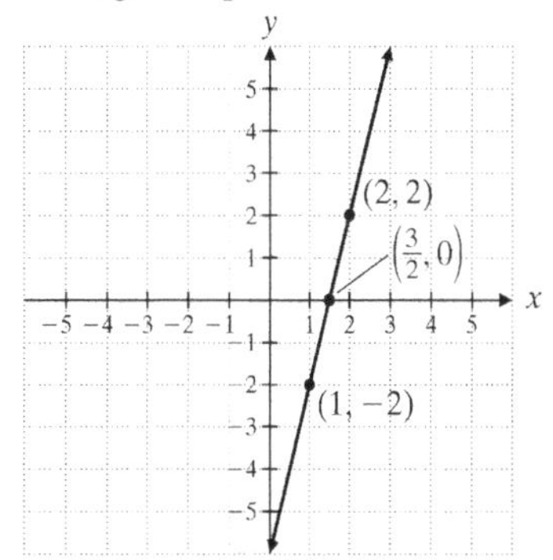

63. $y = -1$

The x-coordinate can be any number as long as y is -1.

$(-5, -1)$, $(2, -1)$, $(5, -1)$

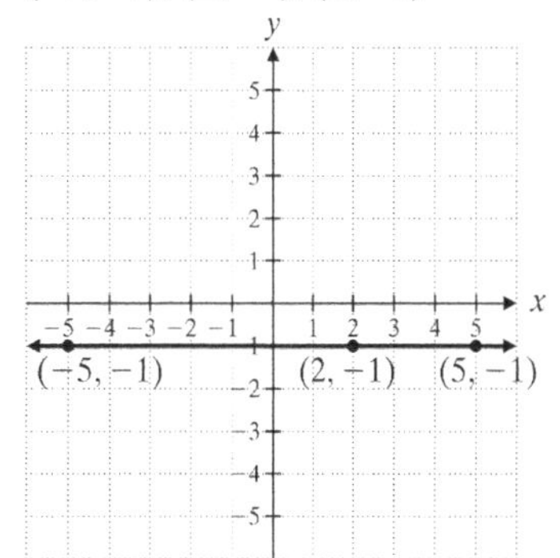

How Am I Doing? Chapter 9 Test

1. The largest region represents the age group 18–20.

2. If we add 44 and 33, we get 77. Thus, 77% of the students are between 18 and 24.

3. If we add 6 and 44, we get 50. Thus, 50% of the voters are 20 or younger.

Copyright © 2012 Pearson Education, Inc.

4. 7% of 5000 = 0.07 × 5000 = 350
350 students are over age 27.

5. In Michigan, the water in Grand Traverse Bay freezes approximately 11 days later in the year and thaws 12 days earlier in the year than it did 100 years ago.

6. In Finland, the water in Nasijarvi Lake freezes approximately 6 days later in the year and thaws 9 days earlier in the year than it did 100 years ago.

7. According to this graph, the body of water where there was the least amount of change in the time of the year that the water freezes is Otsego Lake, NY.

8. According to this graph, the body of water where there was the least amount of change in the time of the year that the water thaws is Baikal Lake, Russia.

9. Honda Accord and Toyota Camry do not have a roadside assistance plan.

10. The Volkswagen Jetta drive train is under warranty for 5 years.

11. 6 − 5 = 1; the Lexus 350 drive train is under warranty 1 more year than the Honda Accord drive train.

12. a. The basic warranty bar for the Lexus ES350 is taller than the other two basic warranty bars.

b. The drive train bar for the Lexus ES350 is taller than any of the other three drive train bars.

c. The rust bar for the Volkswagen Jetta is taller than any of the other three rust bars.

d. The roadside assistance bar for the Lexus ES350 is taller than the other roadside assistance bar.

13. Lexus: $\frac{4+6+6+4}{4} = \frac{20}{4} = 5$ yr

Honda: $\frac{3+5+5}{3} = \frac{13}{3} = 4.33$ yr

14. Toyota: $\frac{3+5+5}{3} = \frac{13}{3} = 4.33$ yr

Volkswagen: $\frac{3+5+12+3}{4} = \frac{23}{4} = 5.75$ yr

15.
$$\frac{89+46+85+91+83+90}{6} = \frac{484}{6} = 80.66... \approx 80.7$$

16. 46, 83, 85, 89, 90, 91
↑

median $= \frac{85+89}{2} = \frac{174}{2} = 87$

17. To plot (3, 5), or $x = 3$, $y = 5$, we start at the origin and move 3 units in the positive x-direction followed by 5 units in the positive y-direction.

18. To plot (0, 0), or $x = 0$, $y = 0$, we start at the origin and move 0 units in the x-direction followed by 0 units in the y-direction.

19. To plot (2, −1), or $x = 2$, $y = -1$, we start at the origin and move 2 units in the positive x-direction followed by 1 unit in the negative y-direction.

20. To plot (−3, 0), or $x = -3$, $y = 0$, we start at the origin and move 3 units in the negative x-direction and followed by 0 units in the y-direction.

17–20.

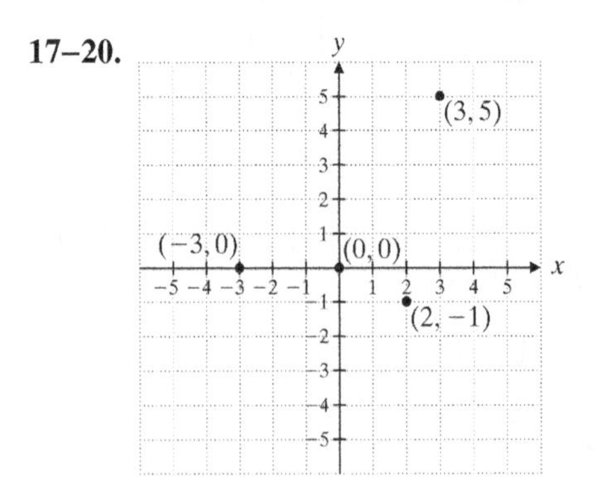

21. $A = (4, 2)$

22. $B = (0, -2)$

23. $C = (3, -1)$

24. $D = (-1, -3)$

Copyright © 2012 Pearson Education, Inc.

25. (1, 2), (−3, 2), (0, 2), (4, 2)

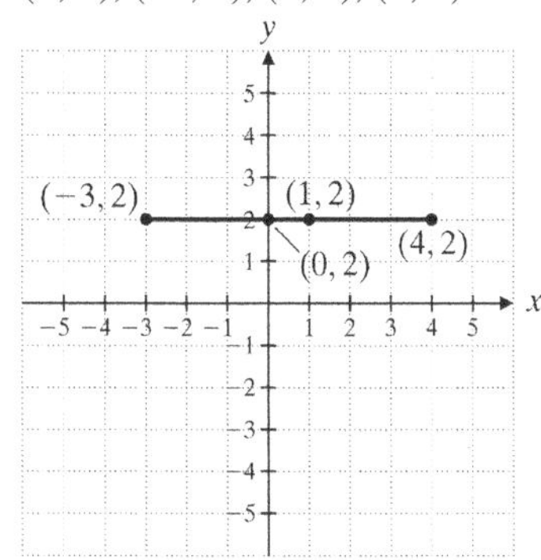

26. $y = 3$
The x-coordinate can be any number as long as y is 3.
(−3, 3), (0, 3), (3, 3)

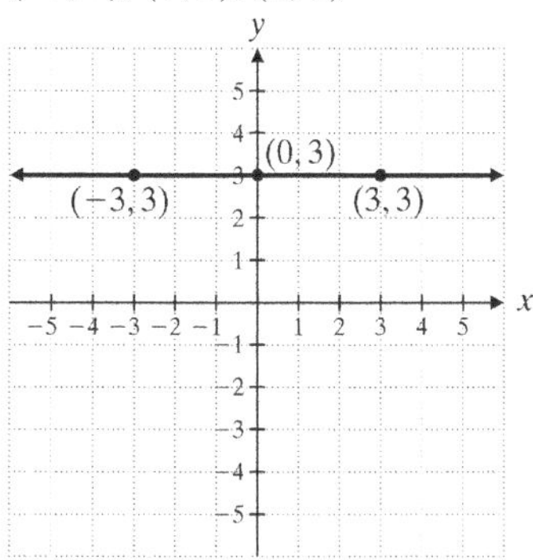

27. $y = 4x - 2$

We choose three values for x: 0, 1, and $\frac{1}{2}$.

$y = 4(0) - 2$
$y = 0 - 2$
$y = -2$

$y = 4(1) - 2$
$y = 4 - 2$
$y = 2$

$y = 4\left(\frac{1}{2}\right) - 2$
$y = 2 - 2$
$y = 0$

(x, y)
(0, −2)
(1, 2)
$\left(\frac{1}{2}, 0\right)$

Plot these ordered pairs and draw a straight line through the points.

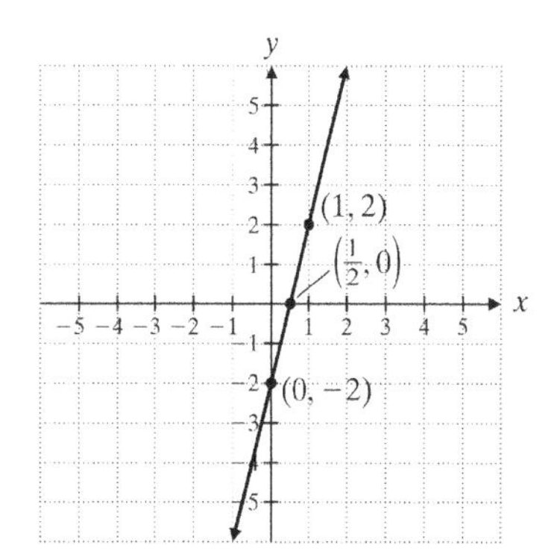

28. $y = 3x + 1$
We choose three values for x: −1, 0, and 1.

$y = 3(-1) + 1$
$y = -3 + 1$
$y = -2$

$y = 3(0) + 1$
$y = 0 + 1$
$y = 1$

$y = 3(1) + 1$
$y = 3 + 1$
$y = 4$

(x, y)
(−1, −2)
(0, 1)
(1, 4)

Plot these ordered pairs and draw a straight line through the points.

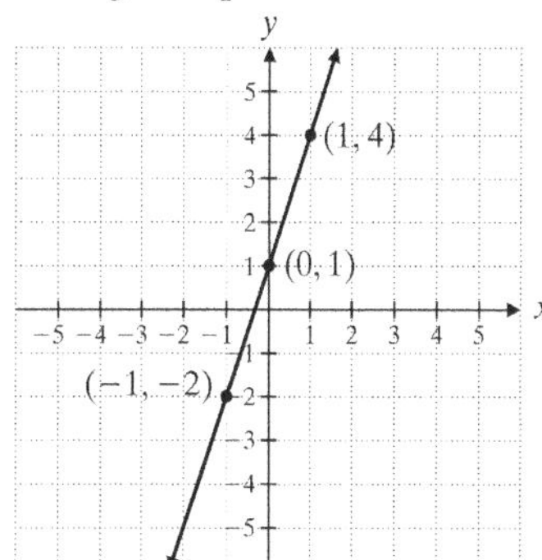

Chapter 10

10.1 Exercises

1. 1 foot = 12 inches

3. 2 pints = 1 quart

5. 1 ton = 2000 pounds

7. 4 quarts = 1 gallon

9. 7 days = 1 week

11. 60 seconds = 1 minute

13. 21 ft = ? yd

$$21\text{ ft}\times\frac{1\text{ yd}}{3\text{ ft}} = 21\cancel{\text{ ft}}\times\frac{1\text{ yd}}{3\cancel{\text{ ft}}} = 21\times\frac{1}{3}\text{ yd} = 7\text{ yd}$$

15. 7920 ft = ? mi

$$\begin{aligned}7920\text{ ft}\times\frac{1\text{ mi}}{5280\text{ ft}} &= 7920\cancel{\text{ ft}}\times\frac{1\text{ mi}}{5280\cancel{\text{ ft}}}\\ &= 7920\times\frac{1}{5280}\text{ mi}\\ &= 1.5\text{ mi}\end{aligned}$$

17. 69 in. = ? ft

$$\begin{aligned}69\text{ in.}\times\frac{1\text{ ft}}{12\text{ in.}} &= 69\cancel{\text{ in.}}\times\frac{1\text{ ft}}{12\cancel{\text{ in.}}}\\ &= 69\times\frac{1}{12}\text{ ft}\\ &= 5.75\text{ ft}\end{aligned}$$

19. 13 tons = ? pounds

$$\begin{aligned}13\text{ tons}\times\frac{2000\text{ lb}}{1\text{ ton}} &= 13\cancel{\text{ tons}}\times\frac{2000\text{ lb}}{1\cancel{\text{ ton}}}\\ &= 13\times 2000\text{ lb}\\ &= 26{,}000\text{ lb}\end{aligned}$$

21. 7 gal = ? qt

$$7\text{ gal}\times\frac{4\text{ qt}}{1\text{ gal}} = 7\cancel{\text{ gal}}\times\frac{4\text{ qt}}{1\cancel{\text{ gal}}} = 7\times 4\text{ qt} = 28\text{ qt}$$

23. 11 days = ? hr

$$\begin{aligned}11\text{ days}\times\frac{24\text{ hr}}{1\text{ day}} &= 11\cancel{\text{ days}}\times\frac{24\text{ hr}}{1\cancel{\text{ day}}}\\ &= 11\times 24\text{ hr}\\ &= 264\text{ hr}\end{aligned}$$

25. Change feet to inches.

$$218\cancel{\text{ ft}}\times\frac{12\text{ in.}}{1\cancel{\text{ ft}}} = 218\times 12\text{ in.} = 2616\text{ in.}$$

Find the total number of inches.
2616 in. + 10 in. = 2626 in.
The javelin flew 2626 inches.

27. Change the number of days to a decimal.

$$2\frac{1}{2}\text{ days} = 2.5\text{ days}$$

Change days to hours.

$$2.5\cancel{\text{ days}}\times\frac{24\text{ hr}}{1\cancel{\text{ day}}} = 60\text{ hr}$$

Find the total charge for parking.

$$60\cancel{\text{ hr}}\times\frac{\$2.25}{1\cancel{\text{ hr}}} = \$135$$

The stockbroker paid $135 for parking.

29. Change ounces to pounds.

$$24\cancel{\text{ oz}}\times\frac{1\text{ lb}}{16\cancel{\text{ oz}}} = 1.5\text{ lb}$$

Find the total cost of the cheese.

$$1.5\cancel{\text{ lb}}\times\frac{\$4}{1\cancel{\text{ lb}}} = \$6$$

Phoebe will pay $6 for the cheese.

31. 50 cm = ? mm
Move the decimal point 1 place to the right.
50 cm = 500 mm

33. 3.6 km = ? m
Move the decimal point 3 places to the right.
3.6 km = 3600 m

35. 2.43 kL = ? mL
Move the decimal point 6 places to the right.
2.43 kL = 2,430,000 mL

37. 1834 mL = ? kL
Move the decimal point 6 places to the left.
1834 mL = 0.001834 kL

39. 0.78 g = ? kg
Move the decimal point 3 places to the left.
0.78 g = 0.00078 kg

41. 5.9 kg = ? mg
Move the decimal point 6 places to the right.
5.9 kg = 5,900,000 mg

Copyright © 2012 Pearson Education, Inc.

43. 7 mL = ? L = ? kL
Move the decimal point 3 places to the left each time.
7 mL = 0.007 L = 0.000007 kL

45. 413 mg = ? g = ? kg
Move the decimal point 3 places to the left each time.
413 mg = 0.413 g = 0.000413 kg

47. 35 mm = ? cm = ? m
Move the decimal point 1 place to the left and then 2 more places to the left.
35 mm = 3.5 cm = 0.035 m

49. 3582 mm = ? m = ? km
Move the decimal point 3 places to the left each time.
3582 mm = 3.582 m = 0.003582 km

51. 0.32 cm = ? m = ? km
Move the decimal point 2 places to the left and then 3 more places to the left.
0.32 cm = 0.0032 m = 0.0000032 km

53. $\frac{\$6.00}{1\ \text{mL}} \times \frac{1000\ \text{mL}}{1\ \text{L}} = \frac{\$6000}{1\ \text{L}}$
It will cost the firm $6000 to produce 1 liter of the vaccine.

55. $\frac{\$850}{1\ \text{mL}} \times \frac{1000\ \text{mL}}{1\ \text{L}} = \frac{\$850{,}000}{1\ \text{L}}$
It costs $850,000 to produce 1 liter.
$0.4\ \text{L} \times \frac{\$850{,}000}{1\ \text{L}} = \$340{,}000$
It will cost the company $340,000 to produce 0.4 liter of the essence.

57. **a.** To change meters to centimeters, move the decimal point 2 places to the right.
4818 m = 481,800 cm
The track is 481,800 centimeters high.

b. To change meters to kilometers, move the decimal point 3 places to the left.
4818 m = 4.818 km
The track is 4.818 kilometers high.

59. 5280 ft = ? yd

$$5280\ \text{ft} \times \frac{1\ \text{yd}}{3\ \text{ft}} = 5280\ \text{ft} \times \frac{1\ \text{yd}}{3\ \text{ft}} = 5280 \times \frac{1}{3}\ \text{yd} = 1760\ \text{yd}$$

61. 4 mi = ? ft

$$4\ \text{mi} \times \frac{5280\ \text{ft}}{1\ \text{mi}} = 4\ \text{mi} \times \frac{5280\ \text{ft}}{1\ \text{mi}} = 4 \times 5280\ \text{ft} = 21{,}120\ \text{ft}$$

63. 9 tons = ? pounds

$$9\ \text{tons} \times \frac{2000\ \text{lb}}{1\ \text{ton}} = 9\ \text{tons} \times \frac{2000\ \text{lb}}{1\ \text{ton}} = 9 \times 2000\ \text{lb} = 18{,}000\ \text{lb}$$

65. 14.6 kg = ? g
Move the decimal point 3 places to the right.
14.6 kg = 14,600 g

67. 3.22 kL = ? L
Move the decimal point 3 places to the right.
3.22 kL = 3220 L

69. 7183 mg = ? g = ? kg
Move the decimal point 3 places to the left each time.
7183 mg = 7.183 g = 0.007183 kg

71. 528 megabytes = ? bytes
Move the decimal point 6 places to the right.
(1 million $= 10^6$)
528 megabytes = 528,000,000 bytes

73. 24.9 gigabytes = ? bytes
Move the decimal point 9 places to the right.
(1 billion $= 10^9$)
24.9 gigabytes = 24,900,000,000 bytes

Cumulative Review

74. 14 out of 70 is what percent?
$\frac{14}{70} = 0.2 = 20\%$
14 out of 70 is 20%.

75. What is 23% of 250?
23% of 250 = 0.23 × 250 = 57.5
57.5 is 23% of 250.

Copyright © 2012 Pearson Education, Inc.

76. What is 1.7% of \$18,900?
1.7% of $\$18,900 = 0.017 \times \$18,900 = \$321.30$
\$321.30 is 1.7% of \$18,900.

77. What is 8% of \$8960?
8% of $\$8960 = 0.08 \times \$8960 = \$716.80$
The salesperson will earn \$716.80.

Quick Quiz 10.1

1. Change feet to miles.
$$14,496\ \cancel{\text{ft}} \times \frac{1\text{ mi}}{5280\ \cancel{\text{ft}}} \approx 2.75\text{ mi}$$
Mount Whitney is approximately 2.75 miles high.

2. 500 mm = ? m
Move the decimal point 3 places to the left.
500 mm = 0.5 m

3. 9 kg = ? mg
Move the decimal point 6 places to the right.
9 kg = 9,000,000 mg

4. Answers may vary. Multiply 240 ounces by the unit fraction $\frac{1\text{ pound}}{16\text{ ounces}}$.

10.2 Exercises

1. $4\ \cancel{\text{ft}} \times \frac{0.305\text{ m}}{1\ \cancel{\text{ft}}} = 1.22\text{ m}$

3. $14\ \cancel{\text{m}} \times \frac{1.09\text{ yd}}{1\ \cancel{\text{m}}} = 15.26\text{ yd}$

5. $15\ \cancel{\text{km}} \times \frac{0.62\text{ mi}}{1\ \cancel{\text{km}}} = 9.3\text{ mi}$

7. $24\ \cancel{\text{yd}} \times \frac{0.914\text{ m}}{1\ \cancel{\text{yd}}} = 21.94\text{ m}$

9. $82\ \cancel{\text{mi}} \times \frac{1.61\text{ km}}{1\ \cancel{\text{mi}}} = 132.02\text{ km}$

11. $25\ \cancel{\text{m}} \times \frac{3.28\text{ ft}}{1\ \cancel{\text{m}}} = 82\text{ ft}$

13. $17.5\ \cancel{\text{cm}} \times \frac{0.394\text{ in.}}{1\ \cancel{\text{cm}}} \approx 6.90\text{ in.}$

15. $5\ \cancel{\text{gal}} \times \frac{3.79\text{ L}}{1\ \cancel{\text{gal}}} = 18.95\text{ L}$

17. $4.5\ \cancel{\text{L}} \times \frac{1.06\text{ qt}}{1\ \cancel{\text{L}}} = 4.77\text{ qt}$

19. $7\ \cancel{\text{oz}} \cdot \frac{28.35\text{ g}}{1\ \cancel{\text{oz}}} = 198.45\text{ g}$

21. $11\text{ kg} \times \frac{2.2\text{ lb}}{1\text{ kg}} = 24.2\text{ lb}$

23. $126\ \cancel{\text{g}} \times \frac{0.0353\text{ oz}}{1\ \cancel{\text{g}}} \approx 4.45\text{ oz}$

25. $4\ \cancel{\text{kg}} \times \frac{2.2\ \cancel{\text{lb}}}{\cancel{\text{kg}}} \times \frac{16\text{ oz}}{1\ \cancel{\text{lb}}} = 140.8\text{ oz}$

27. $230\ \cancel{\text{cm}} \times \frac{1\ \cancel{\text{in.}}}{2.54\ \cancel{\text{cm}}} \times \frac{1\text{ ft}}{12\ \cancel{\text{in.}}} = 7.55\text{ ft}$

29. $16.5\ \cancel{\text{ft}} \times \frac{12\ \cancel{\text{in.}}}{\cancel{\text{ft}}} \times \frac{2.54\text{ cm}}{\cancel{\text{in.}}} = 502.92\text{ cm}$

31. $\frac{50\ \cancel{\text{km}}}{\text{hr}} \times \frac{0.62\text{ mi}}{1\ \cancel{\text{km}}} = 31\text{ mi/hr}$

33. $\frac{45\text{ mi}}{\text{hr}} \times \frac{1.61\text{ km}}{1\text{ mi}} = 72.5\text{ km/hr}$

35. Move the decimal point one place to the left.
13 mm = 1.3 cm
Now multiply by a unit fraction.
$$1.3\text{ cm} \times \frac{0.394\text{ in.}}{1\text{ cm}} \approx 0.51\text{ in.}$$
The wire is about 0.51 inch wide.

37. $F = 1.8 \times C + 32 = 1.8 \times 0 + 32 = 0 + 32 = 32°\text{F}$

39.
$$\begin{aligned} F &= 1.8 \times C + 32 \\ &= 1.8 \times 85 + 32 \\ &= 153 + 32 \\ &= 185°\text{F} \end{aligned}$$

 Copyright © 2012 Pearson Education, Inc.

41. $C = \frac{5 \times F - 160}{9}$
$= \frac{5 \times 168 - 160}{9}$
$= \frac{840 - 160}{9}$
$= \frac{680}{9}$
$\approx 75.56°C$

43. $C = \frac{5 \times F - 160}{9}$
$= \frac{5 \times 86 - 160}{9}$
$= \frac{430 - 160}{9}$
$= \frac{270}{9}$
$= 30°C$

45. $9 \text{ in.} \times \frac{2.54 \text{ cm}}{1 \text{ in.}} = 22.86 \text{ cm}$

47. $32.2 \text{ m} \times \frac{1.09 \text{ yd}}{1 \text{ m}} \approx 35.10 \text{ yd}$

49. $19 \text{ L} \times \frac{0.264 \text{ gal}}{1 \text{ L}} \approx 5.02 \text{ gal}$

51. $32 \text{ lb} \times \frac{0.454 \text{ kg}}{1 \text{ lb}} \approx 14.53 \text{ kg}$

53. $F = 1.8 \times C + 32$
$= 1.8 \times 12 + 32$
$= 21.6 + 32$
$= 53.6°F$

55. $C = \frac{5 \times F - 160}{9}$
$= \frac{5 \times 68 - 160}{9}$
$= \frac{340 - 160}{9}$
$= \frac{180}{9}$
$= 20°C$

57. $635 \text{ kg} \times \frac{2.2 \text{ lb}}{1 \text{ kg}} = 1397 \text{ lb}$

59. First convert 67 miles to kilometers.
$67 \text{ mi} \times \frac{1.61 \text{ km}}{1 \text{ mi}} = 107.87 \text{ km}$
Now find the sum.
107.87 km + 36 km = 143.87 km
The Westons traveled a total of 143.87 kilometers.

61. First convert 15 gallons to liters.
$15 \text{ gal} \times \frac{3.79 \text{ L}}{1 \text{ gal}} = 56.85 \text{ L}$
Now find the difference.
56.85 L − 38 L = 18.85 L
Pierre had 18.85 liters of gas left.

63. First convert 32°C to Fahrenheit.
$F = 1.8 \times C + 32$
$= 1.8 \times 32 + 32$
$= 57.6 + 32$
$= 89.6°F$
Now find the difference.
89.6°F − 80°F = 9.6°F
The difference in temperatures is 9.6°F.

65. 4 o'clock: $F = 1.8 \times C + 32$
$= 1.8 \times 19 + 32$
$= 34.2 + 32$
$= 66.2°F$
7 o'clock: $F = 1.8 \times C + 32$
$= 1.8 \times 45 + 32$
$= 81 + 32$
$= 113°F$

67. $0.768 \text{ oz} \times \frac{28.35 \text{ g}}{1 \text{ oz}} \approx 21.773 \text{ g}$

Cumulative Review

69. $2^3 \times 6 - 4 + 3 = 8 \times 6 - 4 + 3 = 48 - 4 + 3 = 47$

70. $5 + 2 - 3 + 5 \times 3^2 = 5 + 2 - 3 + 5 \times 9$
$= 5 + 2 - 3 + 45$
$= 49$

71. $2^2 + 3^2 + 4^3 + 2 \times 7 = 4 + 9 + 64 + 2 \times 7$
$= 4 + 9 + 64 + 14$
$= 91$

72. $5^2 + 4^2 + 3^2 + 3 \times 8 = 25 + 16 + 9 + 3 \times 8$
$= 25 + 16 + 9 + 24$
$= 74$

Copyright © 2012 Pearson Education, Inc.

Quick Quiz 10.2

1. $1.8 \cancel{m} \times \frac{3.28}{1 \cancel{m}} \approx 6$ ft

2. $149 \cancel{kg} \times \frac{2.2 \text{ lb}}{1 \cancel{kg}} \approx 328$ lb

3. $1.5 \cancel{cm} \times \frac{0.394 \text{ in.}}{1 \cancel{cm}} \approx 0.59$ in.

4. Answers may vary. Multiply $\frac{50 \text{ km}}{\text{hr}}$ by the unit fraction $\frac{0.62 \text{ mi}}{1 \text{ km}}$.

10.3 Exercises

1. Vertex: <u>The point at which the two sides of an angle meet</u>.

3. Obtuse angle: <u>An angle whose measure is between 90° and 180°</u>.

5. Adjacent angles: <u>Two angles formed by intersecting lines that share a common side</u>.

7. Line u is called a <u>transversal</u> since it intersects two or more lines at different points.

9. $\angle ABC$ is called <u>a straight angle</u>.

11. $\angle ABD$ is called <u>an obtuse angle</u>.

13. If the sum of two angles is 180° then these angles are called <u>supplementary angles</u>.

15. An angle whose measure is between 0° and 90° is called <u>an acute angle</u>.

17. $\angle a$ and $\angle c$ are called <u>vertical angles</u>.

19. Name the angles that are adjacent to $\angle c$: <u>$\angle d$</u> and <u>$\angle b$</u>.

21. <u>$\angle b$</u> is equal to $\angle d$.

23. $\angle a + \angle b =$ <u>180°</u>

25. $\angle a$ and $\angle d$ are called <u>alternate interior angles</u>.

27. $\angle b$ is equal to the following two angles: <u>$\angle c$</u> and <u>$\angle f$</u>.

29. $\angle f$ is equal to the following two angles: <u>$\angle b$</u> and <u>$\angle c$</u>.

31. $\angle e$ and $\angle d$ are called <u>corresponding angles</u>.

33. $\angle b$ and $\angle f$ are called <u>vertical angles</u>.

35. The middle letter is the vertex: $\angle TSV$ or $\angle VST$.

37. **a.** Let $\angle S$ = the supplement of $\angle M$.
$$\begin{aligned} \angle S + \angle M &= 180° \\ \angle S + 29° &= 180° \\ + \quad -29° &\quad -29° \\ \hline \angle S &= 151° \end{aligned}$$

b. Let $\angle C$ = the complement of $\angle M$.
$$\begin{aligned} \angle C + \angle M &= 90° \\ \angle C + 29° &= 90° \\ + \quad -29° &\quad -29° \\ \hline \angle C &= 61° \end{aligned}$$

39. **a.** Let $\angle S$ = the supplement of $\angle X$.
$$\begin{aligned} \angle S + \angle X &= 180° \\ \angle S + 45° &= 180° \\ + \quad -45° &\quad -45° \\ \hline \angle S &= 135° \end{aligned}$$

b. Let $\angle C$ = the complement of $\angle X$.
$$\begin{aligned} \angle C + \angle X &= 90° \\ \angle C + 45° &= 90° \\ + \quad -45° &\quad -45° \\ \hline \angle C &= 45° \end{aligned}$$

41.
$$\begin{aligned} x + (x + 15°) &= 90° \\ 2x + 15° &= 90° \\ + \quad -15° &\quad -15° \\ \hline 2x &= 75° \\ \frac{2x}{2} &= \frac{75°}{2} \\ x &= 37.5° \end{aligned}$$

43.
$$\begin{aligned} n + (n - 5°) &= 90° \\ 2n - 5° &= 90° \\ + \quad 5° &\quad 5° \\ \hline 2n &= 95° \\ \frac{2n}{2} &= \frac{95°}{2} \\ n &= 47.5° \end{aligned}$$

Copyright © 2012 Pearson Education, Inc.

45. $2m + m = 90°$
$3m = 90°$
$\frac{3m}{3} = \frac{90°}{3}$
$m = 30°$

47. Since $\angle w$ and $\angle y$ share a common side with $\angle z$, they are adjacent to $\angle z$.

49. $\angle w$ and $\angle z$ are adjacent angles of intersecting lines. They are supplementary, and the sum of their measures is 180°.

51. $\angle z = 70°$ since $\angle x$ and $\angle z$ are vertical angles. Since $\angle x$ and $\angle w$ are adjacent angles, they are supplementary.
$\angle x + \angle w = 180°$
$70° + \angle w = 180°$
$+ \ -70° \qquad -70°$
$\angle w = 110°$
$\angle y = 110°$ since $\angle w$ and $\angle y$ are vertical angles.

53. $\angle y = 45°$ since $\angle w$ and $\angle y$ are vertical angles. Since $\angle w$ and $\angle x$ are adjacent angles, they are supplementary.
$\angle w + \angle x = 180°$
$45° + \angle x = 180°$
$+ \ -45° \qquad -45°$
$\angle x = 135°$
$\angle z = 135°$ since $\angle x$ and $\angle z$ are vertical angles.

55. $\angle p = \angle o = 43°$ since $\angle p$ and $\angle o$ are vertical angles.
$\angle o = \angle n = 43°$ since $\angle o$ and $\angle n$ are alternate interior angles.
$\angle n = \angle l = 43°$ since $\angle n$ and $\angle l$ are vertical angles.
Since $\angle m$ and $\angle n$ are adjacent angles, they are supplementary.
$\angle m + \angle n = 180°$
$\angle m + 43° = 180°$
$+ \quad -43° \ -43°$
$\angle m = 137°$

57. Since $\angle a$ and $\angle b$ are adjacent angles, they are supplementary.
$\angle a + \angle b = 180°$
$45° + \angle b = 180°$
$+ -45° \qquad -45°$
$\angle b = 135°$
$\angle a = \angle c = 45°$ since $\angle a$ and $\angle c$ are alternate interior angles.
$\angle c = \angle e = 45°$ since $\angle c$ and $\angle e$ are vertical angles.
$\angle b = \angle d = 135°$ since $\angle b$ and $\angle d$ are corresponding angles.

59. $\angle y = \angle v = \angle u = \angle r = 99°$ since $\angle y$ and $\angle v$ are vertical angles, $\angle v$ and $\angle u$ are alternate interior angles, and $\angle u$ and $\angle r$ are vertical angles. Since $\angle y$ and $\angle x$ are adjacent, they are supplementary.
$\angle y + \angle x = 180°$
$99° + \angle x = 180°$
$+ -99° \qquad -99°$
$\angle x = 81°$
$\angle x = \angle w = \angle t = \angle s = 81°$ since $\angle x$ and $\angle w$ are vertical angles, $\angle w$ and $\angle t$ are alternate interior angles, and $\angle t$ and $\angle s$ are vertical angles.

61. The angles are vertical angles, so their measures are equal.
$2x + 2 = 5x - 10$
$+ -2x \quad -2x$
$2 = 3x - 10$
$+ \quad 10 \qquad 10$
$12 = 3x$
$\frac{12}{3} = \frac{3x}{3}$
$4 = x$

63. $\angle y = \angle x = 33°$ since $\angle y$ and $\angle x$ are vertical angles. Since the triangle is a right triangle, $\angle w + \angle x = 90°$.
$\angle w + 33° = 90°$
$+ \quad -33° - 33°$
$\angle w = 57°$
Since $\angle w$ and $\angle z$ are adjacent angles, $\angle w + \angle z = 180°$.
$57° + \angle z = 180°$
$+ -57° \qquad -57°$
$\angle z = 123°$

Cumulative Review

65. First find the dimensions of the design, in inches.
$70 \text{ squares} \times \frac{1 \text{ in.}}{14 \text{ squares}} = 5 \text{ in.}$
$84 \text{ squares} \times \frac{1 \text{ in.}}{14 \text{ squares}} = 6 \text{ in.}$
Now add the amount needed for framing.
5 in. + 2(2 in.) = 9 in.
6 in. + 2(2 in.) = 10 in.
The dimensions are 9 inches by 10 inches.

Copyright © 2012 Pearson Education, Inc.

66. Divide: 1,391,193 ÷ 109

$$\begin{array}{r} 12763 \\ 109\overline{)1391193} \\ \underline{109} \\ 301 \\ \underline{218} \\ 831 \\ \underline{763} \\ 689 \\ \underline{654} \\ 353 \\ \underline{327} \\ 26 \end{array}$$

Earth's diameter is about 12,760 kilometers.

Quick Quiz 10.3

1. a. Let $\angle C$ = the complement of $\angle A$.

$$\begin{aligned} \angle C + \angle A &= 90° \\ \angle C + 58° &= 90° \\ + \quad -58° &\;\; -58° \\ \hline \angle C &= 32° \end{aligned}$$

b. Let $\angle S$ = the supplement of $\angle A$.

$$\begin{aligned} \angle S + \angle A &= 180° \\ \angle S + 58° &= 180° \\ + \quad -58° &\;\; -58° \\ \hline \angle S &= 122° \end{aligned}$$

2. a. $\angle a = 65°$ since $\angle a$ and $\angle c$ are vertical angles.

b. Since $\angle b$ and $\angle c$ are adjacent angles, they are supplementary.

$$\begin{aligned} \angle b + \angle c &= 180° \\ \angle b + 65° &= 180° \\ + \quad -65° &\;\; -65° \\ \hline \angle b &= 115° \end{aligned}$$

3.
$$\begin{aligned} x + (x + 60) &= 90° \\ 2x + 60° &= 90° \\ + \quad -60° &\;\; -60° \\ \hline 2x &= 30° \\ \frac{2x}{2} &= \frac{30°}{2} \\ x &= 15° \end{aligned}$$

4. Answers may vary. $\angle a$ and $\angle d$ are supplementary. Subtract the measure of $\angle a$ from 180° to find the measure of $\angle b$.

How Am I Doing? Sections 10.1–10.3

1. 21 ft = ? yd

$$21\text{ ft} \times \frac{1\text{ yd}}{3\text{ ft}} = 21\,\cancel{\text{ft}} \times \frac{1\text{ yd}}{3\,\cancel{\text{ft}}} = 21 \times \frac{1}{3}\text{ yd} = 7\text{ yd}$$

2. 6 gal = ? qt

$$6\text{ gal} \times \frac{4\text{ qt}}{1\text{ gal}} = 6\,\cancel{\text{gal}} \times \frac{4\text{ qt}}{1\,\cancel{\text{gal}}} = 6 \times 4\text{ qt} = 24\text{ qt}$$

3. 240 min = ? hr

$$\begin{aligned} 240\text{ min} \times \frac{1\text{ hr}}{60\text{ min}} &= 240\,\cancel{\text{min}} \times \frac{1\text{ hr}}{60\,\cancel{\text{min}}} \\ &= 240 \times \frac{1}{60}\text{ hr} \\ &= 4\text{ hr} \end{aligned}$$

4. 6.3 kg = ? mg
Move the decimal point 6 places to the right.
6.3 kg = 6,300,000 mg

5. 34 mL = ? L = ? kL
Move the decimal point 3 places to the left each time.
34 mL = 0.034 L = 0.000034 kL

6. Change feet to yards.

$$63\,\cancel{\text{ft}} \times \frac{1\text{ yd}}{3\,\cancel{\text{ft}}} = 21\text{ yd}$$

Find the total cost of the fabric.

$$21\,\cancel{\text{yd}} \times \frac{\$4}{1\,\cancel{\text{yd}}} = \$84$$

The fabric will cost $84.

7. $10\,\cancel{\text{in.}} \times \frac{2.54\text{ cm}}{1\,\cancel{\text{in.}}} = 25.4\text{ cm}$

8. $22\,\cancel{\text{lb}} \times \frac{0.454\text{ kg}}{1\,\cancel{\text{lb}}} = 9.99\text{ kg}$

9. $3.5\,\cancel{\text{L}} \times \frac{1.06\text{ qt}}{1\,\cancel{\text{L}}} = 3.71\text{ qt}$

10. $\frac{95\,\cancel{\text{km}}}{\text{hr}} \times \frac{0.62\text{ mi}}{1\,\cancel{\text{km}}} = 58.9\text{ mi/hr}$

11. First convert 35°C to Fahrenheit.
$F = 1.8 \times C + 32 = 1.8 \times 35 + 32 = 63 + 32 = 95°\text{F}$
Now find the difference.
95°F – 89°F = 6°F
The difference in temperatures is 6°F.

Copyright © 2012 Pearson Education, Inc.

12. $\angle b = \angle c = 45°$

13. $\angle x = \angle y = 32°$

14. a. $\angle S$ = the supplement of $\angle D$.

$$\begin{aligned} \angle S + \angle D &= 180° \\ \angle S + 39° &= 180° \\ + \quad -39° & \quad -39° \\ \hline \angle S &= 141° \end{aligned}$$

b. Let $\angle C$ = the complement of $\angle D$.

$$\begin{aligned} \angle C + \angle D &= 90° \\ \angle C + 39° &= 90° \\ + \quad -39° & \quad -39° \\ \hline \angle C &= 51° \end{aligned}$$

15.
$$\begin{aligned} x + (x + 40°) &= 90° \\ 2x + 40° &= 90° \\ + \quad -40° & \quad -40° \\ \hline 2x &= 50° \\ \frac{2x}{2} &= \frac{50°}{2} \\ x &= 25° \end{aligned}$$

16. $\angle z = 55°$ since $\angle z$ and $\angle x$ are vertical angles. Since $\angle w$ and $\angle x$ are adjacent angles, they are supplementary.

$$\begin{aligned} \angle w + \angle x &= 180° \\ \angle w + 55° &= 180° \\ + \quad -55° & \quad -55° \\ \hline \angle w &= 125° \end{aligned}$$

$\angle y = 125°$ since $\angle y$ and $\angle w$ are vertical angles.

17. $\angle p = \angle o = 40°$ since $\angle p$ and $\angle o$ are vertical angles.
$\angle p = \angle n = 40°$ since $\angle p$ and $\angle n$ are corresponding angles.
Since $\angle m$ and $\angle n$ are adjacent angles, they are supplementary.

$$\begin{aligned} \angle m + \angle n &= 180° \\ \angle m + 40° &= 180° \\ + \quad -40° & \quad -40° \\ \hline \angle m &= 140° \end{aligned}$$

$\angle m = \angle l = 140°$ since $\angle m$ and $\angle l$ are vertical angles.

10.4 Exercises

1. a. $\sqrt{81}$: What is the square root of 81?

b. $n \cdot n = 81$: What number multiplied by itself equals 81?

c. $n^2 = 81$: What number squared equals 81?

3. 50 is not a perfect square. There is no whole number or fraction that when squared equals 50.

5. 49 is a perfect square because $7^2 = 7 \cdot 7 = 49$.

7. 81 is a perfect square because $9^2 = 9 \cdot 9 = 81$.

9. $\frac{1}{5}$ is not a perfect square. There is no whole number or fraction that when squared equals $\frac{1}{5}$.

11. a. $n^2 = 64 \Leftrightarrow n = \sqrt{64}$

b. $\sqrt{64}$ is the positive square root of 64 $\Leftrightarrow n = \sqrt{64}$

c. $n \cdot n = 64 \Leftrightarrow n = \sqrt{64}$

13. $\sqrt{36} = 6$ since $6 \cdot 6 = 36$.

15. $\sqrt{81} = 9$ since $9 \cdot 9 = 81$.

17. $\sqrt{144} = 12$ since $12 \cdot 12 = 144$.

19. $\sqrt{225} = 15$ and $\sqrt{16} = 4$, so $\sqrt{225} - \sqrt{16} = 15 - 4 = 11$.

21. $\sqrt{9} = 3$ and $\sqrt{49} = 7$, so $\sqrt{9} + \sqrt{49} = 3 + 7 = 10$.

23. $\sqrt{\frac{25}{49}} = \frac{5}{7}$ since $\left(\frac{5}{7}\right)\left(\frac{5}{7}\right) = \frac{25}{49}$.

25. $\sqrt{100} = 10$ and $\sqrt{25} = 5$, so $\sqrt{100} - \sqrt{25} = 10 - 5 = 5$.

27. $\sqrt{121} = 11$ and $\sqrt{81} = 9$, so $\sqrt{121} + \sqrt{81} = 11 + 9 = 20$.

29. $\sqrt{\frac{9}{81}} = \frac{3}{9} = \frac{1}{3}$ since $\left(\frac{3}{9}\right)\left(\frac{3}{9}\right) = \frac{9}{81}$.

31. $\sqrt{\frac{36}{49}} = \frac{6}{7}$ since $\left(\frac{6}{7}\right)\left(\frac{6}{7}\right) = \frac{36}{49}$.

Copyright © 2012 Pearson Education, Inc.

33.
$$A = s^2$$
$$121 = s^2$$
$$\sqrt{121} = s$$
$$11 = s$$
The length of the side of the square is 11 feet.

35.
$$A = s^2$$
$$144 = s^2$$
$$\sqrt{144} = s$$
$$12 = s$$
The length of the side of the square is 12 inches.

37. $\sqrt{44} \approx 6.633$

39. $\sqrt{69} \approx 8.307$

41. $\sqrt{80} \approx 8.944$

43. $\sqrt{90} \approx 9.487$

45.
$$c^2 = a^2 + b^2$$
$$c^2 = 4^2 + 3^2$$
$$c^2 = 16 + 9$$
$$c^2 = 25$$
$$c = \sqrt{25} = 5$$
The length of the unknown side is 5 inches.

47.
$$c^2 = a^2 + b^2$$
$$c^2 = 8^2 + 3^2$$
$$c^2 = 64 + 9$$
$$c^2 = 73$$
$$c = \sqrt{73} \approx 8.544$$
The length of the unknown side is approximately 8.544 inches.

49.
$$c^2 = a^2 + b^2$$
$$16^2 = a^2 + 5^2$$
$$256 = a^2 + 25$$
$$-25 \quad -25$$
$$231 = a^2$$
$$\sqrt{231} = a$$
$$15.199 \approx a$$
The length of the unknown side is approximately 15.199 feet.

51.
$$c^2 = a^2 + b^2$$
$$16^2 = a^2 + 8^2$$
$$256 = a^2 + 64$$
$$-64 \quad -64$$
$$192 = a^2$$
$$\sqrt{192} = a$$
$$a \approx 13.856$$
The length of the unknown side is approximately 13.856 kilometers.

53.
$$c^2 = a^2 + b^2$$
$$c^2 = 11^2 + 6^2$$
$$c^2 = 121 + 36$$
$$c^2 = 157$$
$$c = \sqrt{157} \approx 12.530$$
The length of the unknown side is approximately 12.530 meters.

55.
$$c^2 = a^2 + b^2$$
$$c^2 = 5^2 + 5^2$$
$$c^2 = 25 + 25$$
$$c^2 = 50$$
$$c = \sqrt{50} \approx 7.071$$
The length of the unknown side is approximately 7.071 meters.

57. Area of square: $A = s^2 = (4 \text{ in.})^2 = 16 \text{ in.}^2$

Find base of triangle:
$$c^2 = a^2 + b^2$$
$$5^2 = a^2 + 4^2$$
$$25 = a^2 + 16$$
$$-16 \quad -16$$
$$9 = a^2$$
$$3 = a$$

Area of triangle: $A = \frac{bh}{2} = \frac{(3 \text{ in.})(4 \text{ in.})}{2} = 6 \text{ in.}^2$

Sum of areas: $16 \text{ in.}^2 + 6 \text{ in.}^2 = 22 \text{ in.}^2$

Copyright © 2012 Pearson Education, Inc.

59. Area of rectangle:

$A = LW = (12\text{ ft})(4\text{ ft}) = 48\text{ ft}^2$

Find base of triangle:

$$c^2 = a^2 + b^2$$
$$13^2 = a^2 + 12^2$$
$$169 = a^2 + 144$$
$$-144 \quad -144$$
$$25 = a^2$$
$$5 = a$$

Area of triangle: $A = \frac{bh}{2} = \frac{(5\text{ ft})(12\text{ ft})}{2} = 30\text{ ft}^2$

Sum of areas: $48\text{ ft}^2 + 30\text{ ft}^2 = 78\text{ ft}^2$

61.
$$c^2 = a^2 + b^2$$
$$c^2 = 8^2 + 15^2$$
$$c^2 = 64 + 225$$
$$c^2 = 289$$
$$c = \sqrt{289} = 17$$

The wire is 17 feet long.

63.
$$c^2 = a^2 + b^2$$
$$c^2 = 3^2 + 4^2$$
$$c^2 = 9 + 16$$
$$c^2 = 25$$
$$c = \sqrt{25} = 5$$

Juan is 5 miles from his starting point.

65.
$$c^2 = a^2 + b^2$$
$$20^2 = a^2 + 18^2$$
$$400 = a^2 + 324$$
$$-324 \quad -324$$
$$76 = a^2$$
$$\sqrt{76} = a$$
$$8.7 \approx a$$

The base of the ladder is approximately 8.7 feet from the building.

67. Area of triangle:

$A = \frac{bh}{2} = \frac{(18\text{ ft})(12\text{ ft})}{2} = 108\text{ ft}^2$

Area of front rectangle:

$A = LW = (18\text{ ft})(15\text{ ft}) = 270\text{ ft}^2$

Area of side rectangle:

$A = LW = (30\text{ ft})(15\text{ ft}) = 450\text{ ft}^2$

Sum of areas of all four sides:

$$108\text{ ft}^2 + 270\text{ ft}^2 + 450\text{ ft}^2 + 108\text{ ft}^2 + 270\text{ ft}^2 + 450\text{ ft}^2$$
$$= 1656\text{ ft}^2$$

69. Area of large triangle:

$A = \frac{bh}{2} = \frac{(26\text{ yd})(25\text{ yd})}{2} = 325\text{ yd}^2$

Area of small triangle:

$A = \frac{bh}{2} = \frac{(26\text{ yd})(7\text{ yd})}{2} = 91\text{ yd}^2$

Area of wings: $325\text{ yd}^2 - 91\text{ yd}^2 = 234\text{ yd}^2$

Cost: $234\ \cancel{\text{yd}^2} \cdot \frac{\$90}{1\ \cancel{\text{yd}^2}} = \$21,060$

Cumulative Review

71. $\frac{9x}{22} \cdot \frac{10}{27x^2} = \frac{\cancel{9} \cdot \cancel{x} \cdot \cancel{2} \cdot 5}{\cancel{2} \cdot 11 \cdot 3 \cdot \cancel{9} \cdot \cancel{x} \cdot x} = \frac{5}{33x}$

72. The LCD is 30.

$$\frac{8x}{15} - \frac{3x}{10} = \frac{3x \cdot 2}{15 \cdot 2} - \frac{3x \cdot 3}{10 \cdot 3}$$
$$= \frac{16x}{30} - \frac{9x}{30}$$
$$= \frac{7x}{30}$$

Quick Quiz 10.4

1. a. $\sqrt{\frac{36}{49}} = \frac{6}{7}$ since $\left(\frac{6}{7}\right)\left(\frac{6}{7}\right) = \frac{36}{49}$.

b. $\sqrt{4} = 2$ and $\sqrt{9} = 3$, so $\sqrt{4} + \sqrt{9} = 2 + 3 = 5$.

2. $\sqrt{8} \approx 2.83$

3.
$$c^2 = a^2 + b^2$$
$$8^2 = a^2 + 4^2$$
$$64 = a^2 + 16$$
$$-16 \quad -16$$
$$48 = a^2$$
$$\sqrt{48} = a$$
$$6.93 \approx a$$

The other leg is approximately 6.93 centimeters.

Copyright © 2012 Pearson Education, Inc.

4. Answers may vary. The square of the hypotenuse of a right triangle is equal to the sum of the squares of the legs. The Pythagorean Theorem can be used to find the length of the third side of a right triangle given the lengths of two of the sides.

10.5 Exercises

1. The distance around a circle is called the circumference.

3. The diameter is two times the radius of the circle.

5. $r = \frac{d}{2} = \frac{45 \text{ yd}}{2} = 22.5 \text{ yd}$

7. $r = \frac{d}{2} = \frac{3.8 \text{ cm}}{2} = 1.9 \text{ cm}$

9. $C = \pi d \approx 3.14(18 \text{ cm}) \approx 56.5 \text{ cm}$

11. $C = 2\pi r \approx 2(3.14)(11 \text{ in.}) \approx 69.1 \text{ in.}$

13. distance $= C = \pi d \approx 3.14(28) \approx 87.9$ in.

15. distance $= 5C = 5\pi d \approx 5 \cdot 3.14(15) \approx 235.5$ in.

17. $A = \pi r^2 \approx (3.14)(6 \text{ yd})^2 = (3.14)(36 \text{ yd}^2)$
$A \approx 113.0 \text{ yd}^2$

19. $r = \frac{44 \text{ cm}}{2} = 22 \text{ cm}$
$A = \pi r^2 \approx (3.14)(22 \text{ cm})^2 = (3.14)(484 \text{ cm}^2)$
$A \approx 1519.8 \text{ cm}^2$

21. $A = \pi r^2 \approx (3.14)(10 \text{ ft})^2 = (3.14)(100 \text{ ft}^2)$
$A \approx 314 \text{ ft}^2$
The area watered is approximately 314 ft^2.

23. $r = \frac{90 \text{ mi}}{2} = 45 \text{ mi}$
$A = \pi r^2 \approx (3.14)(45 \text{ mi})^2 = (3.14)(2025 \text{ mi}^2)$
$A \approx 6358.5 \text{ mi}^2$
The radio station reaches 6358.5 square miles.

25. Find the circumference of the window.
$C = \pi d \approx (3.14)(2 \text{ ft}) = 6.28 \text{ ft}$
The insulating strip is approximately 6.28 feet long.

27. The distance traveled is 35 times the circumference of the tire.
$35C = 35(2\pi r) \approx 35(2)(3.14)(14 \text{ in.}) = 3077.2 \text{ in.}$
convert this distance to feet.
$3077.2 \text{ in.} \times \frac{1 \text{ ft}}{12 \text{ in.}} \approx 256.43 \text{ ft}$
The car travels approximately 256.43 feet.

29. The distance traveled in one revolution equals the circumference of the tire.
$C = 2\pi r \approx 2(3.14)(16 \text{ in.}) = 100.48 \text{ in.}$
Now divide.
$20{,}096 \div 100.48 = 200$
The wheels completed 200 revolutions.

31. $r = \frac{d}{2} = \frac{64 \text{ in.}}{2} = 32 \text{ in.}$
$A = \pi r^2 \approx (3.14)(32 \text{ in.})^2 = (3.14)(1024 \text{ in.}^2)$
$A \approx 3215.36 \text{ in.}$
The area of the sign is approximately 3215.36 square inches.

33. First find the area of the tabletop in square yards.
$r = \frac{d}{2} = \frac{6 \text{ ft}}{2} = 3 \text{ ft} = 1 \text{ yd}$
$A = \pi r^2 \approx (3.14)(1 \text{ yd})^2 = (3.14)(1 \text{ yd}^2)$
$A \approx 3.14 \text{ yd}^2$
Now multiply.
$\frac{\$72}{1 \text{ yd}^2} \cdot 3.14 \text{ yd}^2 = \$226.08$
The tabletop costs \$226.08.

35. Find the area of the rectangle.
$A = LW = (120 \text{ yd})(40 \text{ yd}) = 4800 \text{ yd}^2$
Together, the two semicircles have an area equal to the area of one circle with diameter 40 yards. Find this area.
$r = \frac{d}{2} = \frac{40 \text{ yd}}{2} = 20 \text{ yd}$
$A = \pi r^2 \approx (3.14)(20 \text{ yd})^2 = (3.14)(400 \text{ yd}^2)$
$A \approx 1256 \text{ yd}^2$
Add the areas.
$4800 \text{ yd}^2 + 1256 \text{ yd}^2 = 6056 \text{ yd}^2$
Find the cost.
$\frac{\$0.20}{1 \text{ yd}^2} \times 6056 \text{ yd}^2 = \$1211.20$
It will cost \$1211.20 to fertilize the field.

 Copyright © 2012 Pearson Education, Inc.

37. a. $\frac{\$6}{8 \text{ slices}} = \$0.75$ per slice

$\frac{1}{8}\cdot\pi r^2 \approx \frac{1}{8}\cdot 3.14\left(\frac{15}{2}\right)^2 \approx 22.08 \text{ in.}^2$

b. $\frac{\$4}{6 \text{ slices}} = \$0.67$ per slice

$\frac{1}{6}\cdot\pi r^2 \approx \frac{1}{6}\cdot 3.14\left(\frac{12}{2}\right)^2 \approx 18.84 \text{ in.}^2$

c. 12 in.: $\frac{\$0.67}{18.8 \text{ in.}^2} \approx \$0.036 \text{ per in.}^2$

15 in.: $\frac{\$0.75}{22.1 \text{ in.}^2} \approx \$0.034 \text{ per in.}^2$

The 15 in. pizza is the better buy.

Cumulative Review

39. $V = LWH = 11(5)(6) = 330 \text{ in.}^3$

40. $V = LWH = 8(4)(5) = 160 \text{ in.}^3$

Quick Quiz 10.5

1. a. $C = 2\pi r \approx 2(3.14)(6 \text{ ft}) = 37.68 \text{ ft}$

b. $A = \pi r^2 \approx (3.14)(6 \text{ ft})^2 = (3.14)(36 \text{ ft}^2)$

$A \approx 113.04 \text{ ft}^2$

2. The distance traveled by a point on the rim of the large wheel is 3 times the circumference of the wheel.

$3C = 3(2\pi r) \approx 3(2\times 3.14\times 3.5) = 65.94 \text{ ft}$

The distance traveled in one revolution of the small pulley is $C = 2\pi r \approx 2(3.14)(1.5) = 9.42$ ft.

Now divide.

$65.94 \div 9.42 = 7$

The smaller wheel makes 7 complete revolutions.

3. Find the area of the rug in square yards.

$r = \frac{d}{2} = \frac{12 \text{ ft}}{2} = 6 \text{ ft}\times\frac{1 \text{ yd}}{3 \text{ ft}} = 2 \text{ yd}$

$A = \pi r^2 \approx (3.14)(2 \text{ yd})^2 = (3.14)(4 \text{ yd}^2)$

$A \approx 12.56 \text{ yd}^2$

Find the cost.

$\frac{\$25}{1 \text{ yd}^2}\times 12.56 \text{ yd}^2 = \$314$

The rug costs \$314.

4. Answers may vary. First find the radius of the semicircle using $r = \frac{d}{2}$. Then find the area of a full circle with radius r using $A = \pi r^2$. Finally, divide this area by 2 to find the area of the semicircle.

10.6 Exercises

1. $V = LWH$: box

3. $V = \frac{4\pi r^3}{3}$: sphere

5. $V = \frac{Bh}{3}$: pyramid

7. $V = \pi r^2 h \approx 3.14(3)^2(5) \approx 141.3 \text{ m}^3$

9. $V = \frac{4\pi r^3}{3} \approx \frac{4(3.14)(4)^3}{3} \approx 267.9 \text{ m}^3$

11. $V = \pi r^2 h = 3.14(4)^2(10) = 502.4 \text{ in.}^3$

13. $V = \frac{4}{3}\pi r^3 = \frac{4}{3}\cdot 3.14(3.25)^3$

$V = 143.7 \text{ cm}^3$

15. $V = \frac{1}{2}\cdot\frac{4\pi r^3}{3} \approx \frac{1}{2}\cdot\frac{4(3.14)(9)^3}{3}$

$V \approx 1526.0 \text{ m}^3$

17. $V = \frac{\pi r^2 h}{3} \approx \frac{3.14(9)^2(16)}{3} \approx 1356.5 \text{ cm}^3$

19. $V = \frac{\pi r^2 h}{3} \approx \frac{3.14(5)^2(10)}{3} \approx 261.7 \text{ ft}^3$

21. $V = \frac{Bh}{3} = \frac{3^2(7)}{3} = 21 \text{ m}^3$

23. $V = \frac{Bh}{3} = \frac{6(9)(5)}{3} = 90 \text{ m}^3$

25. $V = \pi(R^2 - r^2)h \approx 3.14(5^2 - 3^2)(20)$

$V \approx 1004.8 \text{ in.}^3$

Copyright © 2012 Pearson Education, Inc.

27. $V = \pi(R^2 - r^2)h = (3.14)(8^2 - 6^2)(30)$
$V = 2637.6$

29. $\frac{\$0.09}{\text{in.}^3} \cdot \left(\frac{3.14(1)^2(6)}{3} + \frac{1}{2} \cdot \frac{4(3.14)(1)^3}{3} \right) = \$0.75$

31. $V = LWH + \pi r^2 h \approx 4(3)(2) + 3.14(1)^2(2)$
$V \approx 30.28 \text{ ft}^3$

33. $\text{cost} = \frac{\$4}{\text{cm}^3} \cdot \frac{3.14(5)^2(9)}{3} = \$942$

35. $SA = 4\pi r^2 \approx 4(3.14)(7)^2 = 615.44 \text{ in.}^2$

Cumulative Review

37.
$$\frac{21}{40} = \frac{x}{120}$$
$$120 \cdot 21 = 40 \cdot x$$
$$2520 = 40x$$
$$\frac{2520}{40} = \frac{40x}{40}$$
$$63 = x$$

38.
$$\frac{18}{x} = \frac{12}{10}$$
$$10 \cdot 18 = x \cdot 12$$
$$180 = 12x$$
$$\frac{180}{12} = \frac{12x}{12}$$
$$15 = x$$

39. $2\frac{1}{4} \times 3\frac{3}{4} = \frac{9}{4} \times \frac{15}{4} = \frac{135}{16}$ or $8\frac{7}{16}$

40.
$$7\frac{1}{2} \div 4\frac{1}{5} = \frac{15}{2} \div \frac{21}{5}$$
$$= \frac{15}{2} \cdot \frac{5}{21}$$
$$= \frac{\cancel{3} \cdot 5 \cdot 5}{2 \cdot \cancel{3} \cdot 7}$$
$$= \frac{25}{14} \text{ or } 1\frac{11}{14}$$

Quick Quiz 10.6

1. $V = \pi r^2 h \approx (3.14)(6 \text{ in.})^2(10 \text{ in.}) = 1130.4 \text{ in.}^3$

2. $V = \frac{\pi r^2 h}{3} \approx \frac{(3.14)(3 \text{ in.})^2(5 \text{ in.})}{3} = 47.1 \text{ in.}^3$

3. The base is a rectangle.
area of base = (6 m)(5 m) = 30 m^2
$V = \frac{Bh}{3} = \frac{(30 \text{ m}^2)(8 \text{ m})}{3} = 80 \text{ m}^3$

4. Answers may vary. Since 5 inches is one-half of 10 inches, divide the volume of the can in Exercise 1 by 2.

10.7 Exercises

1. The corresponding angles of similar triangles are equal.

3. The perimeters of similar triangles have the same ratios as the corresponding sides.

5. The ratio of 3 to 12 is the same as the ratio of 2 to n.
$$\frac{3}{12} = \frac{2}{n}$$
$$3n = 12(2)$$
$$3n = 24$$
$$\frac{3n}{3} = \frac{24}{3}$$
$$n = 8$$
The length of side n is 8 meters.

7. The ratio of 9 to 27 is the same as the ratio of 5 to n.
$$\frac{9}{27} = \frac{5}{n}$$
$$9n = 27(5)$$
$$9n = 135$$
$$\frac{9n}{9} = \frac{135}{9}$$
$$n = 15$$
The length of side n is 15 centimeters.

9. The ratio of 2 to 5 is the same as the ratio of 7 to n.
$$\frac{2}{5} = \frac{7}{n}$$
$$2n = 5(7)$$
$$2n = 35$$
$$\frac{2n}{2} = \frac{35}{2}$$
$$n = 17.5$$
The length of side n is 17.5 centimeters.

Copyright © 2012 Pearson Education, Inc.

11. The ratio of 18 to 8 is the same as the ratio of 5 to n.

$\frac{18}{8} = \frac{5}{n}$

$18n = 8(5)$

$18n = 40$

$\frac{18n}{18} = \frac{40}{18}$

$n = 2.\overline{2} \approx 2.2$

The length of side n is approximately 2.2 inches.

13. Find the perimeter of the smaller triangle.

$10 + 9 + 6 = 25$

Let P = unknown perimeter.

$\frac{25}{P} = \frac{10}{22}$

$10P = (25)(22)$

$10P = 550$

$\frac{10P}{10} = \frac{550}{10}$

$P = 55$

The perimeter of the larger triangle is 55 inches.

15. Find the perimeter of the larger triangle.

$8 + 10 + 12 = 30$

Let P = unknown perimeter.

$\frac{30}{P} = \frac{8}{6}$

$8P = (30)(6)$

$8P = 180$

$\frac{8P}{8} = \frac{180}{8}$

$P = 22.5$

The perimeter of the smaller triangle is 22.5 feet.

17. Let n = unknown length. The ratio of 9 to 90 is the same as the ratio of 5 to n.

$\frac{9}{90} = \frac{5}{n}$

$9n = (90)(5)$

$9n = 450$

$\frac{9n}{9} = \frac{450}{9}$

$n = 50$

The shortest side of the display case will be 50 centimeters.

19. n is to 6 as 24 is to 4.

$\frac{n}{6} = \frac{24}{4}$

$4n = (6)(24)$

$4n = 144$

$\frac{4n}{4} = \frac{144}{4}$

$n = 36$

The flagpole is 36 feet tall.

21. Let n = height of department store wall. n is to 5.5 as 96 is to 6.5.

$\frac{n}{5.5} = \frac{96}{6.5}$

$6.5n = (5.5)(96)$

$6.5n = 528$

$\frac{6.5n}{6.5} = \frac{528}{6.5}$

$n \approx 81$

The department store wall is approximately 81 feet tall.

23. $\frac{20}{n} = \frac{42}{8}$

$42n = (20)(8)$

$42n = 160$

$\frac{42n}{42} = \frac{160}{42}$

$n \approx 3.8$

The length of side n is approximately 3.8 kilometers.

25. $\frac{n}{15} = \frac{14}{9}$

$9n = (15)(14)$

$9n = 210$

$\frac{9n}{9} = \frac{210}{9}$

$n \approx 23.3$

The length of side n is approximately 23.3 centimeters.

27. Let w = width of new kitchen.

$\frac{w}{9} = \frac{20}{12}$

$12w = (9)(20)$

$12w = 180$

$\frac{12w}{12} = \frac{180}{12}$

$w = 15$

The width of the new kitchen is 15 feet.

Copyright © 2012 Pearson Education, Inc.

29. $\left(\frac{2}{3}\right)^2 = \frac{A}{26}$

$\frac{4}{9} = \frac{A}{26}$

$9A = 104$

$\frac{9A}{9} = \frac{104}{9}$

$A \approx 11.6$

The unknown area is approximately 11.6 square yards.

Cumulative Review

31. $2\times3^2+4-2\times5 = 2\times9+4-2\times5$
$= 18+4-10$
$= 12$

32. $300\div(12-2\times3)+2^4 = 300\div(12-6)+2^4$
$= 300\div6+2^4$
$= 300\div6+16$
$= 50+16$
$= 66$

33. $15+9-(4^2+16\div8) = 15+9-(16+16\div8)$
$= 15+9-(16+2)$
$= 15+9-18$
$= 24-18$
$= 6$

34. $108\div6+51\div3+3^3 = 108\div6+51\div3+27$
$= 18+17+27$
$= 35+27$
$= 62$

Quick Quiz 10.7

1. The ratio of 5 to 12 is the same as the ratio of 6 to x.

$\frac{5}{12} = \frac{6}{x}$

$5x = (12)(6)$

$5x = 72$

$\frac{5x}{5} = \frac{72}{5}$

$x = 14.4$

The length of side x is 14.4 centimeters.

2. Let n = height of building. n is to 13.5 as 10 is to 3.

$\frac{n}{13.5} = \frac{10}{3}$

$3n = (13.5)(10)$

$3n = 135$

$\frac{3n}{3} = \frac{135}{3}$

$n = 45$

The building is 45 feet tall.

3. Let L = unknown length.

$\frac{8}{2.4} = \frac{10}{L}$

$8L = (2.4)(10)$

$8L = 24$

$\frac{8L}{8} = \frac{24}{8}$

$L = 3$

The length is 3 feet.

4. Answers may vary. Multiply the length of the tree's shadow earlier in the day, 3 feet, by $\frac{1}{2}$ to find the length of its shadow that afternoon.

Use Math to Save Money

1. Option A: $15{,}000 \text{ mi}\times\frac{1 \text{ gal}}{20 \text{ mi}} = 750 \text{ gal}$

Option B: $15{,}000 \text{ mi}\times\frac{1 \text{ gal}}{32 \text{ mi}} = 468.75 \text{ gal}$

Option C: $15{,}000 \text{ mi}\times\frac{1 \text{ gal}}{60 \text{ mi}} = 250 \text{ gal}$

2. Option A: $750 \text{ gal}\times\frac{\$3.60}{1 \text{ gal}} = \$2700.00$

Option B: $468.75 \text{ gal}\times\frac{\$3.60}{1 \text{ gal}} = \$1687.50$

Option C: $250 \text{ gal}\times\frac{\$3.60}{1 \text{ gal}} = \$900$

3. Savings of Option B compared to Option A: $\$2700 - \$1687.50 = \$1012.50$

4. Number of years to save \$6,610:

$\$6610\times\frac{1 \text{ yr}}{\$1012.50} \approx 6.5 \text{ yr}$

Copyright © 2012 Pearson Education, Inc.

5. Savings after 5 years:
$$\frac{\$1012.50}{\text{yr}} \times 5 \text{ yr} = \$5062.50$$
Savings after 10 years:
$$\frac{\$1012.50}{\text{yr}} \times 10 \text{ yr} = \$10,125.00$$

6. Yearly savings using Option C instead of Option A: \$2700 − \$900 = \$1800

7. Time to save \$2230:
$$\$2230 \times \frac{1 \text{ yr}}{18000} \approx 1.24 \approx 1 \text{ yr, } 3 \text{ months}$$

8. Savings after 5 years: $\frac{\$1800}{\text{yr}} \times 5 \text{ yr} = \$9000$

Savings after 10 years: $\frac{\$1800}{\text{yr}} \times 10 \text{ yr} = \$18,000$

You Try It

1. 35 ft = ? in.
$$35 \cancel{\text{ft}} \times \frac{12 \text{ in.}}{1 \cancel{\text{ft}}} = 35 \times 12 \text{ in.} = 420 \text{ in.}$$

2. a. 5.25 cm = ? km
Move the decimal point 5 places to the left.
5.25 cm = 0.0000525 km

b. 6025 mL = ? dL
Move the decimal point 2 places to the left.
6025 mL = 60.25 dL

3. a. $8 \cancel{\text{mi}} \times \frac{1.61 \text{ km}}{1 \cancel{\text{mi}}} = 12.88 \text{ km}$

b. $25 \cancel{\text{qt}} \times \frac{0.946 \text{ L}}{1 \cancel{\text{qt}}} = 23.65 \text{ L}$

4. a. $12 \cancel{\text{L}} \times \frac{0.264 \text{ gal}}{1 \cancel{\text{L}}} \approx 3.17 \text{ gal}$

b. $\frac{250 \cancel{\text{m}}}{\text{sec}} \times \frac{3.28 \text{ ft}}{1 \cancel{\text{m}}} = 820 \text{ ft/sec}$

5. a. The obtuse angle is the one measuring 150°. It can be named $\angle AOB$, $\angle BOA$, or $\angle O$.

b. The right angle can be named $\angle DEF$, $\angle FED$, or $\angle E$.

c. The straight angle is $\angle XYZ$ or $\angle ZYX$.

6. Let $\angle S$ = the supplement of $\angle x$.
$$\begin{aligned} \angle S + \angle x &= 180^\circ \\ \angle S + 57^\circ &= 180^\circ \\ + \quad -57^\circ & \quad -57^\circ \\ \hline \angle S &= 123^\circ \end{aligned}$$
Let $\angle C$ = the complement of $\angle x$.
$$\begin{aligned} \angle C + \angle x &= 90^\circ \\ \angle C + 57^\circ &= 90^\circ \\ + \quad -57^\circ & \quad -57^\circ \\ \hline \angle C &= 33^\circ \end{aligned}$$

7. Since $\angle x$ and the 105° angle are adjacent angles, they are supplementary.
$$\begin{aligned} \angle x + 105^\circ &= 180^\circ \\ + \quad -105^\circ & \quad -105^\circ \\ \hline \angle x &= 75^\circ \end{aligned}$$
Since $\angle y$ and the 105° angle are vertical angles, $\angle y = 105^\circ$.
Since $\angle x$ and $\angle z$ are vertical angles, $\angle z = 75^\circ$.

8. Since $\angle x$ and $\angle y$ are adjacent angles, they are supplementary.
$$\begin{aligned} \angle x + \angle y &= 180^\circ \\ \angle x + 120^\circ &= 180^\circ \\ + \quad -120^\circ & \quad -120^\circ \\ \hline \angle x &= 60^\circ \end{aligned}$$
$\angle x = \angle z = 60^\circ$ since $\angle x$ and $\angle z$ are vertical angles.
$\angle x = \angle v = 60^\circ$ since $\angle x$ and $\angle v$ are alternate interior angles.
$\angle w = \angle y = 120^\circ$ since $\angle w$ and $\angle y$ are corresponding angles.

9. $\sqrt{49} = 7$ since $7 \cdot 7 = 49$.

10.
$$\begin{aligned} c^2 &= a^2 + b^2 \\ 12^2 &= a^2 + 5^2 \\ 144 &= a^2 + 25 \\ -25 & \quad\; -25 \\ \hline 119 &= a^2 \\ \sqrt{119} &= a \\ 10.9 &\approx a \end{aligned}$$

11. $A = \frac{bh}{2} = \frac{(4 \text{ ft})(4.56 \text{ ft})}{2} = 9 \text{ ft}^2$

12. $C = 2\pi r \approx 2(3.14)(5 \text{ m}) = 31.4 \text{ m}$

Copyright © 2012 Pearson Education, Inc.

13. $A = \pi r^2 \approx 3.14(5\text{ m})^2 = 78.5\text{ m}^2$

14. $V = \pi r^2 h \approx 3.14(4\text{ ft})^2(6\text{ ft}) = 301.44\text{ ft}^3$

15. $V = \frac{4}{3}\pi r^3 \approx \frac{4}{3}(3.14)(5\text{ ft})^3 \approx 523.33\text{ ft}^3$

16. $V = \frac{\pi r^2 h}{3} \approx \frac{3.14(5\text{ m})^2(8\text{ m})}{3} \approx 209.33\text{ m}^3$

17. $V = \frac{8h}{3} = \frac{(8\text{ ft}\times 10\text{ ft})(5\text{ ft})}{3} \approx 133.33\text{ ft}^3$

18.
$$\frac{12}{4} = \frac{x}{3}$$
$$4x = (12)(3)$$
$$4x = 36$$
$$\frac{4x}{4} = \frac{36}{4}$$
$$x = 9\text{ ft}$$

19. Find the perimeter of the smaller triangle.
4 m + 7 m + 5 m = 16 m
Set up the proportion and solve for *P*.
$$\frac{16}{P} = \frac{4}{7}$$
$$7\times 16 = 4P$$
$$112 = 4P$$
$$\frac{112}{4} = \frac{4P}{4}$$
$$28 = P$$
The perimeter of the larger triangle is 28 meters.

Chapter 10 Review Problems

1. right angle—an angle that measures 90°

2. supplementary angles—two angles whose sum is 180°

3. complementary angles—two angles whose sum is 90°

4. vertical angles—two angles formed by intersecting lines that are opposite each other

5. adjacent angles—two angles formed by intersecting lines that share a common side

6. alternate interior angles—two angles that are on opposite sides of the transversal and between the two parallel lines

7. corresponding angles—two angles that are on the same side of the transversal and are both above or below the other two lines

8. right triangle—a triangle with a 90° angle

9. Pythagorean Theorem—in a right triangle the square of the longest side is equal to the sum of the squares of the other two sides

10. radius—the length of a line segment from the center to a point on the circle

11. diameter—the length of a line segment across the circle that passes through the center

12. circumference—the distance around a circle

13. area of a triangle: $A = \frac{1}{2}bh$

14. circumference of a circle: $C = \pi d = 2\pi r$

15. area of a circle: $A = \pi r^2$

16. volume of a cylinder: $V = \pi r^2 h$

17. volume of a sphere: $V = \frac{4}{3}\pi r^3$

18. volume of a cone: $V = \frac{1}{3}\pi r^2 h$

19. volume of a pyramid: $V = \frac{1}{3}Bh$

20. 33 ft = ? yd
$$33\text{ ft}\times\frac{1\text{ yd}}{3\text{ ft}} = 33\ \cancel{\text{ft}}\times\frac{1\text{ yd}}{3\ \cancel{\text{ft}}} = 33\times\frac{1}{3}\text{ yd} = 11\text{ yd}$$

21. 1500 sec = ? min
$$1500\text{ sec}\times\frac{1\text{ min}}{60\text{ sec}} = 1500\ \cancel{\text{sec}}\times\frac{1\text{ min}}{60\ \cancel{\text{sec}}}$$
$$= 1500\times\frac{1}{60}\text{ min}$$
$$= 25\text{ min}$$

 Copyright © 2012 Pearson Education, Inc.

22. 78 in. = ? ft

$$78 \text{ in.} \times \frac{1 \text{ ft}}{12 \text{ in.}} = 78 \cancel{\text{in.}} \times \frac{1 \text{ ft}}{12 \cancel{\text{in.}}} = 78 \times \frac{1}{12} \text{ ft} = 6.5 \text{ ft}$$

23. 15,840 ft = ? mi

$$15{,}840 \text{ ft} \times \frac{1 \text{ mi}}{5280 \text{ ft}} = 15{,}840 \cancel{\text{ft}} \times \frac{1 \text{ mi}}{5280 \cancel{\text{ft}}} = 15{,}840 \times \frac{1}{5280} \text{ mi} = 3 \text{ mi}$$

24. 3 tons = ? lb

$$3 \text{ tons} \times \frac{2000 \text{ lb}}{1 \text{ ton}} = 3 \cancel{\text{tons}} \times \frac{2000 \text{ lb}}{1 \cancel{\text{ton}}} = 3 \times 2000 \text{ lb} = 6000 \text{ lb}$$

25. 15 gal = ? qt

$$15 \text{ gal} \times \frac{4 \text{ qt}}{1 \text{ gal}} = 15 \cancel{\text{gal}} \times \frac{4 \text{ qt}}{1 \cancel{\text{gal}}} = 15 \times 4 \text{ qt} = 60 \text{ qt}$$

26. 92 oz = ? lb

$$92 \text{ oz} \times \frac{1 \text{ lb}}{16 \text{ oz}} = 92 \cancel{\text{oz}} \times \frac{1 \text{ lb}}{16 \cancel{\text{oz}}} = 92 \times \frac{1}{16} \text{ lb} = 5.75 \text{ lb}$$

27. 31 pt = ? qt

$$31 \text{ pt} \times \frac{1 \text{ qt}}{2 \text{ pt}} = 31 \cancel{\text{pt}} \times \frac{1 \text{ qt}}{2 \cancel{\text{pt}}} = 31 \times \frac{1}{2} \text{ qt} = 15.5 \text{ qt}$$

28. 59 mL = ? L
Move the decimal point 3 places to the left.
59 mL = 0.059 L

29. 8 cm = ? mm
Move the decimal point 1 place to the right.
8 cm = 80 mm

30. 2598 mm = ? cm
Move the decimal point 1 place to the left.
2598 mm = 259.8 cm

31. 778 mg = ? g
Move the decimal point 3 places to the left.
778 mg = 0.778 g

32. 6.3 m = ? cm
Move the decimal point 2 places to the right.
6.3 m = 630 cm

33. 5 km = ? m
Move the decimal point 3 places to the right.
5 km = 5000 m

34. 15 kL = ? L
Move the decimal point 3 places to the right.
15 kL = 15,000 L

35. 473 m = ? km
Move the decimal point 3 places to the left.
473 m = 0.473 km

36. 196 kg = ? g
Move the decimal point 3 places to the right.
196 kg = 196,000 g

37. 721 kg = ? g
Move the decimal point 3 places to the right.
721 kg= 721,000 g

38. Change 4 liters to millimeters.
4 L = 4000 mL
Divide.

$$\frac{4000 \text{ mL}}{24 \text{ jars}} \approx 166.67 \text{ mL per jar}$$

39. a. $P = 7\frac{2}{3} + 4\frac{1}{3} + 5 = 17 \text{ ft}$

b. $P = 17 \text{ ft} \times \frac{12 \text{ in.}}{1 \text{ ft}} = 204 \text{ in.}$

40. $14 \text{ kg} \times \frac{2.2 \text{ lb}}{1 \text{ kg}} = 30.8 \text{ lb}$

41. $20 \text{ lb} \frac{0.454 \text{ kg}}{1 \text{ lb}} = 9.08 \text{ kg}$

42. $15 \text{ ft} \times \frac{0.305 \text{ m}}{1 \text{ ft}} \approx 4.58 \text{ m}$

43. $2.4 \text{ ft} \times \frac{12 \text{ in.}}{1 \text{ ft}} \times \frac{2.54 \text{ cm}}{1 \text{ in.}} \approx 73.15 \text{ cm}$

Copyright © 2012 Pearson Education, Inc.

44. $13 \text{ oz} \times \frac{28.35 \text{ g}}{1 \text{ oz}} = 368.55 \text{ g}$

45. $F = 1.8C + 32 = 1.8(32) + 32 = 89.6°\text{F}$

46. $14 \text{ cm} \times \frac{0.394 \text{ in.}}{1 \text{ cm}} \approx 5.52 \text{ in.}$

47. $C = \frac{5F - 160}{9} = \frac{5(32) - 160}{9} = 0°\text{C}$

48. $200 \text{ mi} - 90 \frac{\text{km}}{\text{hr}} \times 3 \text{ hr} \times \frac{0.62 \text{ mi}}{\text{km}} = 32.6 \text{ mi}$

49. $\frac{\$0.14}{1 \text{ oz}} \times 450 \text{ g} \times \frac{0.0353 \text{ oz}}{1 \text{ g}} = \$2.22$

50. $\angle MNO$, $\angle ONM$

51. $\angle PNO$, $\angle ONP$

52. a. Let $\angle S$ = the supplement of $\angle M$.

$$\begin{aligned} \angle S + \angle M &= 180° \\ \angle S + 39° &= 180° \\ + \quad -39° & \quad -39° \\ \hline \angle S &= 141° \end{aligned}$$

b. Let $\angle C$ = the complement of $\angle C$.

$$\begin{aligned} \angle C + \angle M &= 90° \\ \angle C + 39° &= 90° \\ + \quad -39° & \quad -39° \\ \hline \angle C &= 51° \end{aligned}$$

53.

$$\begin{aligned} m + (m + 10°) &= 90° \\ 2m + 10° &= 90° \\ + \quad -10° & \quad -10° \\ \hline 2m &= 80° \\ \frac{2m}{2} &= \frac{80°}{2} \\ m &= 40° \end{aligned}$$

The measure of $\angle ABD$ is 40°.

54. $\angle w$, $\angle y$

55. $\angle w$ and $\angle z$ are adjacent angles of intersecting lines. They are supplementary, and the sum of their measures is 180°.

56. $\angle x = 65°$ since $\angle x$ and $\angle z$ are vertical angles. Since $\angle z$ and $\angle w$ are adjacent angles, they are supplementary.

$$\begin{aligned} \angle z + \angle w &= 180° \\ 65° + \angle w &= 180° \\ + -65° & \quad -65° \\ \hline \angle w &= 115° \end{aligned}$$

$\angle y = 115°$ since $\angle y$ and $\angle w$ are vertical angles.

57. $\angle v = \angle w = 45°$ since $\angle v$ and $\angle w$ are vertical angles.
$\angle u = \angle w = 45°$ since $\angle u$ and $\angle w$ are corresponding angles.
$\angle s = \angle u = 45°$ since $\angle s$ and $\angle u$ are vertical angles.
Since $\angle t$ and $\angle u$ are adjacent angles, they are supplementary.

$$\begin{aligned} \angle t + \angle u &= 180° \\ \angle t + 45° &= 180° \\ + \quad -45° & \quad -45° \\ \hline \angle t &= 135° \end{aligned}$$

58. $\sqrt{144} = 12$ since $12 \cdot 12 = 144$.

59. $\sqrt{\frac{64}{81}} = \frac{8}{9}$ since $\left(\frac{8}{9}\right)\left(\frac{8}{9}\right) = \frac{64}{81}$.

60. $\sqrt{45} \approx 6.708$

61. $\sqrt{10} \approx 3.162$

62. $\sqrt{64} - \sqrt{25} = 8 - 5 = 3$

63. $\sqrt{121} + \sqrt{16} = 11 + 4 = 15$

64. Area of square: $A = s^2 = (5.1 \text{ m})^2 = 26.01 \text{ m}^2$
Area of triangle:
$A = \frac{1}{2}bh = \frac{1}{2}(9.6 \text{ m})(5.1 \text{ m}) = 24.48 \text{ m}^2$
Find the total area:
$26.01 \text{ m}^2 + 24.48 \text{ m}^2 \approx 50.5 \text{ m}^2$

Copyright © 2012 Pearson Education, Inc.

65.
$$c^2 = a^2 + b^2$$
$$13^2 = a^2 + 12^2$$
$$169 = a^2 + 144$$
$$\underline{-144 \qquad -144}$$
$$25 = a^2$$
$$\sqrt{25} = a$$
$$5 = a$$
The length of the unknown side is 5 yards.

66.
$$c^2 = a^2 + b^2$$
$$7^2 = a^2 + 6^2$$
$$49 = a^2 + 36$$
$$\underline{-36 \qquad -36}$$
$$13 = a^2$$
$$\sqrt{13} = a$$
$$3.61 \approx a$$
The width of the door is approximately 3.61 feet.

67. $C = \pi d \approx 3.14(12 \text{ in.}) \approx 37.68 \text{ in.}$

68. $C = 2\pi r \approx 2(3.14)(7 \text{ in.}) \approx 43.96 \text{ in.}$

69. $A = \pi r^2 \approx 3.14(6 \text{ m})^2 = 113.04 \text{ m}^2$

70. $r = \dfrac{d}{2} = \dfrac{16 \text{ ft}}{2} = 8 \text{ ft}$

$A = \pi r^2 \approx 3.14(8 \text{ ft})^2 \approx 200.96 \text{ ft}^2$

71. $\dfrac{\$30}{1 \text{ yd}^2} \times \dfrac{1 \text{ yd}^2}{9 \text{ ft}^2} = \dfrac{\$30}{9 \text{ ft}^2} = \$3\frac{1}{3} \text{ per ft}^2$

Find the area of the rug.

$r = \dfrac{d}{2} = \dfrac{10 \text{ ft}}{2} = 5 \text{ ft}$

$A = \pi r^2 \approx (3.14)(5 \text{ ft})^2 = 78.5 \text{ ft}^2$

Find the cost.

$78.5 \text{ ft}^2 \times \dfrac{\$3\frac{1}{3}}{\text{ft}^2} = 78.5 \times \$3\frac{1}{3} \approx \$261.67$

The rug will cost \$261.67.

72. $V = \dfrac{4\pi r^3}{3} \approx \dfrac{4(3.14)(1.2)^3}{3} \approx 7.23 \text{ ft}^3$

73. a. $V = LWH - \pi r^2 h$

$V \approx 4(3)(7) - 3.14(1)^2(7) \approx 62.02 \text{ m}^3$

b. $\text{cost} = 62.02 \text{ m}^3 \cdot \dfrac{\$1.20}{\text{m}^3} = \$74.42$

74. $V = \pi r^2 h + \dfrac{1}{2} \cdot \dfrac{4\pi r^3}{3}$

$V \approx 3.14(2)^2(9) + \dfrac{1}{2} \cdot \dfrac{4(3.14)(2)^3}{3}$

$V \approx 129.79 \text{ cm}^3$

75. $V = \dfrac{Bh}{3} = \dfrac{16(18)(18)}{3} = 1728 \text{ m}^3$

76. $V = \dfrac{\pi r^2 h}{3} \approx \dfrac{3.14(17)^2(30)}{3} = 9074.6 \text{ yd}^3$

77. The ratio of 3 to 45 is the same as the ratio of 2 to n.

$\dfrac{3}{45} = \dfrac{2}{n}$

$3n = (45)(2)$

$3n = 90$

$\dfrac{3n}{3} = \dfrac{90}{3}$

$n = 30$

The length of side n is 30 meters.

78. Find the perimeter of the smaller figure.

5 + 18 + 5 + 26 = 54 cm

Let P = unknown perimeter.

$\dfrac{54}{P} = \dfrac{18}{108}$

$18P = (54)(108)$

$18P = 5832$

$\dfrac{18P}{18} = \dfrac{5832}{18}$

$P = 324$

The perimeter of the larger figure is 324 centimeters.

How Am I Doing? Chapter 10 Test

1. a. $145 \text{ oz} \times \dfrac{1 \text{ lb}}{16 \text{ oz}} = 145 \cancel{\text{oz}} \times \dfrac{1 \text{ lb}}{16 \cancel{\text{oz}}}$

$= 145 \times \dfrac{1}{16} \text{ lb}$

$= 9.0625 \text{ lb}$

Copyright © 2012 Pearson Education, Inc.

b. $5\text{ yd}\times\frac{36\text{ in.}}{1\text{ yd}} = 5\ \not{\text{yd}}\times\frac{36\text{ in.}}{1\ \not{\text{yd}}}$
$= 5\times 36\text{ in.}$
$= 180\text{ in.}$

2. a. Move the decimal point 3 places to the left.
162 g = 0.162 kg

b. Move the decimal point 1 place to the right.
2.66 cm = 26.6 mm

3. a. Move the decimal point 3 places to the right.
2 kg = 2000 g

b. $2\text{ kg}\times\frac{2.2\text{ lb}}{\text{kg}} = 4.4\text{ lb}$

4. $F = 1.8C + 32 = 1.8(200) + 32 = 392°\text{F}$

5. $14\text{ km} + 4\text{ mi}\cdot\frac{1.61\text{ km}}{\text{mi}} = 20.44\text{ km}$

6. a. Let $\angle S$ = supplement of $\angle T$.

$$\begin{aligned} \angle S + \angle T &= 180° \\ \angle S + 41° &= 180° \\ + \quad -41° & \quad -41° \\ \hline \angle S &= 139° \end{aligned}$$

b. Let $\angle C$ = complement of $\angle T$.

$$\begin{aligned} \angle C + \angle T &= 90° \\ \angle C + 41° &= 90° \\ + \quad -41° & \quad -41° \\ \hline \angle C &= 49° \end{aligned}$$

7.

$$\begin{aligned} m + (m + 8°) &= 90° \\ 2m + 8° &= 90° \\ + \quad -8° & \quad -8° \\ \hline 2m &= 82° \\ \frac{2m}{2} &= \frac{82°}{2} \\ m &= 41° \end{aligned}$$

8. a. $\angle c = \angle a = 70°$ since $\angle c$ and $\angle a$ are vertical angles.

b. $\angle d$ and $\angle a$ are adjacent angles, so they are supplementary.

$$\begin{aligned} \angle d + \angle a &= 180° \\ \angle d + 70° &= 180° \\ + \quad -70° & \quad -70° \\ \hline \angle d &= 110° \end{aligned}$$

9. $\angle u = \angle w = 43°$ since $\angle u$ and $\angle w$ are vertical angles. $\angle v$ and $\angle w$ are adjacent angles, so they are supplementary.

$$\begin{aligned} \angle v + \angle w &= 180° \\ \angle v + 43° &= 180° \\ + \quad -43° & \quad -43° \\ \hline \angle v &= 137° \end{aligned}$$

$\angle t = \angle w = 43°$ since $\angle t$ and $\angle w$ are corresponding angles. $\angle s = \angle v = 137°$ since $\angle s$ and $\angle v$ are corresponding angles.

10. $81 = 9^2$, yes

11. $4^2 = 16,\ 5^2 = 25$
24 is not a perfect square.

12. $\frac{1}{2^2} = \frac{1}{4},\ \frac{1}{3^2} = \frac{1}{9}$
$\frac{1}{8}$ is not a perfect square.

13. $A = s^2 = 64 \Rightarrow s = \sqrt{64} = 8\text{ ft}$

14. $\sqrt{49} = 7$ since $7\cdot 7 = 49$.

15. $\sqrt{\frac{9}{16}} = \frac{3}{4}$ since $\left(\frac{3}{4}\right)\left(\frac{3}{4}\right) = \frac{9}{16}$.

16. $\sqrt{6} \approx 2.45$

17. $\sqrt{25} + \sqrt{36} = 5 + 6 = 11$

18. $A = \frac{1}{2}bh = \frac{1}{2}(10)(12) = 60\text{ cm}^2$

19. Area of rectangle: $A = LW = (20)(30) = 600\text{ ft}^2$

Area of triangle: $A = \frac{1}{2}bh = \frac{1}{2}(20)(10) = 100\text{ ft}^2$

Total area: $600\text{ ft}^2 + 100\text{ ft}^2 = 700\text{ ft}^2$

Copyright © 2012 Pearson Education, Inc.

20. $c^2 = a^2 + b^2$
$7^2 = a^2 + 3^2$
$49 = a^2 + 9$
$-9 \quad -9$
$40 = a^2$
$\sqrt{40} = a$
$6.32 \approx a$
The length is approximately 6.32 centimeters.

21. $c^2 = a^2 + b^2$
$c^2 = 14^2 + 11^2$
$c^2 = 196 + 121$
$c^2 = 317$
$c = \sqrt{317} \approx 17.8 \text{ ft}$
The ladder is approximately 17.8 feet long.

22. $C = \pi d \approx 3.14(5.15) = 16.171 \text{ m}$

23. $2\pi R = n\pi r \Rightarrow 2(2.1) = n(0.75)$
$n = 5.6 \text{ rev}$

24. $A = \pi r^2 \approx 3.14(1.2)^2 \approx 4.5 \text{ cm}^2$

25. Find the area in square yards.
$r = \frac{d}{2} = \frac{6 \text{ ft}}{2} = 3 \text{ ft} \times \frac{1 \text{ yd}}{3 \text{ ft}} = 1 \text{ yd}$
$A = \pi r^2 \approx (3.14)(1 \text{ yd})^2 = 3.14 \text{ yd}^2$
Find the cost.
$3.14 \text{ yd}^2 \times \frac{\$20}{1 \text{ yd}^2} = \$62.80$
The tablecloth will cost $62.80.

26. $V = \pi r^2 h \approx 3.14(3)^2(7.1) \approx 200.65 \text{ in}^3$

27. $V = \frac{4\pi r^3}{3} \approx \frac{4(3.14)(8)^3}{3} \approx 2143.57 \text{ in.}^3$

28. $V = \frac{\pi r^2 h}{3} \approx \frac{3.14(8)^2(12)}{3} = 803.84 \text{ cm}^3$

29. $V = \frac{Bh}{3} = \frac{10(7)(12)}{3} = 280 \text{ m}^3$

30. The ratio of 4 to 18 is the same as the ratio of 13 to n.
$\frac{4}{18} = \frac{13}{n}$
$4n = (18)(13)$
$4n = 234$
$\frac{4n}{4} = \frac{234}{4}$
$n = 58.5$
The length of side n is 58.5 inches.

31. Let n = the height of flagpole.
$\frac{n}{20} = \frac{2.5}{2}$
$2n = (20)(2.5)$
$2n = 50$
$\frac{2n}{2} = \frac{50}{2}$
$n = 25$
The flagpole is 25 feet tall.

32. $\frac{1}{w} = \frac{4.5}{36}$
$4.5w = (1)(36)$
$\frac{4.5w}{4.5} = \frac{36}{4.5}$
$w = 8$
The width of the larger rectangle is 8 centimeters.

Copyright © 2012 Pearson Education, Inc.

Practice Final Examination

1. The number to the right of the hundred thousands place, 4, is less than 5.
   7,543,876 = 7,500,000 to the nearest hundred thousand

2. A number decreased by 5: $x - 5$

3. $$\begin{aligned}(2+5+x)+4 &= (7+x)+4 \\ &= (x+7)+4 \\ &= x+(7+4) \\ &= x+11\end{aligned}$$

4. Replace x with 9 and y with 12.
   $x + y = 9 + 12 = 21$

5. The unlabeled horizontal side is 6 ft + 15 ft = 21 ft, and the unlabeled vertical side is 5 ft + 4 ft = 9 ft.
   $P = 6 + 5 + 15 + 4 + 21 + 9 = 60$ ft

6. $$\begin{array}{r} 751 \\ -\ 482 \\ \hline 269 \end{array}$$
   Check: $269 + 482 \stackrel{?}{=} 751,\ 751 = 751$

7. $$\begin{array}{r} 5282 \\ \times\ \ 806 \\ \hline 31692 \\ 42256 \\ \hline 4257292 \end{array} \rightarrow 4,257,292$$

8. 812,869 ÷ 743 = 1094 R 27
   $$\begin{array}{r} 1094 \\ 743\overline{)812869} \\ \underline{743} \\ 6986 \\ \underline{6687} \\ 2999 \\ \underline{2972} \\ 27 \end{array}$$

9. $$\begin{aligned}7+15\div 3-3^2 &= 7+15\div 3-9 \\ &= 7+5-9 \\ &= 12-9 \\ &= 3\end{aligned}$$

10. $$\begin{aligned}y+4y+3y &= 16 \\ 8y &= 16 \\ y &= 2\end{aligned}$$

11. \$30 + \$30 + \$70 + \$190 = \$320

12. $-5 < -3$
    Since −5 is to the left of −3 on a number line.

13. $-|-13| = -(-(-13)) = -(13) = -13$

14. $$\begin{aligned}7+(-3)+9+(-4) &= (7+9)+((-3)+(-4)) \\ &= 16+(-7) \\ &= 9\end{aligned}$$

15. $-9-7+8 = -9+(-7)+8 = -16+8 = -8$

16. $$\begin{aligned}(-6)(2)(-1)(5)(-3) &= (-12)(-5)(-3) \\ &= 60(-3) \\ &= -180\end{aligned}$$

17. $64 \div (-8) = -8$

18. $$\begin{aligned}-9+5y+7-3y &= 5y-3y-9+7 \\ &= (5-3)y+(-9+7) \\ &= 2y-2\end{aligned}$$

19. Replace a with −4 and b with 7.
    $a^2 - b = (-4)^2 - 7 = 16 - 7 = 9$

20. $-3(a-2) = -3a - (-3)(2) = -3a + 6$

21. $$\begin{aligned}5(3-7) &= x-3 \\ 5(-4) &= x-3 \\ -20 &= x-3 \\ +\quad 3 &\quad\ \ 3 \\ \hline -17 &= x\end{aligned}$$
    Check: $$\begin{aligned}5(3-7) &= x-3 \\ 5(3-7) &\stackrel{?}{=} -17-3 \\ 5(-4) &\stackrel{?}{=} -20 \\ -20 &= -20\ \checkmark\end{aligned}$$

22. $$\begin{aligned}6(3x) &= 54 \\ 18x &= 54 \\ \frac{18x}{18} &= \frac{54}{18} \\ x &= 3\end{aligned}$$
    Check: $$\begin{aligned}6(3x) &= 54 \\ 6(3\cdot 3) &\stackrel{?}{=} 54 \\ 6(9) &\stackrel{?}{=} 54 \\ 54 &= 54\ \checkmark\end{aligned}$$

Copyright © 2012 Pearson Education, Inc.

23. a. The length is five times the width: $L = 5W$

b. Replace L with 40.

$$40 = 5W$$
$$\frac{40}{5} = \frac{5W}{5}$$
$$8 = W$$

The width is 8 feet.

24. The shape can be divided into three rectangles by extending the two inner vertical edges.

Area of leftmost rectangle:

$A = LW = (21 \text{ in.})(10 \text{ in.}) = 210 \text{ in.}^2$

Area of middle rectangle:

$L = 21 \text{ in.} - 14 \text{ in.} = 7 \text{ in.}$

$A = LW = (7 \text{ in.})(9 \text{ in.}) = 63 \text{ in.}^2$

Area of rightmost rectangle:

$A = LW = (11 \text{ in.})(7 \text{ in.}) = 77 \text{ in.}^2$

Total area:

$210 \text{ in.}^2 + 63 \text{ in.}^2 + 77 \text{ in.}^2 = 350 \text{ in.}^2$

25.

$$A = bh$$
$$108 = 9b$$
$$\frac{108}{9} = \frac{9b}{9}$$
$$12 = b$$

The base is 12 meters.

26.

$$V = LWH$$
$$200 = (10)(W)(5)$$
$$200 = 50W$$
$$\frac{200}{50} = \frac{50W}{50}$$
$$4 = W$$

The width is 4 feet.

27. a. $(21 \text{ ft})(15 \text{ ft}) \cdot \frac{1 \text{ yd}^2}{9 \text{ ft}^2} = 35 \text{ yd}^2$

b. $35 \text{ yd}^2 \cdot \frac{\$18}{\text{yd}^2} = \$630$

28. $(5y)(z^6)(5y^4)(3z^2) = 5 \cdot 5 \cdot 3y^{4+1}z^{6+2} = 75y^5z^8$

29. $V = LWH = (2y^4)(3y^2)(5y) = 30y^7$

30.

$$7\overline{)57} = 8 \text{ R } 1, \quad 56, \quad 1$$

$$\frac{57}{7} = 8\frac{1}{7}$$

31. $3\frac{1}{5} = \frac{5 \cdot 3 + 1}{5} = \frac{15+1}{5} = \frac{16}{5}$

32. $\frac{72y}{117yz} = \frac{72\cancel{y}}{117\cancel{y}z} = \frac{\cancel{9} \cdot 8}{\cancel{9} \cdot 13z} = \frac{8}{13z}$

33. $\frac{27}{-45} = \frac{9 \cdot 3}{9 \cdot (-5)} = \frac{3}{-5} = -\frac{3}{5}$

34. $\frac{18x^7}{42x^{10}} = \frac{\cancel{6} \cdot 3\cancel{x^7}}{\cancel{6} \cdot 7\cancel{x^7} \cdot x^3} = \frac{3}{7x^3}$

35. $(3^3)^2 = 3^{3 \cdot 2} = 3^6 = 729$

36. $\frac{21}{49} = \frac{\cancel{7} \cdot 3}{\cancel{7} \cdot 7} = \frac{3}{7}$

37. $\frac{156 \text{ ft}}{12 \text{ hr}} = 13 \text{ ft per hr}$

38. a. $\frac{42}{14} = 3$ secretaries per lawyer

b. $\frac{12}{3} = 4$ paralegals per lawyer

c.

$$\frac{12}{3} = \frac{n}{80}$$
$$3n = 960$$
$$\frac{3n}{3} = \frac{960}{3}$$
$$n = 320 \text{ paralegals}$$

39.

$$\frac{24}{x} = \frac{8}{5}$$
$$8x = (24)(5)$$
$$8x = 120$$
$$\frac{8x}{8} = \frac{120}{8}$$
$$x = 15$$

Copyright © 2012 Pearson Education, Inc.

Check: $\frac{24}{x} = \frac{8}{5}$

$\frac{24}{15} \stackrel{?}{=} \frac{8}{5}$

$(24)(5) \stackrel{?}{=} (15)(8)$

$120 = 120$ ✓

40. $\frac{1}{90} = \frac{5}{x}$

$x = (90)(5)$

$x = 450$

5 inches represent 450 miles.

41. $\frac{8y}{15} \cdot \frac{30}{24y^2} = \frac{\cancel{8} \cdot \cancel{y} \cdot \cancel{15} \cdot 2}{\cancel{15} \cdot \cancel{8} \cdot 3 \cdot \cancel{y} \cdot y} = \frac{2}{3y}$

42. $\frac{9}{14} \div \frac{-72}{21} = -\frac{\cancel{9}}{\cancel{7} \cdot 2} \cdot \frac{\cancel{7} \cdot 3}{\cancel{9} \cdot 8} = -\frac{3}{16}$

43. $\frac{1}{3} \div 4 = \frac{1}{3} \cdot \frac{1}{4} = \frac{1}{12}$ lb in each container

44. $6x = 2 \cdot 3 \cdot x,\ 18x = 2 \cdot 3^2 x,\ \ 27 = 3^3$

LCM $= 2 \cdot 3^3 x = 54x$

45. $\frac{8}{y} - \frac{4}{y} = \frac{8-4}{y} = \frac{4}{y}$

46. $\frac{-5}{16} \cdot \frac{3}{3} + \frac{9}{24} \cdot \frac{2}{2} = \frac{-15+18}{48} = \frac{3}{48} = \frac{1}{16}$

47. $\frac{5x}{18} \cdot \frac{5}{5} - \frac{7x}{45} \cdot \frac{2}{2} = \frac{25x - 14x}{90} = \frac{11x}{90}$

48. $12\frac{1}{7} \cdot \frac{3}{3} - 7\frac{13}{21} = 11\frac{24}{21} - 7\frac{13}{21} = 4\frac{11}{21}$

49. $3\frac{3}{7} \div 6\frac{2}{3} = \frac{24}{7} \cdot \frac{3}{20} = \frac{6 \cdot \cancel{4} \cdot 3}{7 \cdot \cancel{4} \cdot 5} = \frac{18}{35}$

50. $\left(\frac{3}{4}\right)^2 + \frac{1}{9} \div \frac{1}{3} = \frac{9}{16} + \frac{1}{9} \cdot \frac{3}{1}$

$= \frac{9}{16} \cdot \frac{3}{3} + \frac{1}{3} \cdot \frac{16}{16}$

$= \frac{27+16}{48}$

$= \frac{43}{48}$

51. $\frac{3}{4}x = 27$

$\frac{4}{3}\left(\frac{3}{4}x\right) = \frac{4}{3}(27)$

$x = 36$

52. Subtract.

$1\frac{1}{3} - \frac{2}{3} = \frac{4}{3} - \frac{2}{3} = \frac{2}{3}$

The blade must be lowered $\frac{2}{3}$ inch.

53. $(-4y^2 - 5) - (2y^2 + 3y + 2) + (-y^2 - 3)$

$= -4y^2 - 5 - 2y^2 - 3y - 2 - y^2 - 3$

$= -7y^2 - 3y - 10$

54. $(-3x)(x-7) = (-3x)(x) + (-3x)(-7)$

$= -3x^2 + 21x$

55. $(z^5)(z^6 - 3z - 5) = z^5 \cdot z^6 - z^5 \cdot 3z - z^5 \cdot 5$

$= z^{11} - 3z^6 - 5z^5$

56. $(y+7)(4y^2 - 4y + 6)$

$= 4y^3 - 4y^2 + 6y + 28y^2 - 28y + 42$

$= 4y^3 + 24y^2 - 22y + 42$

57. Let $\angle b = x$.

The measure of $\angle a$ is 40° more than the measure of $\angle b$: $\angle a = x + 40°$.

The measure of $\angle c$ is three times the measure of $\angle b$: $\angle c = 3x$.

58. $(y+5)(y-3) = y^2 - 3y + 5y - 15 = y^2 + 2y - 15$

59. $(x-4)(4x+3) = 4x^2 + 3x - 16x - 12$

$= 4x^2 - 13x - 12$

60. $4z + 12x = 4 \cdot z + 4 \cdot 3x = 4(z + 3x)$

61. $9x^2y - 3xy = 3xy \cdot 3x - 3xy \cdot 1$

$= 3xy(3x - 1)$

62. $4x - 8y + 16 = 4 \cdot x - 4 \cdot 2y + 4 \cdot 4$

$= 4(x - 2y + 4)$

Copyright © 2012 Pearson Education, Inc.

63.
$$
\begin{aligned}
-15-24 &= 2x+15-3x\\
-39 &= 2x+15-3x\\
-39 &= -x+15\\
-15 & \quad\; -15\\
\hline
-54 &= -x\\
\frac{-54}{-1} &= \frac{-x}{-1}\\
54 &= x
\end{aligned}
$$

Check:
$$
\begin{aligned}
-15-24 &= 2x+15-3x\\
-15-24 &\stackrel{?}{=} 2(54)+15-3(54)\\
-39 &\stackrel{?}{=} 108+15-162\\
-39 &= -39 \checkmark
\end{aligned}
$$

64.
$$
\begin{aligned}
-\frac{16}{4} &= -3(6z)-5z\\
-4 &= -3(6z)-5z\\
-4 &= -18z-5z\\
-4 &= -23z\\
\frac{-4}{-23} &= \frac{-23z}{-23}\\
\frac{4}{23} &= z
\end{aligned}
$$

Check:
$$
\begin{aligned}
-\frac{16}{4} &= -3(6z)-5z\\
-\frac{16}{4} &\stackrel{?}{=} -3\left(6\cdot\frac{4}{23}\right)-5\cdot\frac{4}{23}\\
-4 &\stackrel{?}{=} -3\left(\frac{24}{23}\right)-\frac{20}{23}\\
-4 &\stackrel{?}{=} \frac{-72}{23}-\frac{20}{23}\\
-4 &\stackrel{?}{=} \frac{-92}{23}\\
-4 &= -4 \checkmark
\end{aligned}
$$

65.
$$
\begin{aligned}
-z &= \frac{1}{4}\\
-1(-z) &= -1\left(\frac{1}{4}\right)\\
z &= -\frac{1}{4}
\end{aligned}
$$

Check:
$$
\begin{aligned}
-z &= \frac{1}{4}\\
-\left(-\frac{1}{4}\right) &\stackrel{?}{=} \frac{1}{4}\\
\frac{1}{4} &= \frac{1}{4} \checkmark
\end{aligned}
$$

66.
$$
\begin{aligned}
5y-9-9y &= 23\\
-4y-9 &= 23\\
+\qquad 9 & \quad 9\\
\hline
-4y &= 32\\
\frac{-4y}{-4} &= \frac{32}{-4}\\
y &= -8
\end{aligned}
$$

Check:
$$
\begin{aligned}
5y-9-9y &= 23\\
5(-8)-9-9(-8) &\stackrel{?}{=} 23\\
-40-9+72 &\stackrel{?}{=} 23\\
23 &= 23 \checkmark
\end{aligned}
$$

67.
$$
\begin{aligned}
7a-5+3a &= -9+6a\\
10a-5 &= 6a-9\\
-6a & \quad -6a\\
\hline
4a-5 &= -9\\
5 & \quad 5\\
\hline
4a &= -4\\
\frac{4a}{4} &= \frac{-4}{4}\\
a &= -1
\end{aligned}
$$

Check:
$$
\begin{aligned}
7a-5+3a &= -9+6a\\
7(-1)-5+3(-1) &\stackrel{?}{=} -9+6(-1)\\
-7-5-3 &\stackrel{?}{=} -9-6\\
-15 &= -15 \checkmark
\end{aligned}
$$

68.
$$
\begin{aligned}
-2(y-5) &= 6(y+5)-12\\
-2y+10 &= 6y+30-12\\
-2y+10 &= 6y+18\\
2y & \quad 2y\\
\hline
10 &= 8y+18\\
-18 & \quad -18\\
\hline
-8 &= 8y\\
\frac{-8}{8} &= \frac{8y}{8}\\
-1 &= y
\end{aligned}
$$

69.
$$
\begin{aligned}
\frac{x}{2}+\frac{x}{3} &= 10\\
6\left(\frac{x}{2}+\frac{x}{3}\right) &= 6(10)\\
6\left(\frac{x}{2}\right)+6\left(\frac{x}{3}\right) &= 6(10)\\
3x+2x &= 60\\
5x &= 60\\
\frac{5x}{5} &= \frac{60}{5}\\
x &= 12
\end{aligned}
$$

Copyright © 2012 Pearson Education, Inc.

70. a. L = fall students
$L + 95$ = spring students
$L - 75$ = summer students

b. $L + (L + 95) + (L - 75) = 386$

c. $3L + 20 = 386 \Rightarrow 3L = 366$
$L = 122$ fall students
$L + 95 = 217$ spring students
$L - 75 = 47$ summer students

d. $122 + (122 + 95) + (122 - 75) \stackrel{?}{=} 386$
$386 = 386$

71. $751.7596 = 751.76$ to the nearest hundredth

72. $0.588 + 75 = 8.59 = 84.178$

73. $0.042 \times 0.04 = 0.00168$

74. $40(9.50) + 12(13.65) = \$543.80$

75. $-4.75 \div 3.2 = -1.484375 = -1.48$ to the nearest hundredth

76.

Fraction	Decimal	Percent
$\frac{1}{4}$	0.25	25%
$\frac{14}{25}$	0.56	56%
$3\frac{1}{2}$	3.5	350%

77. 79 is 25% of what number?
$79 = 0.25 \times n$
$$\frac{79}{0.25} = \frac{0.25n}{0.25}$$
$316 = n$
79 is 25% of 316.

78. Find 30% of 90.
$n = 0.30 \times 90$
$n = 27$
30% of 90 is 27.

79. What percent of \$28.50 is \$5.75?
$n \times 28.50 = 5.75$
$$\frac{28.50n}{28.50} = \frac{5.75}{28.50}$$
$n \approx 0.20 = 20\%$
The tip was approximately 20% of the bill.

80. Since 60% should be filled during the first week, 100% − 60% or 40% should be available after the first week.

81. a. There is one more symbol next to Biology than is next to Chemistry. Since each symbol represents 50 students, there are 50 more students in Biology than in Chemistry.

b. There are four symbols next to Physics and seven symbols next to Biology, for a total of 11 symbols. Since each symbol represents 50 students, the combined total enrollment is 11(50) or 550 students.

82. a. 40% + 30% = 70%

b. 10% of \$4600 = $0.10 \times \$4600 = \$460$

83.

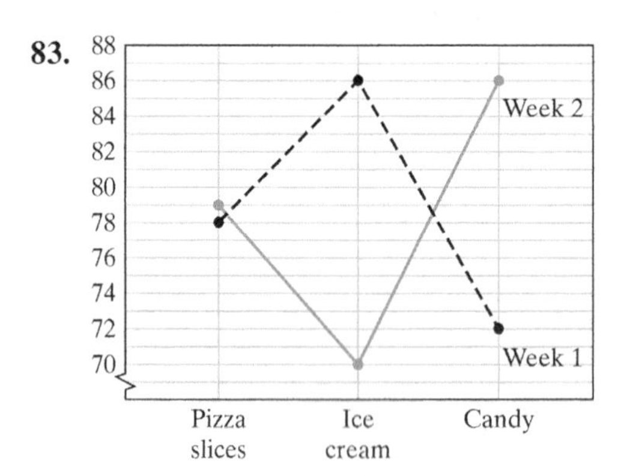

84. 59, 69, 72, 77, 78, 83, 87, 92, 95
↑
median = 78

85. $$\frac{89 + 87 + 92 + 97 + 85 + 85 + 95}{7} = 90°\text{F}$$

86. 2, 7, 7, 8, 8, 9, 9, 9, 10
mode (9, 9, 9)

87. a. $R(-3, 5)$

b. $T(0, 3)$

c. $S(-4, -4)$

d. $U(6, 2)$

Copyright © 2012 Pearson Education, Inc.

88. $y = 2x - 4$
Let $y = -6$.
$-6 = 2x - 4$
$-2 = 2x$
$-1 = x$
$(-1, -6)$ is a solution.
Let $y = -4$.
$-4 = 2x - 4$
$0 = 2x$
$0 = x$
$(0, -4)$ is a solution.
Let $y = -8$.
$-8 = 2x - 4$
$-4 = 2x$
$-2 = x$
$(-2, -8)$ is a solution.

89. $y = 3x - 2$
The slope is 3, and the y-intercept is -2. Plot $(0, -2)$ and then move 1 unit right and 3 units up to $(1, 1)$. Draw the line through $(0, -2)$ and $(1, 1)$.

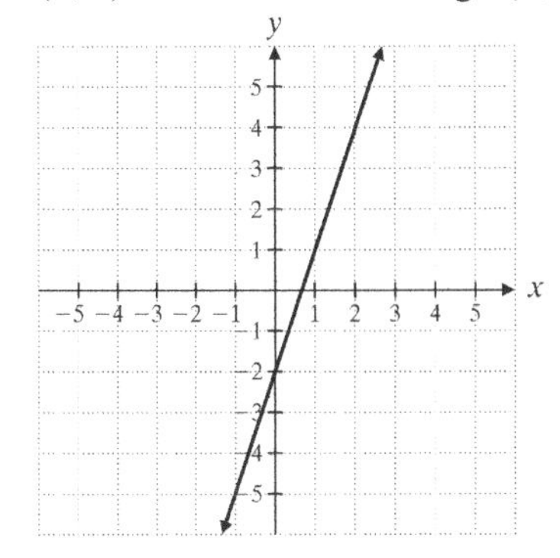

90. $y = -3$
The line is horizontal with y-intercept -3.

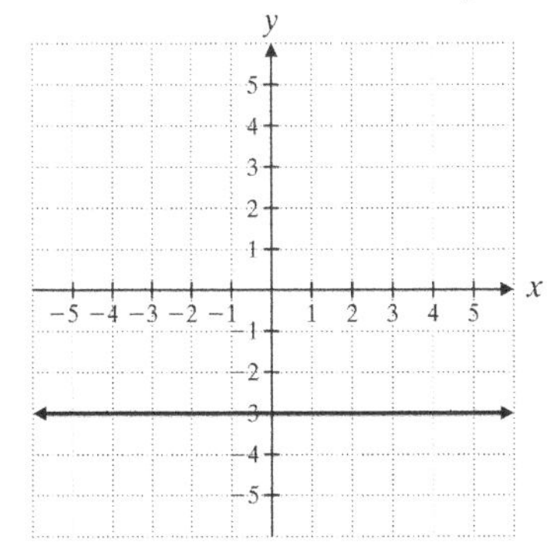

91. $14 \text{ gal} \times \frac{4 \text{ qt}}{\text{gal}} = 56 \text{ qt}$

92. Move the decimal point 3 places to the right.
8 km = 8000 m

93. $13 \text{ ft} \times \frac{0.305 \text{ m}}{1 \text{ ft}} = 3.965 \text{ m}$

94. $\sqrt{81} = 9$ since $9 \cdot 9 = 81$.

95. $\sqrt{121} + \sqrt{16} = 11 + 4 = 15$

96.
$$c^2 = a^2 + b^2$$
$$26^2 = a^2 + 24^2$$
$$676 = a^2 + 576$$
$$100 = a^2$$
$$\sqrt{100} = a$$
$$10 = a$$
The unknown side is 10 inches.

97. $C = \pi d \approx 3.14(13) = 40.82 \text{ ft}$

98. $r = \frac{d}{2} = \frac{18}{2} = 9 \text{ m}$
$A = \pi r^2 \approx 3.14(9)^2 = 254.34 \text{ m}^2$

99. $V = \frac{Bh}{3} = \frac{8(10)(9)}{3} = 240 \text{ ft}^3$

100. The ratio of 7 to 49 is the same as the ratio of 3 to n.
$$\frac{7}{49} = \frac{3}{n}$$
$$7n = (49)(3)$$
$$7n = 147$$
$$\frac{7n}{7} = \frac{147}{7}$$
$$n = 21$$
The length of side n is 21 meters.

Copyright © 2012 Pearson Education, Inc.